BK 621.381 H523E
ELECTRONIC SYSTEMS AND INSTRUMENTATION
/HENRY, RIC
C1978 18.95 FV

3000 487512 30017
St. Louis Community College

W9-DIU-142

621.381 H523e

HENRY
 ELECTRONIC SYSTEMS AND
 INSTRUMENTATION

FV

18.95

WITHDRAWN

St. Louis Community
College

Library

5801 Wilson Avenue
St. Louis, Missouri 63110

Electronic Systems and Instrumentation

Electronic Systems and Instrumentation

Richard W. Henry
Bucknell University

John Wiley & Sons
New York
Santa Barbara
Chichester
Brisbane
Toronto

Copyright © 1978, by John Wiley & Sons, Inc.

All rights reserved. Published simultaneously in Canada.

No part of this book may be reproduced by any means, nor transmitted, nor translated into a machine language without the written permission of the publisher.

Library of Congress Cataloging in Publication Data:

Henry, Richard Warfield, 1932—
    Electronic systems and instrumentation.

    Includes index.
    1. Electronics.   2. Electronic apparatus and
appliances.   I. Title.
TK7815.H43   621.381   77-20170
ISBN 0-471-02487-2

Printed in the United States of America

10 9 8 7 6 5 4 3 2 1

Preface

This book is intended as a brief introduction to electronics for the user of electronic instrumentation systems, both linear and digital. For a user to be able to connect together commercially made electronic building blocks, he or she must have a thorough understanding of their terminal properties (e.g., impedances, frequency-response, noise behavior). However, even a thorough understanding of commercial instruments is not enough. Both in the linear and the digital areas it is often desirable to build part or all of a system from readily available "components." This need may arise because one may not find or be able to afford a commercial instrument to do the job or because one needs the circuit "right now." The increase in sophistication of components available for low-cost, off-the-shelf delivery that has followed the development of integrated circuits has markedly expanded the user's ability to design and construct very complex systems, usually at a parts cost that is one or two orders of magnitude below that of a commercial system.

The text is designed for use in a first or second course in electronics for students with a mathematical preparation that includes *at least* two semesters of calculus. Prior exposure to Fourier series is *not* assumed. Instead, my starting point in the description of systems is the concept of *impulse response* and the technique of *convolution*. This approach allows me to introduce the concept of e^{st} as an eigenfunction, in Chapter 2, and then to proceed directly to Laplace and Fourier transforms. Fourier series are treated as the Fourier transforms of periodic functions.

Chapter 4 introduces the three basic passive components and the concepts of complex impedance and admittance, again using the e^{st} notation, along with pole-zero and Bode plots. In addition, the h-parameter representation for two-port networks is described, and various input configurations of amplifiers are discussed. Frequency-domain techniques are used to describe systems with feedback in Chapter 5 and to investigate the criteria for stability.

Chapter 6 is the only one that deals directly with the underlying physics of electronic devices. The treatment of the transistor and development of circuit models along with the earlier discussion of two-port networks allows me to deal with biasing and amplification in

Chapter 7. Chapter 8 extends the treatment of differential amplifiers to op-amps and their applications.

Chapters 9 and 10 are concerned with combinational and sequential logic circuits. Karnaugh maps are introduced and used as a design tool. The discussion includes a brief description of microprocessors and the basic operation and organization of read-only, random-access, and programmable read-only semiconductor memories.

Chapters 11 and 12 deal with systems that require an understanding of both linear and digital techniques. D-to-A and A-to-D conversion methods, as well as sampling and multiplication and their applications are described.

Chapter 13 investigates thermal and shot noise and a model for the introduction of noise by two-port networks. Much of the material is based on statistical concepts, such as spectral density, developed in Chapters 1 and 3. A more practical approach to noise is taken in Chapter 14 where I discuss shielding, grounding, and systems for the extraction of signals from noise.

The final chapter provides a very brief introduction to discrete systems. The approach parallels the earlier treatment of continuous systems; the discrete impulse response, discrete convolution, eigensequences, z transforms, and discrete Fourier transforms are analogous to impulse response, convolution, eigenfunctions, and Laplace and Fourier transforms introduced in Chapters 1 to 3. I conclude with an introduction to the Fast Fourier Transform.

There is more material than most instructors would try to include in a one-semester course. Topics that could be left out without endangering the overall pattern are: (1) the computation of convolution integrals in Chapter 1, (2) applications of Laplace transforms in Chapter 2 (but the section on the eigenfunction concept is essential), (3) all of Chapter 6 (but this means that the transistor model used in Chapter 7 must be taken on faith), (4) Chapters 9 and 10 on logic (but only if similar material is dealt with at an earlier stage in the curriculum), and (5) Chapter 15 on discrete systems. Although some teachers will be inclined to skip the discussions of random signals in Chapters 1 and 3 and the treatment of noise in Chapter 13, it is my opinion that an insufficient understanding of random signals and noise is responsible for much of the naiveté students have concerning the use of electronics when they complete their undergraduate exposure to electronics courses. Indeed, a healthy "philosophy" toward general problems in precision measurement can be gleaned from an understanding of noise in electronic systems.

Although I have made no attempt at a comprehensive treatment of all aspects of electronic instrumentation, I believe there is a sufficient sampling of topics for an introductory course. Also, whereas the typical nonmathematical electronics course makes it arduous for students to proceed on their own to advanced topics, the analytical tools provided here should enable them to go further with a minimum of difficulty.

I firmly believe that problem solving is one of the most effective ways of learning. Problems with a wide range of difficulty follow each chapter. Many of these extend the ideas introduced in the chapter or deal with applications not mentioned in the text.

Special thanks are due to Professors Stephen Becker of Bucknell University and Kenneth Schick of Union College for helpful discussions of certain topics, and to Mrs Annabelle Libby for her patience and accuracy in typing the manuscript.

Richard W. Henry

Contents

CHAPTER 1

Signals and Systems in the Time Domain

1.1. Introduction   My purpose here is to develop both the theoretical background and practical awareness that is necessary for a scientist or engineer to appreciate and to use effectively the various powerful techniques of instrumentation that are employed in modern experimental science. Such techniques may be divided into analog and digital methods; however, more and more instrumentation systems are utilizing both techniques, so that competent scientists will discover that they need more than a superficial knowledge of both areas.

Unfortunately, there is a fair amount of mathematics involved in electronics. To ignore this mathematics is to be a dilettante, a dabbler in the field. We will find it necessary to introduce such mathematical ideas as convolution, Laplace and Fourier transforms, autocorrelation functions, spectral density, and others in which complex number representations of signals are used freely. Most of the necessary mathematics is developed in the first five chapters; the rewards for persevering in the study of the mathematics come mainly later on when we study practical systems such as op-amps, frequency filters, modulators, sampling devices, and systems for the recovery of signals in the presence of noise. The early mathematical approach is tempered by the brief excursion into the underlying physics of semiconductor devices in Chapter 6 and digital circuits and devices in Chapters 9 and 10. We begin by taking a detailed look at the notion of a *system*.

1.2. Linear, time-invariant (LTI) systems   We will use the term "system" to mean any physical device or combination of devices producing an output or outputs that may vary as a function of time (e.g., a voltage or current or oscilloscope deflection in an electrical system; a displacement or velocity in a mechanical system; or a population of molecules in a chemical system) in response to some input or inputs that may vary as a function of time. An arbitrary system with one input $x(t)$ and one output $y(t)$ is shown in Fig. 1.1.

The most general description of a system consists of a list of possible inputs and the corresponding outputs but, in almost every case, we can describe a system more economically than that. For example, if the

1

Fig. 1.   1. A system with one input and one output.

output depends only on the present value of the input and not on past (or future) values, then the system is said to be memoryless, in which case a complete description is afforded by a simple graph of the output variable as a function of the input variable. For a memoryless system, such a graph must be single-valued; that is, there can be only one value of the output for a given value of input. The system for which the input is the current through a resistor and the output is the voltage across the resistor is a memoryless system, but that for which the input is the current in an iron-core electromagnet and the output is the magnetization is not (because of the hysteresis phenomenon). In much of this book we will be concerned with the problem of system analysis; that is, given a certain time function as the input to a system, we wish to compute the output time function.

There is an extremely important class of systems that are not memoryless but are nevertheless amenable to a simple mathematical description. These are systems possessing the two properties of *linearity* and *time invariance*. In words, linearity means that the output of a system (when the input is the sum of two arbitrary time functions) is equal to the sum of the outputs for the two input time functions separately. In equation form, if we denote two arbitrary inputs by $x_1(t)$ and $x_2(t)$ and the corresponding outputs by $y_1(t)$ and $y_2(t)$, respectively, then when the input is $x_1(t) + x_2(t)$, the output must be $y_1(t) + y_2(t)$ if the system is linear. The property of time invariance means simply that the output response to a certain input applied *now* has the same form as the response to the same input applied *later*. That is, if the input $x(t)$ gives the output $y(t)$, then a delayed version of the input, $x(t - \tau)$, will give a delayed version of the output $y(t - \tau)$.

Let us hasten to say that no real system is linear and time invariant; the linearity property will always break down for a large enough input (this is said to be overloading the input), and the time invariant property will break down because of slow aging of components within a system or perhaps even because of a more catastrophic event, such as severe input overload or physical mishandling. However, it is an extremely useful fiction to model real systems as LTI systems, because such systems are subject to description and analysis by a rich and powerful body of mathematical techniques. The LTI models often apply sufficiently accurately and over a wide enough range of conditions to be extremely useful.

Some examples of LTI systems are: (1) the delay function in which the

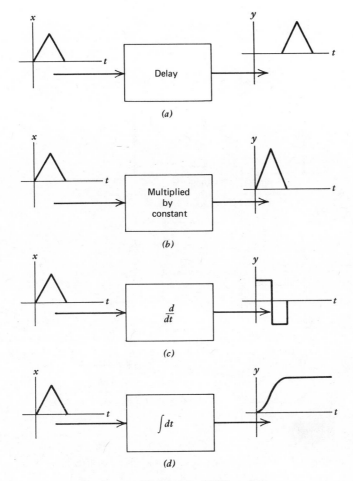

Fig. 1.   2. Examples of LTI systems.

output is simply a delayed version of the input, (2) multiplication by a constant, (3) differentiation with respect to time, and (4) integration with respect to time. These examples are illustrated in Fig. 1.2, where the input is a triangular pulse and the output is as shown. An example of a system that is not linear (though it is time invariant) is a squaring circuit, for which the output at any instant of time is the square of the input at that instant of time.

1.3. The impulse response of an LTI system and the convolution integral   The basic method of attack on the problem of finding the output of an LTI system when the input is some specified function of time is to analyze the input time function into a weighted sum (or superposition) of elementary input time functions. If the response to each

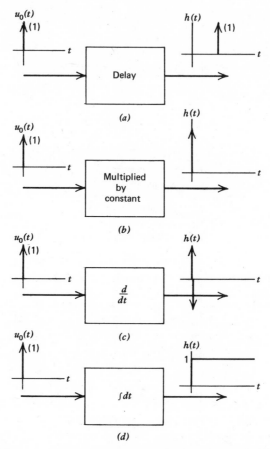

Fig. 1.   3. Impulse responses of systems of Fig. 1.2.

elementary input is known, then, by virtue of the linearity property, the
output of the system is an appropriately weighted sum of the outputs in
response to the elementary inputs. One particularly useful method of
analysis begins by representing the input by a weighted sum of unit
impulses or Dirac delta functions.†ˌThe output of an LTI system when the
input is a unit impulse occurring at time $t = 0$ is called the *impulse
response* of the system. The impulse responses for the four systems of
Fig. 1.2 are sketched in Fig. 1.3. Knowledge of the impulse response of a
system, which is a time function denoted by $h(t)$, is all that one requires

---

† A unit impulse may be thought of, for our purposes, as any very brief pulse that
has unit area beneath it. It may be a rectangular pulse but need not be. In the
diagrams, an impulse will be denoted by a vertical arrow, with the *area* of the impulse
shown next to the arrow in parentheses. The mathematical symbol we'll use for the
unit impulse at time zero is $u_0(t)$. Thus, $u_0(t - a)$ is a unit impulse occurring at $t = a$
(where the argument is zero).

to calculate the output for a known input. This is an exceedingly powerful idea, and we will devote a few pages to showing first, how an arbitrary input can be analyzed or resolved into a weighted sum of unit impulses, and second, how the output can be calculated when the input and the impulse response are known.

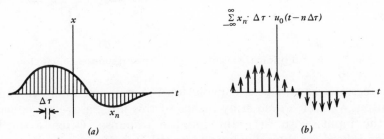

(a)   (b)

Fig. 1.   4. Approximation of a continuous time function by a set of impulses.

Look at Fig. 1.4. Here we have sketched an arbitrary time function $x(t)$. We can approximate this function by a series of rectangular pulses, successively delayed in time; each has width $\Delta\tau$ and height x_n, equal to an ordinate of the original curve at the center of the rectangular pulse. If we make better and better approximations, by using more and more rectangles, each little rectangle of height x_n and width $\Delta\tau$ becomes more and more like an impulse of area $x_n \Delta\tau$. [We should remark that an impulse is best defined mathematically, not as the limit of a narrower and narrower pulse, but instead in terms of its properties within an integral. That is, the value of the integral $\int_{-\infty}^{\infty} x(\tau)u_0(t-\tau)\,d\tau$, where $u_0(t-\tau)$ is a unit impulse located at $t = \tau$, is *defined* to be just $x(t)$. Thus, essentially by definition, an integral containing an impulse is equal to the remainder of the integrand evaluated where the impulse is located.]

By equating the smooth curve $x(t)$ to the limiting sequence of closely spaced impulses, as implied in Fig. 1.4, we are saying that:

$$x(t) = \lim_{\Delta\tau \to 0} \sum_{n=-\infty}^{\infty} x_n \Delta\tau u_0(t - n\,\Delta\tau) = \int_{-\infty}^{\infty} x(\tau)u_0(t-\tau)\,d\tau \quad (1.1)$$

But we see that this equation is just the definition of the impulse function given above.

Suppose we know that the response of a system when the input is a unit impulse at $t = 0$ is the function $h(t)$ of Fig. 1.5. That is, $h(t)$ is the impulse response. The output of the system when the input is the sequence of impulses of Fig. 1.4b must be, because of the LTI property, the sum of a sequence of weighted, delayed pulses, each with the shape of $h(t)$. We can write the limiting sum as an integral:

$$y(t) = \lim_{\Delta\tau \to 0} \sum_{n=-\infty}^{\infty} x_n \Delta\tau h(t - n\,\Delta\tau) = \int_{-\infty}^{\infty} x(\tau)h(t-\tau)\,d\tau \quad (1.2)$$

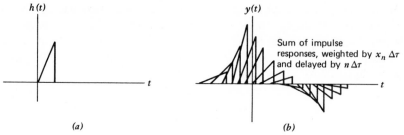

(a)   *(b)*

Fig. 1.   5. Addition of weighted and delayed responses to impulses.

This integral is known as a convolution integral; and when we compute the output $y(t)$ of a system by means of this integral we are convolving $x(t)$, the input, with $h(t)$, the impulse response. Often a shorthand notation is used for the convolution integral; it is written simply as:

$$\int_{-\infty}^{\infty} x(\tau)h(t-\tau)\,d\tau = x(t)*h(t) \tag{1.3}$$

It is important to understand clearly that Eq. 1.2 provides a recipe for calculating the output $y(t)$ of a system when the input $x(t)$ is a known function, and that this recipe can be used for any LTI system provided its impulse response $h(t)$ is a known function.

The process of convolution is commutative, as can be proved by making the substitution $v = t - \tau$, $dv = -d\tau$ in Eq. 1.2, to arrive at:

$$y(t) = \int_{-\infty}^{\infty} x(t-v)h(v)\,dv = h(t)*x(t) \tag{1.4}$$

Therefore, there is a perfect symmetry between the input to a system and the impulse-response of the system in the sense that, as illustrated in Fig. 1.6, if $h(t)$ is the input to a different LTI system with impulse response $x(t)$, the output is still $y(t)$.

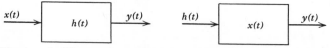

Fig. 1.   6. The symmetry between input and impulse response.

1.4. Evaluation of the convolution integral   Whenever $x(t)$ and $h(t)$ are described by formulas applying over the entire range of time from $-\infty$ to $+\infty$, the calculation of the convolution integral is straightforward; one simply (!) performs the required definite integral. However, it is usually the case that either $x(t)$ or $h(t)$ or both are given by different formulas over different parts of the range. Indeed, for any real system, the impulse response cannot begin before $t = 0$, or else the output would anticipate the input. This is called the realizability property and therefore, for realizable systems, $h(t) = 0$ for $t < 0$. Whenever the formulas for either

$x(t)$ or $h(t)$, or both, are given in pieces, extreme care must be exercised with the limits of the integrals. We illustrate how to handle the limits with an example, which we will solve by a graphical technique.

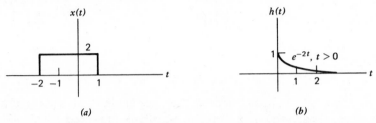

Fig. 1.   7. Example of signals to be convolved.

Example.   Suppose $x(t)$ is the square pulse drawn in Fig. 1.7a and $h(t)$ is the realizable impulse response drawn in Fig. 1.7b. The problem is to compute the output at some known time $t = t_0$.

Our plan of attack is to use Eq. 1.2, but to plot the functions $x(\tau)$ and $h(t_0 - \tau)$ as *functions of* τ for the known time t_0. Then we show the result of multiplying the two functions graphically. The graph will illustrate the limits to be placed on the integral.

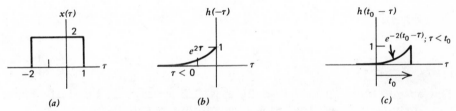

Fig. 1.   8. Steps in the convolution process.

It's easy to plot $x(\tau)$ versus τ; we just redraw $x(t)$ in Fig. 1.8a with the independent variable labeled by τ instead of t. Since it is a little trickier to visualize $h(t_0 - \tau)$ as a function of τ, we arrive at that graph in two steps. First, we draw $h(-\tau)$ in Fig. 1.8b, which is just $h(\tau)$ reflected around the $\tau = 0$ axis. Notice that $h(-\tau)$ is zero for negative values of its argument, $-\tau$, which are positive values of τ. Next we draw $h(t_0 - \tau)$, which is the function $h(-\tau)$ *delayed* (moved to later times τ) by the amount t_0. To check that $h(t_0 - \tau)$ is a delayed rather than an advanced version of $h(-\tau)$, we note that negative values of the argument $t_0 - \tau$ [i.e., where $h(t_0 - \tau) = 0$] correspond to values of τ that are *greater* than t_0.

Finally, in Fig. 1.9a, we plot the product of $x(\tau)$ and $h(t_0 - \tau)$ as a function of τ for the case $t_0 > 1$. The area under this curve is the value of the output $y(t)$ at the time t_0. It is essential to realize that the situation in Fig. 1.9a applies only if we are calculating the output for a time t_0 that is greater than 1. If instead, t_0 lies between -2 and 1, the situation is as

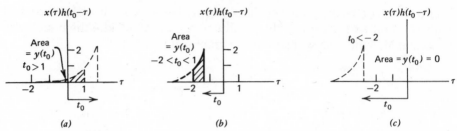

Fig. 1.   9.   Obtaining the output at various times.

drawn in Fig. 1.9b; in other words, the upper limit of the integral is t_0 instead of 1. The remaining case, $t_0 < -2$, is drawn in Fig. 1.9c. Here there is no overlap of the functions $x(\tau)$ and $h(t_0 - \tau)$; the product $x(\tau)h(t_0 - \tau)$ is zero for all τ, and the output is zero. This last result is, of course, expected, since there can be no output of a realizable system before the input begins, which is at $t = -2$ in this example.

Using the graphs of Fig. 1.9 , we compute expressions for the output $y(t)$ for the different ranges of t. (We lose nothing by dropping the subscript on t.)

$$\text{For} \quad t < -2 \qquad y(t) = 0$$

$$\text{For} -2 < t < 1 \qquad y(t) = \int_{-2}^{t} 2e^{-2(t-\tau)}d\tau = 1 - e^{-2(t+2)}$$

$$\text{For} \quad t > 1 \qquad y(t) = \int_{-2}^{1} 2e^{-2(t-\tau)}d\tau = e^{-2t} \cdot (e^2 - e^{-4})$$

$$(1.5)$$

The output function described by Eqs. 1.5 is drawn in Fig. 1.10. There is an exponential rise toward the asymptote, $y = 1$, beginning at $t = -2$ and lasting until $t = 1$, followed by an exponential decay toward $y = 0$.

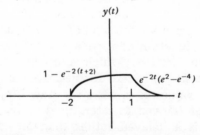

Fig. 1.   10.   Final output signal for the convolution problem illustrated in Fig. 1.7.

Example.   Consider the situation in Fig. 1.11. The input is the same as in the previous example, but this time the impulse response is a square pulse 1 second long and 1 unit high whose onset is delayed by 1 second.

As in the earlier example, the function $h(t - \tau)$ is found by first reflecting $h(\tau)$ about the $\tau = 0$ axis and then shifting it to the right

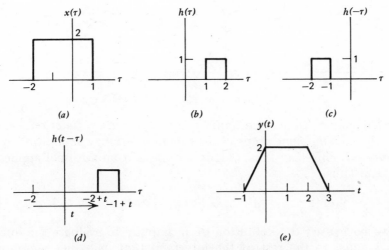

(a)   (b)   (c)

(d)   (e)

Fig. 1.   11. Another example of convolution.

(delaying it) by the amount t. The final output is again the area under the product of $x(\tau)$ and $h(t-\tau)$. As t increases from $-\infty$, there is no overlap until the leading edge of $h(t-\tau)$ at $\tau = -1 + t$ reaches the trailing edge of $x(\tau)$ at $\tau = -2$. This occurs at $-1 + t = -2$ or at $t = -1$. Then there is a linear rise in the output for 1 second while the area of overlap increases, followed by a constant output until the leading edge of $h(t-\tau)$ begins to emerge from under $x(\tau)$ at $-1 + t = 1$, or at $t = 2$. The linear decrease of $y(t)$ then ends at $t = 3$, when the trailing edge of $h(t-\tau)$ at $-2 + t$ reaches the right-hand edge of the input pulse.

To conclude this section we list three important properties of the convolution process.

(a) *The convolution of an impulse with an impulse is another impulse.*   The relation shown in Fig. 1.12 should be clear because the impulse response is the response to a unit impulse at $t = 0$. When the input impulse is multiplied by a constant, A, and delayed in time by t_A, then the output impulse is multiplied by the same constant A and delayed in time by the same amount, t_A.

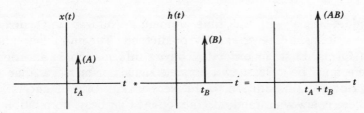

Fig. 1.   12. The convolution of an impulse with an impulse.

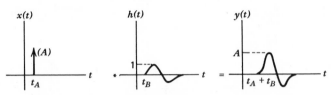

Fig. 1.   13. Convolution with a delayed impulse.

(b) *The convolution of an impulse with an arbitrary function yields an output that has the same shape as the function but may be multiplied by a constant and shifted in time.*   This is evident from a formal application of Eq. 1.2. Thus

$$y(t) = \int_{-\infty}^{\infty} Au_0(\tau - t_A)h(t - \tau)\, d\tau = Ah(t - t_A) \qquad (1.6)$$

Here we have used the definition of an impulse to evaluate the integral, which is equal to the rest of the integrand (not including the impulse) evaluated at $\tau = t_A$. An illustration of this property is shown in Fig. 1.13.

(c) *When two or more LTI systems are cascaded, the order is immaterial.*   We need only prove that the overall impulse response, that is, the output when $x(t)$ is an impulse $u_0(t)$, is the same for the two cases shown in Fig. 1.14. In the first case, when $x(t)$ is a unit impulse, the output of the first system is $h_1(t)$, and the output of the second system is $h_1(t)*h_2(t)$. In the second case the output of the first system is $h_2(t)$, and the output of the second system is $h_2(t)*h_1(t)$. We have already proved that $h_1(t)*h_2(t) = h_2(t)*h_1(t)$; therefore, the two outputs $y(t)$ are identical, as we wanted to show.

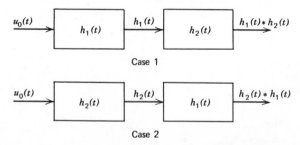

Fig. 1.   14. Cascaded LTI systems: the order is immaterial.

1.5. Random signals in the time domain   Voltages and currents in electronic systems are never exactly predictable. The unpredictability can take two forms. First, in order to convey information in an interval of time, there must be at least two possible signals or messages that can be received. However, in addition to this very useful form of uncertainty of signals, there is always a random component that bears no relation to the information-carrying signal. This random component, called noise, is basically due to fluctuations in the flow of electrons through electronic devices. A signal may be contaminated in various predictable ways too, such as by the addition of 60-Hz interference and its harmonics from power lines, or by radio-frequency interference from man-made sources such as radio stations and automobile ignition systems. Often no distinction is made between random noise and predictable interference; both are simply treated as unwanted signals.

The purpose of many electronic systems is to detect and measure a signal that is, to some extent, predictable when it is accompanied by unpredictable noise. Later we will investigate several such systems and will study some of the mechanisms for the production of noise in electronic systems. But first we must learn how to characterize and model random signals and find how these signals propagate through a system.

Imagine that we have a number of amplifiers, constructed to be as identical as possible. Although the amplifiers are supposed to be identical, nevertheless, if we measure to a sufficient degree of accuracy the output voltages of all the amplifiers at the same time with the same input signal, we obtain a distribution of numbers. Actual experiments reveal that such a distribution is usually bell-shaped or Gaussian.

Fundamentally, the Gaussian distribution results from the fact that a large number of random phenomena are involved in producing the noise in each amplifier. Each phenomenon has its own distribution around some mean value, which may not be Gaussian, but the Central Limit theorem of probability theory tells us that when many random variables are involved, the overall output has a Gaussian distribution.

A Gaussian distribution is characterized by two important parameters. The first is the mean value or, in the case of our ensemble of amplifiers, simply the numerical average of all the output voltages of the various amplifiers at a given time. (Here, averages will be denoted by a bar over a symbol.) The second important parameter provides a measure of the spread of the various values around the average, or the width of the distribution. This parameter, called the variance, is calculated as the mean of the square of all the deviations from the mean. Thus, if the ensemble consisted of just five amplifiers whose output voltages at a certain time were 5, 6, 7, 8, and 9 V, the mean value would be

$$\bar{v} = \frac{5 + 6 + 7 + 8 + 9}{5} = 7 \text{ V}$$

and the variance would be

$$\sigma^2 = \frac{(5-7)^2 + (6-7)^2 + (7-7)^2 + (8-7)^2 + (9-7)^2}{5} = 2.0 \text{ V}^2$$

In this example, the distribution of voltages is not Gaussian but the mean and variance can still give a partial characterization of the distribution.

For a Gaussian distribution, the mean and variance give a nearly complete description.

If the parameters of a distribution remain constant as time goes on, the distribution is said to be stationary. When this is true, the average and the variance can be calculated, provided that the input is recreated anew for each measurement, from a series of measurements of output voltages from just one amplifier taken at various times† instead of the output voltages from all the amplifiers at one time. It should be clear that the distribution of voltages from a real amplifier will not be stationary, because of long-term changes in its properties (called drift) or perhaps owing to changes brought about by a variable ambient temperature. Nevertheless, we will pretend that the time average of the output of one amplifier is indistinguishable from the ensemble average of many amplifiers and will not make a distinction between the two kinds of averages.

An important relation exists between the variance, denoted by $\overline{(\Delta x)^2}$, of a random variable x and the mean value of the variable, $\bar{x}$. From the definition of variance, we have

$$\overline{(\Delta x)^2} = \overline{(x - \bar{x})^2} = \overline{x^2 - 2x\bar{x} + \bar{x}^2} = \overline{x^2} - \bar{x}^2 \qquad (1.7)$$

so that the variance is given by the average of the square minus the square of the average. Notice that, since the variance is always positive (or zero, if all the values of x are the same), the average of the square can never be less than the square of the average.

Even the knowledge that a random process is Gaussian with a certain mean and variance does not provide a complete description of the process, since we still have no idea of the rate at which the variable takes on new values. Look at the two random waveforms in Fig. 1.15. These waveforms might have the same mean and variance but obviously they have a different character. Waveform A varies more rapidly than waveform B. What we need is a quantitative measure of the rate of variation of a waveform. Before dealing with that question, however, we will examine a special case of a random signal.

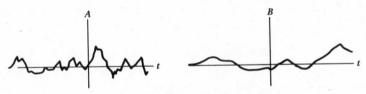

Fig. 1.   15.   Two noise signals.

---

†The measurements must be taken at intervals far enough apart in time that the measurements are not correlated. For example, for a brief time after the output is read to be 5.00 V, we would expect the output to have an unusually high probability of being near 5.00 V.

1.6. Randomly occurring impulses   An important type of random signal is one consisting of impulses all of the same size, but occurring at random times. It turns out that only one parameter is required to describe completely the random nature of this type of signal; that parameter is the average rate of occurrence of the impulses. The mathematical name for a signal consisting of randomly occurring impulses is a "Poisson process." Three examples of Poisson processes in the real world are impacts of raindrops on a roof during a thunderstorm, births of babies in a hospital, and decays of a collection of radioactive nuclei.

Two important questions that can be asked about any Poisson process are: (1) what is the probability that, in a given time interval, exactly n events will be observed (or that n impulses will occur) and (2) what is the variance of the various numbers of events occurring in many intervals of the same duration? The probability of obtaining n events depends, intuitively, upon the number n (it might be more likely that 100 events will be observed during a 1-minute interval than 10 events), and also upon the average number of events that one expects during many repetitions of the time interval (we might expect the probability of obtaining exactly 100 events in 1 minute to be greater if the average number of events per minute is 100 that if it is 10). It can be shown† that the probability of obtaining exactly n events, given that the average number of events expected is a, is given by $P(n)$ where:

$$P(n) = \frac{a^n \cdot e^{-a}}{n!} \tag{1.8}$$

Armed with this relation, we can easily obtain an expression for the variance of the number of events occurring in many intervals of the standard duration. In the notation of Eq. 1.7 the random variable x is now a discrete variable (the number n of events occurring in successive intervals), the average value $\bar{x}$ is given to be a, and we are trying to compute $\overline{(\Delta x)^2}$. Clearly, to do this from Eq. 1.7, we must compute $\overline{x^2}$; that is, the average value of the square of the number of events in an interval. To find this average we multiply each of the infinite number of possible values of n^2 by the probability of obtaining that value (from Eq. 1.8) and sum over all values of n. Thus

$$\overline{x^2} = \overline{n^2} = \sum_{n=0}^{\infty} n^2 P(n) = \sum_{0}^{\infty} \frac{n^2 a^n e^{-a}}{n!} = e^{-a} \sum_{0}^{\infty} \frac{n^2 a^n}{n!} \tag{1.9}$$

The summation in this equation can be evaluated by straightforward

†D. Halliday, *Introductory Nuclear Physics*, Appendix 10, Wiley, New York, 1955; also H. D. Young, *Statistical Treatment of Experimental Data*, pp. 57—59, McGraw-Hill, New York, 1962.

means† to obtain $(a^2 + a) \cdot e^a$ so that $\overline{x^2} = a^2 + a$ and our desired result, the variance, is given by:

$$\overline{(\Delta x)^2} = (a^2 + a) - a^2 = a \tag{1.10}$$

This important result, that the variance of a Poisson process with mean value a is also a, may already be familiar from the statistics of counting a radioactive source. There, a well-known result is that the standard deviation of the number of counts obtained in many counting intervals (the standard deviation is defined as the square root of the variance) is given by $\sqrt{a}$, which of course says the same thing as Eq. 1.10.

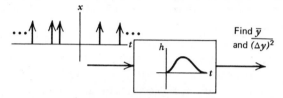

Fig. 1.   16.   A Poisson sequence as the input to an LTI system.

Now an important question to ask is the following. Suppose that the Poisson sequence of impulses is the input to an LTI system with an impulse response $h(t)$. What will be the average value of the output of the system and what will be the variance of the output? The problem is illustrated in Fig. 1.16. The average $y(t)$ is quite easy to compute. If we consider a long time interval T, clearly the average number of input pulses we expect in that time interval is aT. To find the average value of the output, we can calculate the average area under the y versus t curve over the interval T and divide by T. But each input impulse contributes an amount $\int_0^\infty h(t)dt$ to this area, so the general result is

$$\overline{y(t)} = a \cdot \int_0^\infty h(t)\, dt \tag{1.11}$$

It is not so easy to determine the variance of the output. We might be tempted to try first to find the average of the square of the output and then use Eq. 1.7, but the difficulty is that the output pulses can overlap. (This overlap caused no problem when we computed the average because the integral of a sum is equal to the sum of the integrals.) If there were no overlap of the output pulses, the average of the square of the output would be simply $a \cdot \int_0^\infty [h(t)]^2 dt$. To discover what might be involved in

† $\displaystyle\sum_{n=0}^{\infty} \frac{n^2 a^n}{n!} = \sum_0^{\infty} \frac{[n(n-1) + n]a^n}{n!} = \sum_0^{\infty} \frac{n(n-1)a^n}{n!} + \sum_0^{\infty} \frac{na^n}{n!} = \sum_2^{\infty} \frac{a^n}{(n-2)!} + \sum_1^{\infty} \frac{a^n}{(n-1)!}$

$\displaystyle = a^2 \sum_2^{\infty} \frac{a^{n-2}}{(n-2)!} + a \sum_1^{\infty} \frac{a^{n-1}}{(n-1)!} = a^2 \sum_{m=0}^{\infty} \frac{a^m}{m!} + a \sum_{m=0}^{\infty} \frac{a^m}{m!} = a^2 e^a + a e^a$

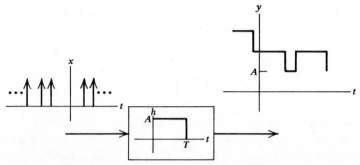

Fig. 1.   17. Another LTI system with Poisson input sequence.

doing it right, look at the special case of Fig. 1.17, where the impulse response is a rectangular pulse of height A and duration T. If we measure the output at any time t, we can understand immediately that the contribution to the output at that time (and, therefore the contribution to the output squared) can come only from input pulses that arrived during the preceding T seconds. If two pulses arrived during that time, the square of the output at time t would be $(2A)^2$. If three pulses arrived, the square of the output would be $(3A)^2$ and if n pulses arrived the square of the output would be $(nA)^2$. To find the average of the square of the output, we simply multiply each possible value by the probability of its occurrence, from the Poisson probability formula (Eq. 1.8), and add all these products. Here, since the average number of pulses expected during T seconds is aT, where a is the average number per second, we find

$$\overline{[y(t)]^2} = \sum_0^\infty (nA)^2 \cdot \frac{(aT)^n \cdot e^{-aT}}{n!} = (a^2 T^2 + aT) \cdot A^2 \qquad (1.12)$$

Note that the sum is the same as in Eq. 1.9 except that a is now replaced by aT, and the whole thing is multiplied by A^2.

One of the problems at the end of the chapter involves a calculation of the average of the square of the output of a system driven by randomly occurring impulses when the impulse response is not a simple rectangular pulse. It will be seen by working this problem that the general case of a continuously varying impulse response represents a considerable mathematical challenge. Rather than continue our attack along these lines we will now introduce a new description of signals in the time domain. This new concept will prove to be so powerful that we will be able to solve the general problem of calculating the variance of an output given the impulse response and the variance of the input; the special case of an input that is a Poisson process will then be very easy to do.

1.7. The autocorrelation function   We now introduce another of what will eventually be a considerable list of transformations of time signals that involve a definite integral. This transformation is called the autocorrelation function of $x(t)$ and is defined by:

$$R_x(\tau) = \overline{x(t) \cdot x(t + \tau)} = \lim_{T \to \infty} \frac{1}{2T} \int_{-T}^{T} x(t)\,x(t + \tau)\,dt \qquad (1.13)$$

Notice that the autocorrelation function $R_x(\tau)$ is a function of a parameter τ. The autocorrelation function for a particular value of this parameter is the average value of the product of the signal $x(t)$ with the same signal shifted *backward* in time by the amount τ. The definition of the autocorrelation function bears more than a superficial relation to the convolution integral. In fact, the autocorrelation function is, except for the normalizing factor $1/2T$, the convolution of the function $x(t)$ with itself turned around in time.†

Although the autocorrelation function is an average, as shown explicitly by its definition, it conveys a different kind of information than do the average and variance of the waveform. Whereas the latter tell something about the distribution of values that the signal can have, the autocorrelation function tells something about the rate at which the function changes. This last point can perhaps be made clearer by looking at the autocorrelation function of a sinusoidal wave of period T_1. (Incidentally, it makes little sense to compute the autocorrelation function for signals of finite duration; the factor $1/2T$ in front of the integral causes the autocorrelation function to be vanishingly small for every value of τ. However, for waves of essentially infinite duration, whether periodic or nonperiodic, the autocorrelation function can be a useful description.)

For the case of a periodic wave it is sufficient to average over just one period of the wave in calculating $R_x(\tau)$. We leave it to one of the problems to show that the average value of a sinusoid multiplied by a time-shifted sinusoid of the same period is given by $(A^2/2) \cdot \cos(2\pi\tau/T_1)$. The original sinusoid and its autocorrelation function are plotted in Fig. 1.18. Two

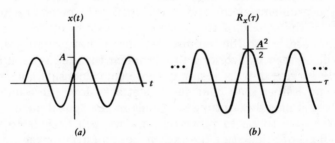

Fig. 1. 18. A sinusoid and its autocorrelation function.

†To see this, write the convolution of $x(t)$ with $x(-t)$ as $\lim_{T \to \infty} \int_{-T}^{T} x(-u) \cdot x(t-u)\,du$ and then make the formal substitutions, $t = \tau$, $u = -v$, to arrive at $2T$ times the autocorrelation function.

features of this example are noteworthy. First, the autocorrelation function is an even function of τ. This is a general property of autocorrelation functions; it results from the fact that it makes no difference to the average whether we shift one of the $x(t)$ functions forward or backward in time. Another important property of the autocorrelation function of a periodic signal is that it is periodic with the same period as the signal itself. The phase information is lost, however, in the sense that the original sinusoidal wave in our example could have any phase and still give precisely the same autocorrelation function.

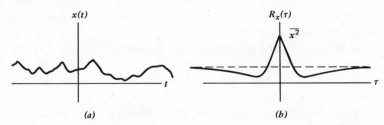

Fig. 1.   19. Typical autocorrelation function for a random signal.

Although autocorrelation functions can be calculated for periodic functions, their real value consists in their ability to characterize random signals. Figure 1.19 suggests what we might expect for the autocorrelation function of a continuous random signal. The maximum value of the autocorrelation function occurs at $\tau = 0$, because it is only when $\tau = 0$ that the integrand in the averaging integral is necessarily positive for all values of t. The value of the autocorrelation function at $\tau = 0$ is the average of the square of the signal as can be seen by setting $\tau = 0$ in the defining equation. As τ departs from zero the autocorrelation function falls off and, for sufficiently large values of τ, approaches a constant value. The reason why $R_x(\tau)$ for random signals approaches a constant for large τ is that values of x that are separated by long intervals of time are completely uncorrelated. That is, when the signal is random, the value of x at one time has no effect on the distribution of possible values of x at another time if those two times are separated sufficiently.

As an example of the calculation of an autocorrelation function for a random signal, we'll do it for a Poisson process. The calculation may be easier to understand if we remember that the autocorrelation function is $1/2T$ times the convolution of the signal with itself turned around in time. We simply have to convolve a lot of impulses, which we already know how to do. Figure 1.20 shows the two sequences of impulses to be convolved. We begin by fixing the limits of integration in the defining integral for the autocorrelation function; at the end we will let T go to infinity. Thus there are approximately $2aT$ impulses in each sequence. Each impulse in one sequence, say $x(t)$, must be convolved with every impulse in the other

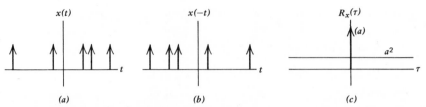

Fig. 1.   20. Autocorrelation function for a Poisson sequence.

sequence. Thus each impulse in $x(t)$ gives rise to $2aT$ impulses spaced randomly over an interval of duration $2T$ except that there is guaranteed to be an impulse at $\tau = 0$. This is true because $x(-t)$ is a mirror image of $x(t)$. Altogether there will be $2aT$ impulses at $\tau = 0$, and the rest of the impulses will be spaced randomly at the average rate of $2a^2 T$ per unit time.

Now to find $R_x(\tau)$, we must divide by $2T$ and take the limit as T becomes very large. When we do this the impulse at the origin remains constant with area a, but since the other impulses are all multiplied by $1/2T$, they all become very small. On the other hand, the number of these other pulses per unit time grows larger in proportion to T. In the limit, this sequence of very many, very small impulses is indistinguishable from a constant function, a^2. Notice that the behavior of $R_x(\tau)$ as a function of τ in this example is an extreme case of what we expect for a continuously random signal. This autocorrelation function is indeed largest for $\tau = 0$ and falls off (very quickly!) to a constant value as τ departs from zero.

1.8. Autocorrelation function of the output of an LTI system†   Finally, we are prepared to ask, and answer, the key question. How can we express the autocorrelation function of the output of an LTI system in terms of the autocorrelation function of the input signal and the impulse response of the system?

We proceed in a formal way, by writing down the convolution integrals for $y(t)$ and $y(t + \tau)$:

$$y(t) = \int_{-\infty}^{\infty} h(u)x(t - u)\, du$$

$$y(t + \tau) = \int_{-\infty}^{\infty} h(v)x(t + \tau - v)\, dv$$

Next, we multiply these together and average over t to obtain $R_y(\tau)$. Thus

$$R_y(\tau) = \overline{\int_{-\infty}^{\infty} h(u) \cdot x(t - u)\, du \cdot \int_{-\infty}^{\infty} h(v)x(t + \tau - v)\, dv}$$

†Sections 8 and 9 of this chapter should be omitted in a first reading.

Since the average involves just the factors that depend on t, we can rearrange the order of integration to give:

$$R_y(\tau) = \iint_{-\infty}^{\infty} h(u) \cdot h(v) \cdot \overline{x(t-u)x(t+\tau-v)}\, du\, dv$$

Now we make the substitution $t' = t - u$. Hence

$$R_y(\tau) = \iint_{-\infty}^{\infty} h(u) \cdot h(v) \cdot \overline{x(t') \cdot x(t'+u-v+\tau)}\, du\, dv$$

The average within the integral is now taken over the variable t', and we can recognize it as the autocorrelation function of the input signal, except that the shift is by $u - v + \tau$ instead of by τ. Thus finally, we have the result in the form we are looking for.

$$R_y(\tau) = \iint_{-\infty}^{\infty} h(u) \cdot h(v) \cdot R_x(u-v+\tau)\, du \cdot dv \qquad (1.14)$$

The autocorrelation function of the output of an LTI system is a *double convolution* of the autocorrelation function of the input with the impulse response of the system.

Instead of attempting to develop an intuitive "feel" for the double convolution integral, we proceed directly to an example, in which the input is a Poisson process. From Fig. 1.20 we see that the autocorrelation function of the input pulse train is given by:

$$R_x(\tau) = a \cdot u_0(\tau) + a^2 \qquad (1.15)$$

Substituting this expression into Eq. 1.14, we have

$$R_y(\tau) = \iint_{-\infty}^{\infty} h(u) \cdot h(v) \cdot [au_0(u-v+\tau) + a^2]\, du\, dv$$

or

$$R_y(\tau) = a \cdot \iint_{-\infty}^{\infty} h(u) \cdot h(v) \cdot u_0(u-v+\tau)\, du\, dv$$

$$+ a^2 \cdot \int_{-\infty}^{\infty} h(u)\, du \cdot \int_{-\infty}^{\infty} h(v)\, dv$$

Next we integrate over v in the double integral, by simply evaluating the integrand where the argument of the impulse is zero — at $v = u + \tau$. The final result for the autocorrelation function of the output when the input is a Poisson process is

$$R_y(\tau) = a \cdot \int_{-\infty}^{\infty} h(u) \cdot h(u+\tau)\, du + \left[a \cdot \int_{-\infty}^{\infty} h(u)\, du \right]^2 \qquad (1.16)$$

We can interpret the first term in this result as the average rate of input pulses times the impulse response convolved with itself turned around in time and the second term as the square of the average value of the output. (Recall the result in Eq. 1.11.)

Equation 1.16 can be applied immediately to the problem of finding the average of the square of the output and the variance of the output by setting $\tau = 0$. We obtain

$$R_y(0) = \overline{[y(t)]^2} = a \cdot \int_{-\infty}^{\infty} h^2(u)\, du + \overline{[y(t)]}^2 \qquad (1.17)$$

for the average of the square and, with the aid of Eq. 1.7,

$$\overline{(\Delta y)^2} = \overline{y^2} - (\bar{y})^2 = a \cdot \int_{-\infty}^{\infty} h^2(u)\, du \qquad (1.18)$$

for the variance.

1.9. The ratemeter   The ratemeter, as its name implies, is an instrument that measures the rate at which events occur. Ratemeters are typically found in radiation survey meters where one is interested in obtaining a quick estimate of the radiation level in an area. All that is required for a ratemeter is an LTI system whose impulse response has nonzero area under it. The input is a sequence of pulses, representing the events to be counted. Anyone who has used a ratemeter in connection with a radioactive source knows that the output of the meter fluctuates. Having obtained formulas for the mean and variance of the output of an LTI system whose input is a Poisson process, we are in a position to predict the average and variance of the output of a ratemeter if we know its impulse response.

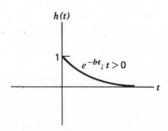

Fig. 1.   21. Impulse response of a ratemeter.

For example, suppose the impulse response of a ratemeter is the function shown in Fig. 1.21. (For reasons that we develop in chapter 3, a system with this impulse response is said to be a first-order, low-pass filter.) The integral of the impulse response and the integral of its square can be evaluated in a straightforward way to give:

$$\int_0^{\infty} h(t)\, dt = \int_0^{\infty} e^{-bt}\, dt = \frac{1}{b} \qquad (1.19)$$

$$\int_0^\infty h^2(t)\, dt = \left| \int_0^\infty e^{-2bt} dt \right| = \frac{1}{2b} \qquad (1.20)$$

We compute the average and the variance of the output of the ratemeter directly from Eqs. 1.11 and 1.18.

$$\overline{y(t)} = \frac{a}{b} \qquad (1.21)$$

$$\overline{(\Delta y)^2} = \frac{a}{2b} \qquad (1.22)$$

A word of caution must be given concerning the use of a ratemeter. In order that Eqs. 1.11 and 1.18 apply to the meter output, the sequence of input pulses must have been entering the ratemeter for an interval of time at least as long as the duration of the impulse response, because the output at a particular time depends on each earlier input pulse in proportion to the value of the impulse response at a time equal to the elapsed time since the pulse occurred. (Indeed, this is just an alternate way of looking at the convolution integral.) Now in the present example, because of the long tail of the impulse response, we would have to leave the ratemeter on for an infinitely long time for it to read correctly. In practice, however, one simply allows a time equal to several "time-constants" of the impulse response to elapse before taking a reading.

As a figure of merit for evaluating the performance of different ratemeters, we might examine the ratio of the average output to the square root of the variance. This can be viewed, in a sense, as the signal-to-noise ratio for the ratemeter, the signal being the average output, which is proportional to the input rate a, and the noise being the fluctuations about this average value. We take the square root of the variance, which we earlier noted is called the standard deviation, so that we are comparing quantities with the same units. The figure of merit, or signal-to-noise ratio for the ratemeter of our example, is therefore $\sqrt{2a/b}$. Note that the signal-to-noise ratio improves in proportion to the square root of the input pulse rate.

Because the ratemeter we have just examined gives greater weight to more recent input pulses in comparison to those in the more distant past, we might expect that it does not give the optimal signal-to-noise ratio. In fact, the optimal signal-to-noise ratio for a ratemeter will occur for one that gives equal weighting to all input pulses that have occurred since it was turned on. If we are willing to wait a time T before reading the output, then the impulse response of such a ratemeter is, effectively, a constant of duration T, because only in that case will all pulses that arrive during the interval 0 to T be given equal weight in the output. In effect, the ratemeter has become an integrator with integration time T. For the integrator, Eq. 1.11 gives, for the average of the output,

$$\overline{y(t)} = aT \qquad (1.23)$$

and Eq. 1.18 gives, for the variance of the output,

$$\overline{(\Delta y)^2} = aT \tag{1.24}$$

The signal-to-noise ratio is therefore $\sqrt{aT}$. Comparing this to the earlier example where the impulse response was the exponential pulse, and realizing that in order for that ratemeter to give a reliable reading in time T we must have $T \gg 1/b$, we see that the integrating type of ratemeter indeed has a better signal-to-noise ratio. Note also that the integrating ratemeter gives exactly the same performance as a counter, for which the corresponding signal-to-noise ratio is $\bar{N}/\sqrt{\bar{N}} = \sqrt{\bar{N}}$, because aT is just the average number $\bar{N}$ of input pulses expected to occur during time T.

1.10. The phase-sensitive detector   As a final example of a system that can be understood by time-domain concepts, we consider the phase-sensitive detector (PSD), sometimes called the lock-in amplifier. The PSD is used to detect and measure the amplitude and phase of a periodic signal, for example, in an ac bridge measurement. Often a significant component of random noise obscures the desired signal. A basic PSD is shown in Fig. 1.22. Here the low-level signal containing the sinusoidal wave to be measured is amplified and then sent to a switching network where it is inverted during alternate half cycles of a reference signal. Following the switching network, it is smoothed (by convolution with a long-duration pulse) and the final output is read on a dc meter.

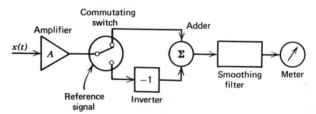

Fig. 1.   22. A phase-sensitive detector.

The operation of the PSD when the input is a pure sinusoid can be understood with the aid of the diagrams in Fig. 1.23. In part a of the figure, the sinusoidal wave is phased so that the positive half cycles coincide exactly with the times the switch is in the up or positive position, and the negative half cycles (shown dotted) coincide exactly with the times the switch is in the down or negative position. The solid wave is the output of the adder; since it clearly has an average value that is positive, the output of the smoothing filter will be positive. In Fig. 1.23b the sinusoidal wave is phased so that it spends equal times being positive and negative during each switching interval; the average output of the smoothing filter will then be zero. In Fig. 1.23c the wave is phased so that

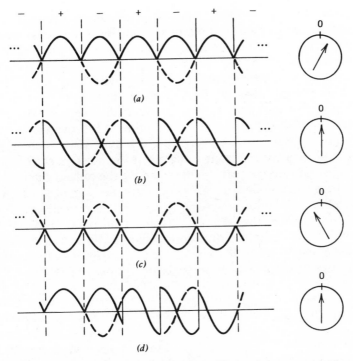

Fig. 1.   23. Response of phase-sensitive detector to sinusoids.

the output of the adder is always negative; the smoothing filter output is therefore negative. Finally in Fig. 1.23d we suggest what happens when the period of an input sinusoid is not exactly equal to that of the switch. The output of the adder averaged over a half cycle of the switch swings back and forth between positive and negative values and, provided the smoothing filter averages over enough cycles, its output will remain nearly zero.

The earliest phase-sensitive detectors used motor driven switches to reverse the polarity of the signal, but modern versions use electronic switching. The signal that controls the switch is called the reference signal; it is usually derived from the original sinusoidal wave that drives the system whose output is the input to the PSD. That system could be an ac bridge, as mentioned earlier; a light-measuring system such as an interferometer or spectrometer, in which case the input light signal would be modulated, usually by a mechanical shutter; or any other measurement system in which the input quantity to be measured can be alternately applied to and removed from the input to the system.

The ability of a PSD to reject signals such as random noise that do not have the same period as the switch depends directly on the averaging time of the smoothing filter. The longer the averaging time the greater is the rejection ability. One modern way of obtaining an accurate numerical estimate of the magnitude of the inphase signal when noise is present is to use an integrator as the smoothing filter. The integration time may be minutes or even hours. At the end of the integration time, the integrator is usually read to a high degree of accuracy with a digital meter. It is important to realize, however, that no matter how good the system, there are random errors inherent in the measurement. Thus in the integrator system we have just described, if the experiment is repeated several times, the integrator output will be slightly different after each experiment because of the slightly different integrals of the noise in each experiment. Whether these differences can be detected by the output meter is another question. Clearly it makes no sense to integrate for hours if, when we integrate for a few minutes, several times, the output meter is unable to discriminate between the slightly different outputs. In that case one experiment of a few minutes duration would clearly suffice. Of course the error in the experiment in this example is not zero; we can say only that it is less than the smallest scale division on the output meter.

Problems for Chapter 1

1.1. The input to an LTI system is the $x(t)$ graphed in Fig. P1.1. Find the output if the impulse response $h(t)$ is identical to $x(t)$.

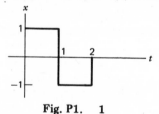

Fig. P1.   1

1.2. The input of problem 1 is fed to an LTI system with impulse response shown in Fig. P1.2. Find and plot the output signal.

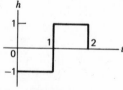

Fig. P1.   2

1.3 The waveforms $x_1(t)$, $x_2(t)$, and $x_3(t)$ shown in Fig. P1.3 all contain the same "energy," $= \int_{-\infty}^{\infty} |x(t)|^2 \, dt$, as the $x(t)$ of problem 1. Find the output of the LTI system of problem 2 for each of these inputs and compare with the output computed in problem 2.

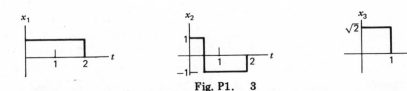

Fig. P1. 3

How can this LTI system be used to "recognize" the presence of the signal $x(t)$ in the presence of other signals with the same energy? (An LTI system whose impulse response is an input signal turned around in time is called the "matched filter" for that input signal.)

1.4. The impulse response of a system is a unit sine function of period T beginning at $t = 0$. Find and graph the response to a square-wave input pulse of duration T in Fig. P1.4.

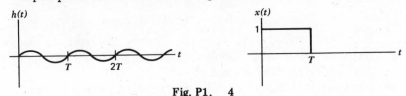

Fig. P1. 4

1.5. The impulse response of a system is a square pulse lasting from $t = 0$ to $t = T$. Find the output waveform if the input is a sine wave of period T, starting at $t = 0$.

1.6. Show by a systems argument that if the impulse response of an LTI system contains zero net area, then the average value of the output is zero, no matter what the input.

1.7. Model a pair of impulses as follows. One is rectangular with duration Δ and area A, the other is rectangular with duration Δ and area B. Now convolve these two "impulses" and sketch and dimension the result. Does your answer agree with the statement that the convolution of an impulse with area A and an impulse with area B is an impulse with area $A \times B$?

1.8. The idea of breaking an input waveform into a sum of more "elementary" waveforms can be useful when a system is LTI, even though the elementary waveforms are *not* impulses. For example, consider a very lightly damped oscillatory system such as a pendulum, or, more practically, a construction crane.
   (a) If the "input" is the horizontal position of the top of the boom and the "output" is the horizontal position of the load, sketch the response of the system to a sudden "step" change in the input.
   (b) A typical problem for a crane operator is to move the load to a new position quickly, but without the load swinging to and fro

after it is moved. Show how the required load movement can be achieved by an appropriate "input" motion of the boom. *Hint:* The time required is half the period of oscillation of the load.

1.9. Find the output of a first-order low-pass filter with impulse response $0 (t < 0)$, $e^{-b\,t} (t \geqslant 0)$ if the input is a single square pulse of height 1 lasting from $t = 0$ to $t = a$.

1.10. The rate of decay of a collection of radioactive nuclei as a function of time may be considered to be the "output" of a system whose "input" is the rate of production of those nuclei.

(a) If the decay constant for a certain species is λ (so that $dN/dt = -\lambda N$), show that the impulse response of this "system" is

$$\lambda e^{-\lambda t}; \qquad t > 0$$
$$0; \qquad t \leqslant 0$$

In other words, this expression gives the rate of decay in response to a unit impulse of production that produces one radioactive nucleus at time 0.

(b) Next suppose that the production rate is a rectangular pulse:

$$R = R_0; \qquad 0 < t < T_0$$
$$0; \qquad t \leqslant 0 \qquad t \geqslant T_0$$

instead of an impulse. Use convolution to sketch and dimension the decay rate of this nuclear species.

(c) Now consider a situation in which the result of each decay is a radioactive daughter nucleus with decay constant λ', so that the rate of decay of the first species becomes the rate of production of the second. Discuss how convolution can be used to find the decay rate of the second species when the *production rate* of the first species is known.

(d) If the rate of production of species 1 is an impulse that results in N_0 nuclei of type 1 at $t = 0$, find a closed form expression for the decay rate of the second species.

1.11. Applications of convolution are not limited to situations where the independent variable is time. As an example, consider a device designed to measure a physical variable such as mass (Fig. P1.11). The measuring device always introduces an uncertainty so that when the input is a definite precise value m_0, as represented by the impulsive probability distribution $P_i(m)$, then different output readings are obtained that have a probability distribution such as $P_0(m)$. Describe how convolution might be used to compute an expected distribution of readings when a set of masses that is distributed in some way around m_0 is measured.

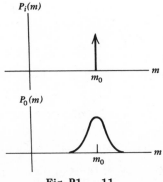

Fig. P1.   11

1.12. Prove that the probability for getting different numbers n of events during a time interval, when summed over all possible values of n, is 1, by summing the series for $P(n)$. If you don't recognize the sum:

$$\sum_{n=0}^{\infty} \frac{a^n}{n!}$$

notice that the derivative of the sum with respect to a is equal, term by term, to the original sum. What function of a do you know that is equal to its derivative?

1.13. The input to a system is a sequence of random unit impulses at average rate a. The impulse response is as shown (Fig. P1.13). Calculate the average, average of the square, and variance of the output.

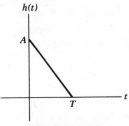

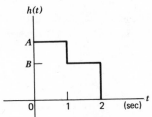

Fig. P1.   13   Fig. P1.   14

1.14. A random sequence of unit impulses, at rate a, is the input to a system with impulse response as shown in Fig. P1.14. At a particular time t, if there have been n_1 input impulses during the preceding second and n_2 input impulses during the second before that, the output will be $n_1 A + n_2 B$, and the square of the output will be $n_1^2 A^2 + n_2^2 B^2 + 2n_1 n_2 AB$. Show, by performing a double sum that is properly weighted by Poisson probabilities, over all possible values of n_1 and n_2, that the average value of $y(t)$ is $a \cdot (A + B)$, and the average value of $[y(t)]^2$ is $a(A^2 + B^2) + a^2(A + B)^2$.

1.15. Find the autocorrelation function of $x(t) = A \sin(2\pi t/T_1)$. *Hint:* expand $\sin[2\pi(t+\tau)/T_1]$ as the sine of the sum of two angles and use the fact that the average value of $\sin^2(2\pi t/T)$ over a period is ½. The average value of $\sin(2\pi t/T_1)\cos(2\pi t/T_1)$ should be intuitively obvious without doing the integral. (Try drawing a picture!)

1.16. Square pulses of height 1 and duration a occurring at random times and at average rate $1/T$ are sent into a system with impulse response $e^{-bt}(t \geqslant 0), 0(t < 0)$. Find the mean of the square of the output by evaluating the autocorrelation function of the output.

1.17. (a) Sketch and dimension the impulse response of a system whose input is the electric field $\vec{E}$ between the vertical deflecting plates of an oscilloscope and whose output is the beam deflection d at the screen. Assume that the electrons have been accelerated through a potential difference of 3000 V before reaching the plates, that the plates have a length of 2 cm, and that the distance from the center of the plates to the screen is 20 cm. (*Hint:* Find the step response to an electric field of 1 V/m and then differentiate to find the impulse response. Note that when the field is suddenly changed, electrons that are already part way through the plate region are not deflected as much as electrons that are just starting to pass between the plates.)

(b) For the oscilloscope deflection system of part (a), at what frequency applied sinusoidal electric field does the deflection first become zero?

1.18. Sketch and dimension the impulse response of a system whose input is the horizontal force F applied to a 1-m simple pendulum of mass 0.1 kg and whose output is the angular deflection θ of the pendulum. (Assume small angles of oscillation. Why?) Include damping, which reduces the amplitude by ½ every minute.

1.19. (a) Sketch and dimension the impulse response of a system whose input is the force exerted on a mass m initially at rest and whose output is the *velocity* of the mass.

(b) Do the same for a system with the same input but with an output that is the position of the mass.

(c) If two force impulses were given the mass at times separated by Δt, could the resulting velocity be obtained by adding the responses of part (a) to the individual impulses? Could the resulting displacement be obtained by adding the responses of part (b) to the individual impulses?

(d) Find, by convolution, the position versus time of a mass of 1 kg that receives a positive impulsive kick of 20 N-sec at $t = 0$, and is then acted upon by a constant negative force of 10 N. (Note

that this approximates the situation of a projectile shot upward
in the gravitational field of the earth.)

(e) Could convolution be used to compute the position of an object
if the force depended on position or velocity rather than time?
Discuss.

REFERENCES

Carlson, A. B., *Communication Systems: An Introduction to Signals and Noise in Electrical Communication*, Chapters 2 and 3, McGraw-Hill, New York, 1968.

Craig, E. J., *Laplace and Fourier Transforms for Electrical Engineers*, Chapter 8, Holt, Rinehart and Winston, New York, 1964.

Mason, S. J., and H. J. Zimmerman, *Electronic Circuits, Signals and Systems*, Chapters 6 and 7, Wiley, New York, 1960.

Schwartz, M., *Information Transmission, Modulation, and Noise*, Chapters 2 and 7, McGraw-Hill, New York, 1959.

CHAPTER 2

Laplace Transforms and LTI Systems

2.1. Introduction The theme of Chapter 1 was that we can compute the output of an LTI system by convolving the input time function with the impulse response of the system. However, as we saw in the examples of section 1.4, convolution can be a fairly involved operation when one or both of the functions are defined by different formulas over different ranges of time. There is an alternative approach to the problem of finding the output time function; this approach involves the concept of a transform (either Laplace or Fourier transform) and is called, generally, the frequency domain approach. This chapter and the next deal with these transform methods of analyzing linear systems.

2.2. The eigenfunction[†] concept We begin by noting a remarkable fact. If we use as the input to our LTI system the function $x(t) = e^{st}$, where s is a real or complex constant, we can calculate the output time function to be:

$$y(t) = \int_{-\infty}^{\infty} h(\tau) x(t - \tau)\, d\tau = \int_{-\infty}^{\infty} h(\tau) e^{s(t - \tau)} d\tau = H(s) \cdot e^{st} \quad (2.1)$$

where

$$H(s) = \int_{-\infty}^{\infty} h(\tau) e^{-s\tau} d\tau$$

provided that the integral exists. In other words, when the input of an LTI system is e^{st}, the output is the *same time function*, except that it is multiplied by the constant $H(s)$, which is determined by integrating $e^{-s\tau}$ times the impulse response from $-\infty$ to $+\infty$. The integral itself is a function of s, the complex constant in the exponent of the input signal; for the impulse responses encountered in electronics, it is always possible to find a range of complex values of s for which the integral exists.

[†] Eigenfunctions occur in several places in theoretical physics. The general idea is that if some function x is operated on (i.e., modified) by some mathematical operation O or by a physical system, then if the result is a constant λ times the original function, that function is said to be an eigenfunction of the operator O. That is, $Ox = \lambda x$ if x is an eigenfunction. The operator is analogous to our *system*, which "operates" on the input time function to give the output time function.

31

An idea of the importance of the eigenfunction concept can be obtained by looking at a sinusoidal input function, such as $x(t) = \cos \omega t$. We can write $\cos \omega t$ as a sum of two e^{st} functions. That is, $\cos \omega t = \frac{1}{2}e^{j\omega t} + \frac{1}{2}e^{-j\omega t}$, so that the two s values are $s_1 = j\omega$ and $s_2 = -j\omega$. To obtain the output $y(t)$, we would just perform the integrals to obtain $H(j\omega)$ and $H(-j\omega)$ and then add the responses to the two exponential inputs.

$$y(t) = \frac{1}{2}H(j\omega)e^{j\omega t} + \frac{1}{2}H(-j\omega)e^{-j\omega t} \tag{2.2}$$

It should be clear that this method could be used to find the response when the input is composed of any finite number of e^{st} functions.

2.3. The Laplace transform   The function $H(s)$ defined by:

$$H(s) = \int_{-\infty}^{\infty} h(\tau)e^{-s\tau}d\tau \tag{2.3}$$

is called the Laplace transform† of the time function $h(\tau)$. A useful notation for the Laplace transform of a time function such as $h(t)$ is $\mathscr{L}[h(t)]$. Although $\mathscr{L}[h(t)]$ is a function of the variable s, the *complex frequency*, it may not be defined for all values of s. For example, consider the exponential pulse function:

$$h(t) = u_{-1}(t) \cdot e^{-at} \tag{2.4}$$

where $u_{-1}(t)$ is the unit step function:

$$u_{-1}(t) = 1 \qquad t \geqslant 0$$
$$= 0 \qquad t < 0$$

and where a is a positive real number. The Laplace transform of the exponential pulse may be found, by direct integration, to be

$$\mathscr{L}[h(t)] = \int_{0}^{\infty} e^{-(a+s)\tau}d\tau = \frac{1}{s+a} \tag{2.5}$$

---

†Equation 2.1 defines, strictly speaking, the bilateral Laplace transform of $x(t)$. Most authors define the Laplace transform with the domain of integration running from 0 to ∞ and, in addition, restrict the functions to be transformed to be zero for $t < 0$. For this class of functions, the definitions are clearly equivalent. The problem is that if the input has been "doing something" prior to $t = 0$ and if the system is not memoryless, then the output at times after $t = 0$ may depend on the earlier history of $x(t)$. A not-too-obvious case of this occurs in the solution of linear differential equations such as $a(d^2y/dt^2) + b(dy/dt) + cy = x(t)$ by means of Laplace transforms. Often there are "initial conditions" to be met; thus, in the second-order equation, the values of y or dy/dt at $t = 0$ may not be zero. These two initial conditions can be viewed as the legacy of the early ($t < 0$) history of the input signal. In a general system an infinite number of parameters, for example, all the values of $x(t)$ over a finite interval, may be necessary for a complete history.

provided that the real part of s is greater than $-a$. That is, if $\text{Re}(s)$ is less than $-a$, the integral does not converge.

The Laplace transform of the impulse response of an LTI system is sometimes called the system function. However, as we will soon see, it is useful to consider the Laplace transforms of input and output time functions as well as impulse responses.

2.4. Laplace transform of the output of an LTI system

The Laplace transform provides a second way of determining the output of an LTI system when the input and impulse response are known functions of time. To see how this is done, we take the Laplace transform of the output, $y(t)$, written as a convolution integral. Thus

$$\mathscr{L}[y(t)] = Y(s) = \int_{-\infty}^{\infty} y(t)e^{-st}dt = \int_{-\infty}^{\infty} e^{-st} \int_{-\infty}^{\infty} h(\tau)x(t-\tau)\, d\tau\, dt \quad (2.6)$$

Now we make the substitution $u = t - \tau$, $du = dt$ and obtain

$$Y(s) = \iint_{-\infty}^{\infty} e^{-s(u+\tau)} h(\tau)x(u)\, d\tau\, du$$

$$= \int_{-\infty}^{\infty} h(\tau)e^{-s\tau}d\tau \cdot \int_{-\infty}^{\infty} x(u)e^{-su}du \quad (2.7)$$

Since each of the integrals on the right of Eq. 2.7 is a Laplace transform, we have

$$Y(s) = H(s)X(s) \quad (2.8)$$

That is, the Laplace transform of the output is given by the product of the Laplace transforms of the input and the impulse response.†

The procedure for finding the output of an LTI system by the Laplace transform method is:

1. Find the Laplace transforms of $x(t)$ and $h(t)$.
2. Multiply these Laplace transforms, as in Eq. 2.8, to obtain the Laplace transform $Y(s)$ of the output time function.
3. Find the function, $y(t)$, whose Laplace transform is $Y(s)$.

Provided that we have a table of Laplace transforms that includes the $Y(s)$ obtained in step 2, the process of convolution of time functions is replaced by multiplication of system functions and simple table lookup.

Before giving examples of the Laplace transform method in action, we show how tables of Laplace transforms are developed.

---

† The only condition on Eq. 2.8 is that the Laplace transforms $H(s)$ and $X(s)$ must be defined for at least overlapping ranges of s.

2.5. Computation of Laplace transforms   Although the Laplace transform is defined in terms of the integral in Eq. 2.3, it is often simpler to obtain Laplace transforms of new functions by performing certain operations on known Laplace transforms. For example, suppose we know the Laplace transform $X(s)$ of a time function $x(t)$ and want to find the Laplace transform of $t \cdot x(t)$.

Starting with the definition:

$$\mathscr{L}[x(t)] = X(s) = \int_{-\infty}^{\infty} x(t)e^{-st}dt \tag{2.9}$$

we differentiate both sides with respect to s, to obtain

$$\frac{dX(s)}{ds} = -\int_{-\infty}^{\infty} t\, x(t)e^{-st}dt \tag{2.10}$$

Comparing Eq. 2.10 with (2.3) or (2.9), we see that $\mathscr{L}[tx(t)] = (-d/ds)\,\mathscr{L}[x(t)]$.

Example.   Find the Laplace transforms of the functions $u_{-1}(t)te^{-t}$ and $u_{-1}(t)t^2 e^{-t}$.

From Eq. 2.5 we know that the Laplace transform of $u_{-1}(t)e^{-t}$ is $1/(s+1)$; $\mathrm{Re}(s) > -1$. From the preceding discussion, since the two new functions just involve multiplication by t,

$$\mathscr{L}[u_{-1}(t)\, te^{-t}] = \frac{-d}{ds}\left(\frac{1}{s+1}\right) = \frac{1}{(s+1)^2};   \mathrm{Re}(s) > -1 \tag{2.11}$$

and

$$\mathscr{L}[u_{-1}(t)\, t^2 e^{-t}] = \frac{-d}{ds}\frac{1}{(s+1)^2} = \frac{2}{(s+1)^3};   \mathrm{Re}(s) > -1 \tag{2.12}$$

Next, suppose that we know the Laplace transform of a function $x(t)$ and wish to find the Laplace transform of a delayed function $x(t-a)$. We start with the defining equation:

$$\mathscr{L}[x(t-a)] = \int_{-\infty}^{\infty} x(t-a)e^{-st}dt \tag{2.13}$$

and make the substitution $u = t - a$. Then

$$\mathscr{L}[x(t-a)] = \int_{-\infty}^{\infty} x(u)e^{-s(u+a)}du$$

$$= e^{-sa}\int_{-\infty}^{\infty} x(u)e^{-su}du = e^{-sa}\,\mathscr{L}[x(t)] \tag{2.14}$$

Thus, the transform of a delayed time function is e^{-sa} times the transform of the original time function, where a is the amount of time delay.

Example.   Find the Laplace transform of the unit step function:

$$u_{-1}(t) = \begin{array}{cc} 1 & t \geqslant 0 \\ 0 & t < 0 \end{array}$$

and of the delayed unit step $u_{-1}(t-2)$.

We find the Laplace transform of the unit step by direct integration:

$$\mathscr{L}[u_{-1}(t)] = \int_0^\infty e^{-st}dt = \frac{1}{s}; \qquad \text{Re}(s) > 0 \qquad (2.15)$$

and use the delay theorem to find the Laplace transform of a step delayed by two time units.

$$\mathscr{L}[u_{-1}(t-2)] = \frac{e^{-2s}}{s}; \qquad \text{Re}(s) > 0 \qquad (2.16)$$

Next we find the Laplace transform of $[dx(t)]/dt$ in terms of the Laplace transform of $x(t)$. We substitute the definition of a derivative:

$$\frac{dx(t)}{dt} = \lim_{a \to 0} \left[\frac{x(t+a) - x(t)}{a} \right]$$

directly into Eq. 2.3,

$$\mathscr{L}\left[\frac{dx}{dt}\right] = \int_{-\infty}^\infty \lim_{a \to 0} \left[\frac{x(t+a) - x(t)}{a} \right] e^{-st}dt$$

$$= \lim_{a \to 0} \frac{1}{a} \left[\int_{-\infty}^\infty x(t+a)e^{-st}dt - \int_{-\infty}^\infty x(t)e^{-st}\,dt \right] \quad (2.17)$$

and use the delay theorem on the transform of $x(t+a)$ to find

$$\mathscr{L}\left[\frac{dx}{dt}\right] = \lim_{a \to 0} \frac{1}{a} \left\{ e^{sa}\,\mathscr{L}[x(t)] - \mathscr{L}[x(t)] \right\}$$

$$= \lim_{a \to 0} \frac{1}{a} \mathscr{L}[x(t)] \cdot (1 + sa + \ldots - 1) = s\,\mathscr{L}[x(t)] \quad (2.18)$$

That is, the Laplace transform of the time derivative is given by s times the Laplace transform of the original time function.†

We can use the time-derivative theorem to find the Laplace transform

---

† This theorem must be modified by the addition of $-x(0)$ to the right-hand side of (2.18) if the Laplace transform is defined by the integral with 0 for the lower limit.

of the important function $u_{-1}(t) \sin \omega t$, where

$$u_{-1}(t) \cdot \sin \omega t = \begin{matrix} \sin \omega t & t \geq 0 \\ 0 & t < 0 \end{matrix} \qquad (2.19)$$

To find $\mathscr{L}[u_{-1}(t) \sin \omega t]$, we take two time derivatives of $u_{-1}(t) \sin \omega t$, noting that the time derivative of the unit step $u_{-1}(t)$ is the unit impulse, $u_0(t)$. Hence,

$$\frac{d}{dt}[u_{-1}(t)\sin \omega t] = \omega u_{-1}(t)\cos \omega t + u_0(t)\sin \omega t = \omega u_{-1}(t)\cos \omega t$$

$$(2.20)$$

and

$$\frac{d^2}{dt^2}[u_{-1}(t)\sin \omega t] = \omega u_0(t)\cos \omega t - \omega^2 u_{-1}(t)\sin \omega t$$

or

$$\frac{d^2}{dt^2}[u_{-1}(t)\sin \omega t] = \omega u_0(t) - \omega^2 u_{-1}(t)\sin \omega t \qquad (2.21)$$

Now we take the Laplace transform of each term in Eq. 2.21:†

$$s^2 \mathscr{L}[u_{-1}(t) \sin \omega t] = \omega - \omega^2 \mathscr{L}[u_{-1}(t) \sin \omega t] \qquad (2.22)$$

and we solve for the Laplace transform of $u_{-1}(t) \sin \omega t$.

$$\mathscr{L}[u_{-1}(t)\sin \omega t] = \frac{\omega}{s^2 + \omega^2} \qquad (2.23)$$

The theorems and results derived above are listed along with some others in Table 2.1.

2.6. Some examples Now that we have built a table of Laplace transforms we are ready to tackle the problem of finding $y(t)$, given $x(t)$ and $h(t)$, by the Laplace transform method. As a first example, let $x(t)$ and $h(t)$ both be the exponential pulse $u_{-1}(t)e^{-at}$. The first step, finding the Laplace transforms of $x(t)$ and $h(t)$, was already done as an example and appears as number 8 in Table 2.1. The second step is to find $Y(s)$ by multiplication of $X(s) \cdot H(s)$. Thus the Laplace transform of the output is

$$Y(s) = \frac{1}{(s + a)^2} \qquad \text{Re}(s) > -a \qquad (2.24)$$

Now to find the time function for which this is the transform requires a

---

† The Laplace transform of the unit impulse is easily seen to be 1 by direct integration:

$$\mathscr{L}[u_0(t)] = \int_{-\infty}^{\infty} u_0(t)e^{-st}dt = e^{-s \cdot 0} = 1$$

TABLE 2.1.   A Short Table of Laplace Transforms

Time Function	Laplace Transform
1. $x(t)$	$\int_{-\infty}^{\infty} x(t)e^{-st}\,dt$
2. $c_1 x_1(t) + c_2 x_2(t)$	$c_1 \mathscr{L}[x(t)] + c_2 \mathscr{L}[x_2(t)]$
3. $x_1(t)*x_2(t)$	$\mathscr{L}[x_1(t)] \cdot \mathscr{L}[x_2(t)]$
4. $u_0(t)$	1
5. $u_{-1}(t) = \begin{cases} 1, & t \geq 0 \\ 0, & t < 0 \end{cases}$	$\dfrac{1}{s},\ \mathrm{Re}(s) > 0$
6. $tx(t)$	$-\dfrac{d}{ds}\mathscr{L}[x(t)]$
7. $\dfrac{dx}{dt}$	$s\mathscr{L}[x(t)]$
8. $u_{-1}(t)e^{-at};\ a \geq 0$	$\dfrac{1}{s+a},\ \mathrm{Re}(s) > -a$
9. $x(t-a)$	$e^{-as}\mathscr{L}[x(t)]$
10. $u_{-1}(t) \cdot t^n$	$\dfrac{n!}{s^{n+1}},\ \mathrm{Re}(s) > 0$
11. $u_{-1}(t)\sin \omega t$	$\dfrac{\omega}{s^2 + \omega^2},\ \mathrm{Re}(s) > 0$
12. $u_{-1}(t)\cos \omega t$	$\dfrac{s}{s^2 + \omega^2},\ \mathrm{Re}(s) > 0$

little ingenuity. We note that since this $Y(s)$ is the derivative with respect to s of $-[1/(s+a)]$, we can use the multiplication by t theorem, number 6 of Table 2.1, to write down the result:

$$y(t) = u_{-1}(t) \cdot te^{-at} \tag{2.25}$$

Although this would not be a difficult problem to do by the convolution method, it is even easier by Laplace transforms.

Now suppose we change the problem a little bit, so that $x(t)$ is an exponential pulse with a different decay constant, say $u_{-1}(t)e^{-bt}$. Now we have

$$Y(s) = \frac{1}{(s+a)(s+b)} \tag{2.26}$$

This can be written, by a partial fraction expansion, as:

$$\frac{1}{(s+a)(s+b)} = \frac{A}{(s+a)} + \frac{B}{(s+b)} \tag{2.27}$$

where A and B are constants. To evaluate A, we multiply both sides of the equation by $s+a$ and set $s = -a$. The result is easily seen to give $A = 1/(b-a)$. A similar process gives $B = 1/(a-b)$; therefore, the Laplace transform of the output is

$$Y(s) = \frac{1/(b-a)}{s+a} + \frac{1/(a-b)}{s+b} \tag{2.28}$$

Now, since each term is seen to be the transform of an exponential pulse,

$$y(t) = u_{-1}(t) \cdot \left(\frac{e^{-at} - e^{-bt}}{b-a} \right) \tag{2.29}$$

This output function is sketched in Fig. 2.1 for the special case $b = 2$, $a = 1$.

As a third example, suppose the output of the first example, $u_{-1}(t)te^{-at}$, is sent into an LTI system with impulse response $u_{-1}(t)e^{-bt}$. We already know that the Laplace transforms of the input and impulse response $1/(s+a)^2$ and $1/(s+b)$, respectively. Therefore, the transform of the final output is $1/(s+a)^2 \cdot 1/(s+b)$. To find the time function of which this is the Laplace transform, we again expand in partial fractions. The coefficient of a second-order polynomial has the general form $As + B$. Hence

$$\frac{1}{(s+a)^2(s+b)} = \frac{As+B}{(s+a)^2} + \frac{C}{(s+b)} \tag{2.30}$$

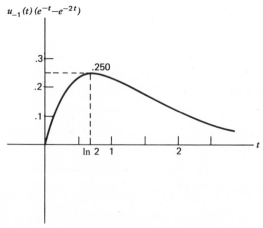

Fig. 2. 1. Output of LTI system: input and impulse response are exponential pulses.

A straightforward way of finding A, B, and C is to obtain a common denominator on the right and then match coefficients of powers of s. Thus

$$\frac{1}{(s+a)^2(s+b)} = \frac{(As+B)(s+b)+C(s+a)^2}{(s+a)^2(s+b)}$$

$$= \frac{s^2(A+C)+s(Ab+B+2Ca)+(Bb+Ca^2)}{(s+a)^2(s+b)} \quad (2.31)$$

so that

$$A + C = 0$$

$$Ab + B + 2Ca = 0 \quad (2.32)$$

and

$$Bb + Ca^2 = 1$$

These equations may readily be solved for A, B, and C in terms of a and b. Finally, the time function is found by examination of the expression:

$$Y(s) = \frac{As}{(s+a)^2} + \frac{B}{(s+a)^2} + \frac{C}{s+b} \quad (2.33)$$

With the help of numbers 6 and 7 of Table 2.1, we find

$$y(t) = u_{-1}(t)\left[A\frac{d}{dt}(te^{-at}) + Bte^{-at} + Ce^{-bt}\right]$$

$$= u_{-1}(t)[Ae^{-at} + (B-Aa)\,te^{-at} + Ce^{-bt}] \quad (2.34)$$

When the solutions for A, B, and C are substituted, the result is

$$y(t) = u_{-1}(t) \cdot \left[\frac{e^{-bt}-e^{-at}}{(b-a)^2} + \frac{te^{-at}}{b-a}\right] \quad (2.35)$$

The great advantage of the Laplace transform method is that convolution, which requires integration and careful bookkeeping on the limits of the integrals, is replaced by a highly mechanized process that requires only algebraic manipulations.

Problems for Chapter 2

2.1. *Domain of convergence* (Fig. P2.1)

(a) The input function e^t (for all time) is applied to a system with impulse response:

$$e^{-2t} \qquad t > 0$$

$$0 \qquad t \leqslant 0$$

Use convolution to find and graph the corresponding output function.

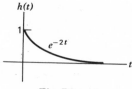

Fig. P2.   1

(b) Repeat part (a) if the input function is e^{-t} for all time.

(c) From the results of parts (a) and (b) show that the system function is $1/(s + 2)$.

(d) Now try to repeat part (a) with the input function e^{-3t}. What catastrophe occurs? Over what range of s values for an input e^{st} function does the convolution process yield an answer?

2.2. In Fig. P2.2 find the system function for a simple delay element (one whose impulse response is a delayed impulse.) *Hint:* Use Eq. 2.3 and the definition of an impulse.

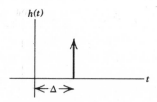

2.3. Show by a "systems" argument that the "area" of the output signal $\int_{-\infty}^{\infty} y(t)dt$ is the product of the "areas" of the input signal and the impulse response. *Hint:* Use the fact that $Y(s) = H(s)X(s)$ for any s and make a wise choice for s.

2.4. Use Laplace transforms to find the output of a system whose impulse response is an exponential pulse, e^{-bt} $(t \geqslant 0)$ and 0 for $t < 0$, when the input is a unit amplitude square pulse of duration a.

2.5. Prove number 12 of Table 2.1. *Hint.* Start with number 11 and use the derivative rule.

2.6. The signal $u_{-1}(t) \sin \omega t$ is sent into an LTI system whose impulse response is the exponential pulse, $u_{-1}(t)e^{-at}$. Find the output time function by Laplace transforms.

2.7. A certain LTI system has an impulse response that is a damped sinusoid, that is, $h(t) = u_{-1}(t)e^{-at}\sin \omega t$. Show, by successive differentiation, that this impulse response satisfies the differential equation:

$$\frac{d^2h}{dt^2} + 2a\,\frac{dh}{dt} + (\omega^2 + a^2)h = \omega u_0(t)$$

and, therefore, that the system function is:

$$H(s) = \frac{\omega}{s^2 + 2as + (\omega^2 + a^2)}$$

REFERENCES

Chirlian, P., *Basic Network Theory*, Chapter 5, McGraw-Hill, New York, 1969.

Craig, E. J., *Laplace and Fourier Transforms for Electrical Engineers*, Chapters 3 and 5, Holt, Rinehart and Winston, New York, 1964.

Kreyszig, E., *Advanced Engineering Mathematics*, Chapter 4, Wiley, New York, 1962.

Kuo, F. F., *Network Analysis and Synthesis*, Chapter 6, Wiley, New York, 1966.

Spiegel, M. R., *Theory and Problems of Laplace Transforms*, Schaum Outline Series, Schaum Publishing Co., 1965.

CHAPTER 3

Signals and Systems in the Frequency Domain

3.1. Introduction We have now become familiar with two methods of characterizing LTI systems so that the output time function can be determined from the input. The first method, convolution, describes the system in terms of its impulse response, $h(t)$. The second, Laplace transforms, describes the system in terms of its system function, $H(s)$, which can be viewed either as the Laplace transform of the impulse response or as the factor by which an e^{st} time function is multiplied as it passes through the system. For most systems, an e^{st} input with pure imaginary s will give a bounded output and, furthermore, for these systems it is usually a straightforward matter to analyze an arbitrary input waveform into sums of these exponential inputs with imaginary exponents. This is the realm of Fourier transforms.

Although the Laplace transform is more general than the Fourier transform in the sense that not every function that has a Laplace transform has a Fourier transform, the latter should be studied separately, for it can often provide us with an illuminating new view of LTI systems.

3.2. The Fourier transform The Fourier transform is another integral transformation of time signals that has important applications in the analysis of instrumentation systems. It is defined† by:

$$X(f) \equiv \int_{-\infty}^{\infty} x(t)e^{-j2\pi ft}\,dt \qquad f \text{ real} \tag{3.1}$$

This integral, which exists for any of the functions $x(t)$ that represent signals of interest in electronics, can be seen to be a special case of the Laplace transform, as we defined it in Eq. 2.1, in which s has the pure imaginary value $j2\pi f$. The real variable f is called the frequency of the signal. Because of the factor $e^{-j2\pi ft}$ in the integrand, even if $x(t)$ is real, the function $X(f)$ is, in general, complex. Notice that if s is not purely imaginary, so that $s = a + j2\pi f$, then the Laplace transform of a function

---

† Several other definitions appear in the literature. These differ in the placement of the factor 2π. If it does not appear in the exponent it appears in a normalizing factor in front of the integral in either Eq. 3.1 or Eq. 3.2, or both.

$x(t)$ can be interpreted as the Fourier transform of the modified function $x(t) \cdot e^{-at}$. The weighting factor e^{-at} is sometimes necessary to allow the integral to give a finite result.

The Fourier transform satisfies the Laplace transform property in Eq. 2.8, or in number 3 of Table 2.1. That is, the Fourier transform $Y(f)$ of the output of an LTI system is the product of the Fourier transforms of the input $X(f)$ and the impulse response $H(f)$, and therefore it affords an alternative method for analyzing such systems. One advantage of the Fourier transform over the Laplace transform is that we will find it easier to determine the time function when its Fourier transform is known, such as when we have found $Y(f)$ by multiplication of $X(f)$ and $H(f)$, than to do the same operation on a Laplace transform. Instead of referring to a list of Laplace transforms of known functions, we can use the Fourier inversion theorem, which states that

$$x(t) = \int_{-\infty}^{\infty} X(f)e^{j2\pi ft}\,df \qquad (3.2)$$

Note that this equation differs from the defining Eq. 3.1, in that the integration is over the variable f rather than t and the sign in the exponent is positive rather than negative. We can interpret this inverse Fourier transform relation as expressing $x(t)$ by an infinite linear superposition of e^{st} functions with $s = j2\pi f$; that is, it decomposes $x(t)$ into complex functions of frequency.

Equations 3.1 and 3.2 state that if we are given $x(t)$ and compute $X(f)$ by means of Eq. 3.1, then we can plug this $X(f)$ into the integral in Eq. 3.2 and obtain the original $x(t)$ back again. Instead of attempting a general proof of this reciprocal relation for an arbitrary $x(t)$ and its Fourier transform, we will show it explicitly in two important examples.

Example 1.   Find the Fourier transform $X(f)$ of the symmetrical square pulse of Fig. 3.1a and show that the square pulse is recreated when this $X(f)$ is inserted into Eq. 3.2.

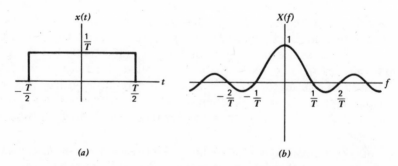

(a)   (b)

Fig. 3.   1. A square pulse and its Fourier transform.

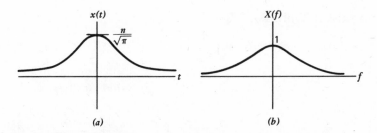

Fig. 3.   2. The Fourier transform of a Gaussian pulse is also a Gaussian pulse.

Solution:

$$X(f) = \int_{-T/2}^{T/2} \frac{1}{T} \cdot e^{-j2\pi ft} dt = \frac{1}{-j2\pi fT} e^{-j2\pi ft} \Big|_{-T/2}^{T/2} = \frac{e^{j\pi fT} - e^{-j\pi fT}}{2\pi fjT} = \frac{\sin \pi fT}{\pi fT}$$

Now, put this $X(f)$ into Eq. 3.2:

$$\int_{-\infty}^{\infty} \frac{\sin \pi fT}{\pi fT} e^{j2\pi ft} df = \int_{-\infty}^{\infty} \frac{\sin \pi fT \cos 2\pi ft \, df}{\pi fT}^{\ddagger} + j \int_{-\infty}^{\infty} \frac{\sin \pi fT}{\pi fT} \cdot \sin 2\pi ft \, df^{\dagger}$$

$$= \begin{cases} \dfrac{1}{T}, & |t| < T/2 \\[2mm] \dfrac{1}{2T}, & |t| = T/2 \\[2mm] 0, & |t| > T/2 \end{cases}$$

Example 2.   Find the Fourier transform of the Gaussian pulse $x(t) = (n/\sqrt{\pi})e^{-n^2 t^2}$, Fig. 3.2a, and show that it satisfies Eq. 3.2.

Solution:

$$X(f) = \frac{n}{\sqrt{\pi}} \int_{-\infty}^{\infty} e^{-n^2 t^2} e^{-j2\pi ft} \, dt$$

$$= \frac{n}{\sqrt{\pi}} \int_{-\infty}^{\infty} e^{-[n^2 t^2 + j2\pi ft + (j^2 \pi^2 f^2 / n^2)]} \cdot e^{-(\pi^2 f^2 / n^2)} \, dt$$

$$= \frac{n}{\sqrt{\pi}} e^{-(\pi^2 f^2 / n^2)} \int_{-\infty}^{\infty} e^{-[nt + (j\pi f / n)]^2} \, dt$$

$$= \frac{n}{\sqrt{\pi}} e^{-(\pi^2 f^2 / n^2)} \int_{-\infty}^{\infty} e^{-v^2} \frac{dv}{n} = e^{-(\pi^2 f^2 / n^2)}$$

†The imaginary part is zero because it is an odd function of f integrated over symmetric limits.

‡This integral appears in tables, such as *Mathematical Tables from Handbook of Chemistry and Physics*, Ninth Edition, No. 342, p. 273, Chemical Rubber Publishing Co., Cleveland, 1952.

Now put this $X(f)$ into Eq. 3.2.

$$\int_{-\infty}^{\infty} e^{-(\pi^2 f^2/n^2)} e^{j2\pi ft} df = \int_{-\infty}^{\infty} e^{-[(\pi^2 f^2/n^2)-j2\pi ft+j^2 n^2 t^2]} \cdot e^{-n^2 t^2} df$$

$$= e^{-n^2 t^2} \int_{-\infty}^{\infty} e^{-[(\pi f/n)-jnt]^2} df$$

$$= \frac{n}{\pi} e^{-n^2 t^2} \int_{-\infty}^{\infty} e^{-u^2} du = \frac{n}{\sqrt{\pi}} e^{-n^2 t^2}$$

These two examples are important because either one can be used as the basis of a theory of the unit impulse. Note that the area under the time functions in Figs. 3.1 and 3.2 is 1,[†] independent of the value of T in example 1, or n in example 2. Also, as $T \to 0$ in example 1 or as $n \to \infty$ in example 2, the time functions each become very narrow and very tall. The corresponding Fourier transforms shown in Figs. 3.1*b* and 3.2*b* each become very wide, while the value near $f = 0$ remains at 1. In the limit, the time functions approach a unit impulse, while the corresponding transforms approach the constant value 1. *Thus, the time function $u_0(t)$ and the frequency function 1 form a Fourier transform pair.* Although we have not given a rigorous proof of this statement, example 2 can be made the basis for such a proof.[‡]

3.3. Fourier transform of periodic functions: Fourier series   Next we address the question of whether periodic functions, that is, functions that look the same when advanced or delayed in time by a certain amount,

---

[†]To prove this for the pulse in Fig. 3.2*a*, write $I = \int_{-\infty}^{\infty} e^{-n^2 t^2} dt$. Then

$$I^2 = \int_{-\infty}^{\infty} e^{-n^2 x^2} dx \cdot \int_{-\infty}^{\infty} e^{-n^2 y^2} dy = \iint_{-\infty}^{\infty} e^{-n^2(x^2+y^2)} dx\, dy$$

Now change to polar coordinates:

$$I^2 = \int_0^{\infty} \int_0^{2\pi} e^{-n^2 r^2} r d\theta\, dr = 2\pi \int_0^{\infty} e^{-n^2 r^2} r dr$$

and let $u = r^2$, $du = 2r\, dr$:

$$I^2 = \pi \int_0^{\infty} e^{-n^2 u} du = \frac{\pi}{n^2}.$$

Thus

$$I = \frac{\sqrt{\pi}}{n}$$

[‡]See M. J. Lighthill, *Introduction to Fourier Analysis and Generalized Functions*, Chapter 2, University Press, Cambridge, 1958.

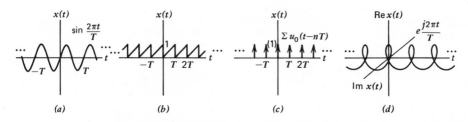

Fig. 3.   3. Examples of periodic time functions.

such as sines and cosines, and square waves, have Fourier transforms in the sense of Eqs. 3.1 and 3.2. Several examples of periodic time functions are shown in Fig. 3.3. Notice particularly the complex time function, $e^{j2\pi t/T}$, which is sketched as a function of t with the real part plotted vertically and the imaginary part plotted on an axis "out of the page." Every periodic function with period T can be seen to have also period $2T, 3T, \ldots, nT$; that is, each one may be displaced by any integral multiple of T and remains indistinguishable from the original function.

We can expect, intuitively, that if a periodic function can be composed of a superposition of functions of the form $e^{j2\pi ft}$, as in Eq. 3.2, then the f values ought to be restricted to integral multiples of $1/T$ so that every one of the $e^{j2\pi ft}$ functions will become $e^{j2\pi(nt/T)}$ and have period T. In order to so restrict these exponential functions within the notational framework of Eq. 3.2, we must restrict $X(f)$ to be a set of impulses, possibly with complex areas, at the frequency values n/T (see Fig. 3.4). Then if the integral in Eq. 3.2 is performed, that integral is converted into a sum of discrete terms. There may be (but need not be) an infinite number of these terms, so that in general we have, when $x(t)$ has period T,

$$x(t) = \sum_{n=-\infty}^{\infty} x_n\, e^{j2\pi(nt/T)} \tag{3.3}$$

Although it is easy to suggest that a periodic function can be represented by such a series, which is one form of Fourier series, it is not easy to find a rigorous mathematical proof that the sum is indeed equivalent to the function for a broad class of periodic functions,† including all those functions that we will encounter in our study of electronic systems. We will not reproduce such a proof here. However, if Eq. 3.3 is true in some sense, then we can give a prescription for finding the coefficient x_n when the time function is given. This prescription makes use of the "orthogonality" property of the functions $e^{j2\pi(nt/T)}$. Thus, to find a particular coefficient, say x_{n_1}, we multiply both sides of Eq. 3.3 by $e^{-j(2\pi n_1 t/T)}$ and then integrate over one period of t, for example, from $t = -T/2$ to $t = T/2$. On the right side we have an infinite number of terms in the

†See Lighthill, op. cit., Chapter 5, for a modern theory of Fourier series.

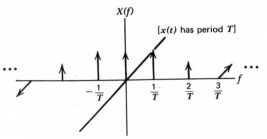

Fig. 3.   4. The Fourier transform of a periodic time signal.

integrand of the form $e^{j2\pi(t/T)(n-n_1)}$. All of the terms with $n \neq n_1$ integrate to zero by straightforward means but the one term with $n = n_1$ becomes $x_{n_1} \cdot \int_{-T/2}^{T/2} e^{j \cdot 0} dt = x_{n_1} \cdot T$. The result is that the coefficient x_{n_1} is given explicitly by the formula:

$$x_{n_1} = \frac{1}{T} \int_{\text{period}} x(t)\, e^{-j2\pi n_1 t/T}\, dt \tag{3.4}$$

We illustrate the calculation of the "Fourier coefficients" x_n by only one example, that of the infinite sequence of real unit impulses of Fig. 3.5a. We will show presently that the transforms of all other periodic functions can be obtained from the transform of the impulse train.

Example.   Compute the x_n for the infinite impulse train.

Using Eq. 3.4, we see that the range of integration covers only one of the impulses. Thus

$$x_{n_1} = \frac{1}{T} \int_{-T/2}^{T/2} u_0(t)\, e^{-j2\pi n_1 t/T}\, dt = \frac{1}{T} \cdot e^{-j2\pi n_1 \cdot 0/T} = \frac{1}{T} \tag{3.5}$$

and we have the simple result that all the Fourier coefficients for the periodic train of impulses have the same value, $1/T$. In other words, the Fourier transform of a train of equal impulses in the time domain is a train of equal impulses in the frequency domain (see Fig. 3.5b).

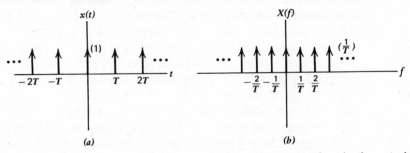

Fig. 3.   5. The Fourier transform of a periodic train of impulses is also a train of impulses.

3.4. Construction of a table of Fourier transforms   Although it must be possible to use the definition, Eq. 3.1, to calculate the Fourier transform of a time function if the Fourier transform exists, it is often more convenient to use one or more of a number of theorems and relations concerning Fourier transforms. By using a relatively brief library of basic transforms and applying the theorems, it is possible to build up extensive tables of Fourier transforms.

A basic library of Fourier transforms appears in Table 3.1. Note that we have already proved all but numbers 3, 4, 7, and 8 (number 2 is just T times the Fourier transform pair of example 1 in the last section). To prove numbers 3 and 4, we can start with the transform as a pair of impulses and use the inversion theorem to obtain a pair of complex exponential time functions at frequencies $\pm f_0$. Then apply the fundamental relation, $e^{jx} = \cos x + j \sin x$, with $x = \pm 2\pi f_0 t$, to obtain the sine and cosine functions in the time domain. To prove number 7 we simply do the indicated integration in the defining equation, noting that the integral in this case runs from $t = 0$ to ∞. Number 8 is the limiting case of 7 as a approaches zero; since the total "area" under the frequency function in 7 can be shown[†] to be ½, independent of a, as a approaches zero, there is a real impulse at $f = 0$ as well as a pair of imaginary hyperbolas, $1/j2\pi f$.

Table 3.2 is a compilation of theorems and properties that we will find useful not only in obtaining new transforms but in learning to "visualize" the time function when its transform is known or vice versa. Proofs of the theorems will be left as exercises for the student. Here we will be content merely to draw attention to the various theorems in the table with a few comments.

Theorems 1, 7, 8, and 11 deal with symmetry properties of the time function $x(t)$ or its transform $X(f)$. For example, theorem 1 states that if we know the Fourier transform $X(f)$ of a real $x(t)$ for positive frequencies, we can find the Fourier transform for negative frequencies simply by changing the sign of f and taking the complex conjugate. And theorem 11 states that if a periodic time function is such that when we delay it by half a period it is an inverted version of the original function, then the even Fourier coefficients are all zero. An example would be a symmetric triangular wave (Fig. 3.6).

Theorem 3 shows that the Fourier transform is a linear operation, that is, that superposition holds in both time and frequency domains. Theorem 9, the duality theorem, is a consequence of the near symmetry

†

$$\int_{-\infty}^{\infty} \frac{df}{a + j2\pi f} = \int_{-\infty}^{\infty} \frac{(a - j2\pi f)\,df}{a^2 + 4\pi^2 f^2} = \int_{-\infty}^{\infty} \frac{a\,df}{a^2 + 4\pi^2 f^2} = \frac{1}{2}$$

See, for example, *Mathematical Tables from Handbook of Chemistry and Physics*, Ninth Edition, No. 16, Chemical Rubber Publishing Co., Cleveland, 1952.

TABLE 3.1   A Basic Library of Fourier Transform Pairs

	Time Function	Frequency Function		
1.	$u_0(t)$	1		
2.	1: $-T/2 < t < T/2$ 0 otherwise	$\dfrac{\sin \pi f T}{\pi f}$		
3.	$\cos 2\pi f_0 t$	$\frac{1}{2}[u_0(f - f_0) + u_0(f + f_0)]$		
4.	$\sin 2\pi f_0 t$	$\dfrac{j}{2}[u_0(f + f_0) - u_0(f - f_0)]$		

5. $\displaystyle\sum_{-\infty}^{\infty} u_0(t-nT)$ $\displaystyle\frac{1}{T} \cdot \sum_{n=-\infty}^{\infty} u_0\left(f-\frac{n}{T}\right)$

6. $\displaystyle\frac{n}{\sqrt{\pi}}\, e^{-n^2 t^2}$ $e^{-(\pi^2 f^2/n^2)}$

7. $e^{-at}; \qquad t>0$ $\displaystyle\frac{1}{a+j2\pi f}$

8. $u_{-1}(t)\begin{cases}1; & t>0 \\ 0; & t<0\end{cases}$ $\displaystyle\frac{1}{2}\cdot u_0(f) \qquad f=0$

    $\displaystyle\frac{1}{j2\pi f} \qquad f\neq 0$

51

TABLE 3.2   Some Theorems on Fourier Transforms

Time Function	Frequency Function	Comments
1. Real $x(t) = x^*(t)$	$X(f) = X^*(-f)$	$X(f)$ has "conjugate" symmetry
2. $x(t-a)$	$e^{-j2\pi fa} \cdot X(f)$	Time shifting theorem
3. $x_1(t) + x_2(t)$	$X_1(f) + X_2(f)$	Linearity
4. $\dfrac{dx}{dt}$	$-j2\pi f \cdot X(f)$	Transform of time derivative
5. $x_1(t) * x_2(t)$	$X_1(f) \cdot X_2(f)$	Convolution of time functions; product of FTs
6. $x_1(t) \cdot x_2(t)$	$X_1(f) * X_2(f)$	Product of time functions; convolution of FTs
7. Even function $x(t) = x(-t)$	Real $X(f) = X^*(f)$	Even time function; real FT
8. Odd function $x(t) = -x(-t)$	Imaginary $X(f) = -X^*(f)$	Odd time function; imaginary FT
9. $X(t)$	$x(-f)$	Duality
10. $t \cdot x(t)$	$\dfrac{j}{2\pi} \cdot \dfrac{dX(f)}{df}$	Multiplication by t
11. For periodic $x(t)$ $x(t-T/2) = -x(t)$	$x_n = 0$ for n even	"Odd harmonic" time function

between the Fourier transform and the inverse Fourier transform. Theorems 4 and 10 also exhibit this close symmetry. Theorem 5 is our primary motivation for introducing transforms; that is, in the frequency domain we need only multiply transforms to obtain the transform of the

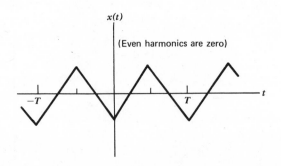

Fig. 3.   6. Example of an odd-harmonic time function.

output of an LTI system, while in the time domain we have to perform the more difficult process of convolution to find the output. Theorem 6, a sort of dual of theorem 5, says that if we must multiply two time functions (an operation that has widespread application in communications, such as radio and television, and also in modern digital data processing systems), we can view the multiplication process in the time domain as convolution in the frequency domain. And last, but not least, theorem 2 deals with the important time-delay operation.

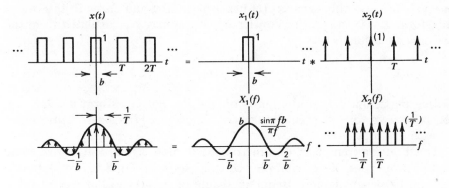

Fig. 3.   7. Obtaining the Fourier transform of a square wave.

Example.   Use the basic library and the theorems to find the Fourier transform of a square wave of period T, consisting of square pulses, symmetric around $t = nT$ of duration b (see Fig. 3.7).

The square wave may be viewed as the convolution of a single square pulse of duration b, with the periodic impulse train. This decomposition into two signals is shown in the top row of Fig. 3.7, which should be read from left to right. Since we convolve these two time signals to get the original $x(t)$, then according to number 5 of Table 3.2 we must multiply their transforms to get the transform $X(f)$ of the original $x(t)$. The transforms, from numbers 2 and 5 of Table 3.1, are shown in the bottom line of the figure, which should be read from right to left. Thus, the Fourier transform of the square wave is a sequence of impulses at frequencies equal to multiples of $1/T$, with the area of the nth impulse being given by $(\sin \pi nb/T)/\pi n$. An important special case, called a 50% duty-cycle square wave, occurs for $b = T/2$. The areas of the impulses are now $\sin(n\pi/2)/\pi n$, and we note that this gives zero for all even, nonzero values of n, while the odd values of n give $1/\pi$, $-(1/3\pi)$, $1/5\pi$, etc.

3.5. Random signals in the frequency domain: spectral density   In Chapter 1 we saw that it was useful to characterize random signals in the time domain with two parameters — the mean and variance — and also with an integral transform — called the autocorrelation function (from which the mean and variance can be obtained) — which shows how rapidly a random time function is changing. Let us now examine random signals from the frequency domain point of view.

We cannot, of course, know the details of a random signal $x(t)$ for all time, but we can measure and record the instantaneous values of the signal over a finite time, say, of duration T. We now *define* the function $x_T(t)$ to be equal to the signal $x(t)$ during the interval T and zero outside the interval. Because the signal $x_T(t)$ has a beginning and an end, its Fourier transform $X_T(f)$ exists; the Fourier transform may be calculated from the defining equation:

$$X_T(f) = \int_{-\infty}^{\infty} x_T(t)\, e^{-j2\pi ft}\, dt \qquad (3.6)$$

Notice particularly that the limits on the Fourier transform integral are, formally, $-\infty$ to $+\infty$, even though the function $x_T(t)$ is zero outside the interval of duration T.

The Fourier transform of the truncated time signal, $x_T(t)$, has some useful properties. For example, setting $f = 0$ in Eq. 3.6, we obtain a relation between the zero frequency value of $X_T(f)$ and the average value of the waveform $x_T(t)$ over the interval T. Thus

$$X_T(0) = \int_{-\infty}^{\infty} x_T(t)\, dt = \int_{-T/2}^{T/2} x_T(t)\, dt = T \cdot \overline{x_T(t)}$$

$$(3.7)$$

or

$$\overline{x_T(t)} = \frac{1}{T} \cdot X_T(0)$$

We can also obtain a relation between the Fourier transform $X_T(f)$ and the average of the square of the time signal over the interval T. However, to do this, we must first prove a theorem known as Parseval's theorem. In its general form, this theorem deals with two time functions, say $x_1(t)$ and $x_2(t)$, and their Fourier transforms (assumed to exist), $X_1(f)$ and $X_2(f)$. Parseval's theorem, which is more useful in proving certain other theorems than in its own right, states that

$$\int_{-\infty}^{\infty} x_1(t)x_2{}^*(t)\, dt = \int_{-\infty}^{\infty} X_1(f)X_2{}^*(f)\, df \qquad (3.8)$$

This may be proved immediately by substituting for $x_1(t)$ on the left side in terms of its Fourier transform:

$$x_1(t) = \int_{-\infty}^{\infty} X_1(f)\, e^{j2\pi ft}\, df \qquad (3.9)$$

and by substituting for $X_2(f)$ on the right side in terms of its inverse Fourier transform:

$$X_2(f) = \int_{-\infty}^{\infty} x_2(t)e^{-j2\pi ft} \, dt \qquad (3.10)$$

Each side of Parseval's theorem becomes a double integral with identical integrands and, provided the order of integration is immaterial, the two sides are equal and we have proved the theorem.

We now apply Parseval's theorem to obtain a frequency domain method of calculating the average value of the square of a random signal. We deal again with the truncated signal $x_T(t)$. The average of its square over the interval T is

$$\overline{|x_T(t)|^2} = \frac{1}{T} \cdot \int_{-T/2}^{T/2} |x_T(t)|^2 dt = \frac{1}{T} \int_{-\infty}^{\infty} |x_T(t)|^2 dt \qquad (3.11)$$

Now, if we let $x_1(t) = x_2(t) = x_T(t)$ in Parseval's theorem, we arrive at:

$$\overline{|x_T(t)|^2} = \frac{1}{T} \int_{-\infty}^{\infty} X_T(f) \cdot X_T{}^*(f) \, df = \frac{1}{T} \int_{-\infty}^{\infty} |X_T(f)|^2 df \qquad (3.12)$$

In the limit, as $T \to \infty$, the quantity $(1/T)\,|X_T(f)|^2$, a real function of frequency, is called the *spectral density* of the signal $x(t)$ and is sometimes given a special symbol, $W_x(f)$. That is,

$$W_x(f) \equiv \lim_{T \to \infty} \frac{1}{T} \cdot |X_T(f)|^2 \qquad (3.13)$$

The spectral density, like the autocorrelation function, conveys information about how rapidly a random waveform is varying. In addition, both the spectral density and the autocorrelation function allow us to calculate the average of the square of the waveform; we do this by setting $\tau = 0$ in the autocorrelation function or by integrating the spectral density over all frequencies. That is,

$$\overline{|x(t)|^2} = R_x(0) = \int_{-\infty}^{\infty} W_x(f) \, df \qquad (3.14)$$

Thus, there is an intimate relation between the autocorrelation function and the spectral density. As a matter of fact, they form a Fourier transform pair! The spectral density is the Fourier transform of the autocorrelation function, and the autocorrelation function is the inverse Fourier transform of the spectral density. We can demonstrate this rather neatly with a "systems" argument rather than a formal mathematical proof. Consider the system of Fig. 3.8, in which the signal $x_T(t)$ is the input to a system whose impulse response is $x_T(-t)$; that is, the signal itself turned around in time. We have already seen in Chapter 1 that the output of this system, which is the convolution of $x_T(t)$ with itself turned

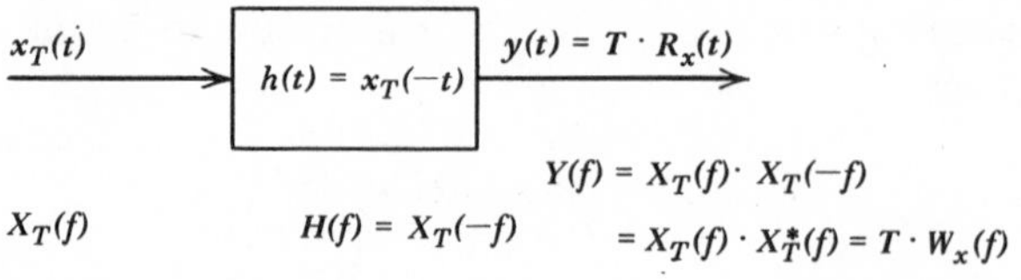

$x_T(t)$

$h(t) = x_T(-t)$ $y(t) = T \cdot R_x(t)$

$X_T(f)$ $H(f) = X_T(-f)$ $Y(f) = X_T(f) \cdot X_T(-f)$

$= X_T(f) \cdot X_T^*(f) = T \cdot W_x(f)$

Fig. 3.   8. Showing that the spectral density is the Fourier transform of the autocorrelation function.

around in time, is, in the limit as $T \to \infty$, the autocorrelation function of $x(t)$ multiplied by T. But the Fourier transform of the output is the product of the Fourier transform of the input and of the impulse response, which are $X_T(f)$ and $X_T(-f)$,† respectively and, since $x(t)$ is real, $X_T^*(f) = X_T(-f)$ by theorem 1 of Table 3.2. Therefore, the Fourier transform of the output is, in the limit as $T \to \infty$, equal to T times the spectral density. The argument is summarized in Fig. 3.8, and the results in Eqs. 3.15.

$$R_x(\tau) = \int_{-\infty}^{\infty} W_x(f)\, e^{j2\pi f \tau}\, df$$

$$W_x(f) = \int_{-\infty}^{\infty} R_x(\tau)\, e^{-j2\pi f \tau}\, d\tau$$

(3.15)

3.6. The spectral density of the output of an LTI system   In Chapter 1 we learned how to compute the autocorrelation function of the output of an LTI system in terms of the autocorrelation function of the input and the impulse response of the system. That method, embodied in Eq. 1.14, involved the rather difficult process of double convolution. If we could obtain the spectral density of the output of the system, which we can denote by $W_y(f)$, then we could, by the results of the last section, simply take the inverse Fourier transform to obtain $R_y(\tau)$. That is,

$$R_y(\tau) = \int_{-\infty}^{\infty} W_y(f)\, e^{j2\pi f \tau}\, df$$

(3.16)

It is a simple matter to calculate $W_y(f)$, since the Fourier transform of

† That the Fourier transform of $x(-t)$ is $X(-f)$ is easily proved by writing:

$$x(t) = \int_{-\infty}^{\infty} X(f)\, e^{j2\pi f t}\, df$$

and substituting $-t$ for t and $-f$ for f. Thus

$$x(-t) = \int_{-\infty}^{\infty} X(-f)\, e^{j2\pi f t}\, df$$

and we recognize that $X(-f)$ is the Fourier transform of $x(-t)$.

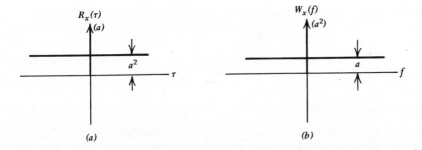

Fig. 3.   9. Autocorrelation function and spectral density of a Poisson process.

the output of the system is the product of the transforms of the input and the impulse response by theorem 5 of Table 3.2. Thus

$$Y_T(f) = X_T(f) \cdot H(f) \tag{3.17}$$

If we take the square of the magnitude of both sides of this equation, we have

$$|Y_T(f)|^2 = |X_T(f)|^2 \cdot |H(f)|^2 \tag{3.18}$$

and we divide by T and take the limit as $T \to \infty$ to arrive at:

$$W_y(f) = W_x(f) |H(f)|^2 \tag{3.19}$$

Once again we may take the Poisson process as an example of a random waveform. Figure 3.9 shows the spectral density of the Poisson process, obtained as the Fourier transform of the autocorrelation function. Figure 3.10 suggests how the spectral density of the output of an LTI system is obtained when its impulse response is known and its input is the Poisson sequence at average rate a.

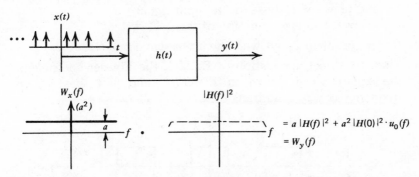

Fig. 3.   10. Obtaining the output spectral density when the input is a Poisson sequence.

The output spectral density is seen to be given by:

$$W_y(f) = a \mid H(f) \mid^2 + a^2 \mid H(0) \mid^2 \cdot u_0(f) \qquad (3.20)$$

This equation is seen to be the term by term Fourier transform of Eq. 1.16.

Problems for Chapter 3

3.1. (a) Find the Fourier coefficients x_n by direct application of Eq. 3.4 for the periodic sequence of impulses in Fig. P3.1. Plot the x_n as impulses in frequency on a Fourier transform graph similar to Fig. 3.4 in the text.

(b) Check your answer to part (a) by considering $x(t)$ to be a sum of two periodic functions, one consisting of the positive impulses and the other of the negative impulses. Use the delay property of Fourier transforms to find the transform of the negative sequence.

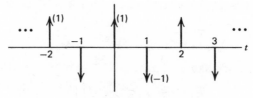

Fig. P3.   1

3.2. The input signal $x(t) = \cos \omega t$ is applied to an LTI system with system function $H(s) = 1/(s + 2)$.

(a) Find and plot the Fourier transform $X(f)$ of the input signal as a pair of impulses on a graph like Fig. 3.4.

(b) If $\omega = 1$, find the Fourier transform $Y(f)$ of the output signal by multiplying $X(f)$ by $H(s)$ with s set equal to $j\omega = j$.

(c) What time function does $Y(f)$ represent? Compare its magnitude and phase with those of the input signal.

(d) Is $\cos \omega t$ an eigenfunction of this LTI system?

3.3. Repeat problem 2, but with the input signal $x(t) = \sin \omega t$.

3.4 Find, by direct application of Eq. 3.4, the Fourier coefficients for the square wave shown in Fig. P3.4, and plot these x_n versus frequency as a Fourier transform.

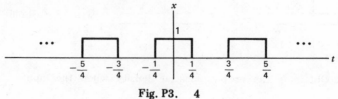

Fig. P3.   4

3.5. Find the *period* of the periodic time function whose Fourier transform is shown in Fig. P3.5.

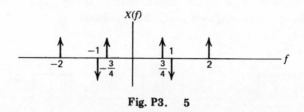

Fig. P3. 5

3.6. A square wave that spends equal time at ±1 and has period 1.0 msec is the input to an LTI system whose system function $H(f)$ is pure imaginary and given by Fig. P3.6. Find the output time function.

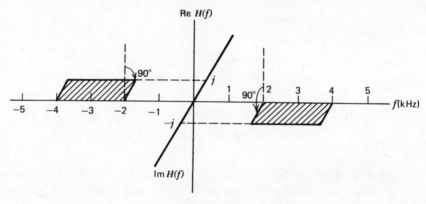

Fig. P3. 6

3.7. The expression $\cos^2 2\pi f t$ can be converted to a sum of sinusoids by a trigonometric identity. An easy way of deriving this identity is to compute the Fourier transform of $\cos^2 2\pi f t$, by convolving the Fourier transform of $\cos 2\pi f t$ with itself. In this way, prove that $\cos^2 2\pi f t = \frac{1}{2} + \frac{1}{2} \cos 4\pi f t$.

3.8. Prove that $\sin^2 2\pi f t = \frac{1}{2} - \frac{1}{2} \cos 4\pi f t$ by the method of Fourier transforms outlined in problem 7.

3.9. Prove that $\sin 2\pi f t \cdot \cos 2\pi f t = \frac{1}{2} \sin 4\pi f t$ by means of Fourier transforms.

3.10. Find the Fourier transform of $\cos^3 2\pi f t$ by doing two successive convolutions of the transform of the cosine function. Since $\cos^3 2\pi f t$ is an even function of time, what can you tell about its transform before working it out?

3.11. Find the Fourier transform of the periodic square wave shown in Fig. P3.11 by multiplying the transforms of a single square pulse and a periodic train of impulses. Which, if any, of the harmonics of the fundamental frequency $1/T$ are missing?

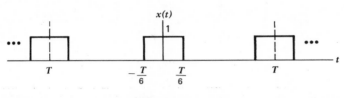

Fig. P3.   11

3.12. Given the full-wave rectified cosine wave in Fig. P3.12a,

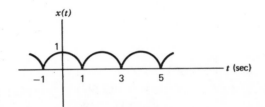

Fig. P3.   12a

(a) Tell whether its Fourier transform $X(f)$ is real or imaginary, impulsive or continuous, an even or odd function of f, periodic or nonperiodic in f.
(b) Find its lowest nonzero frequency component.
(c) Find its zero-frequency component.
(d) The answers to the above can be verified by treating $x(t)$ as the product of a cosine wave and a square wave $s(t)$, which switches between ±1. The transform of $s(t)$ is indicated in Fig. P3.12b. Sketch and dimension $X(f)$ in the range $-1 \leqslant f \leqslant 1$ Hz.

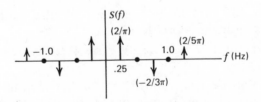

Fig. P3.   12b

3.13. A single square pulse is shown in Fig. P3.13. Its Fourier transform is $Q(f)$ and the Fourier transform of its derivative, $p(t) \equiv dq/dt$ is $P(f)$.
(a) Sketch and dimension a graph of $p(t)$.
(b) Find $P(f)$ directly from the definition $P(f) = \int_{-\infty}^{\infty} p(t)e^{-j2\pi ft}dt$ by doing the integral.

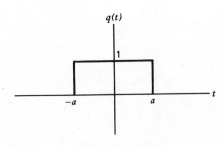

Fig. P3.   13

(c) Use the derivative rule for transforms to express $P(f)$ in terms of $Q(f)$.
(d) By equating the answers to (b) and (c), find $Q(f)$ as an explicit function of f and simplify to the level of a trigonometric function or functions.

3.14. Find the Fourier transform of the triangular pulse and of the periodic wave obtained by repeating this pulse every 2 seconds (see Fig. P3.14). Then use the fact that $\int_{-\infty}^{\infty} X(f)df = x(0)$ to show that

$$\left(1 + \frac{1}{9} + \frac{1}{25} + \frac{1}{49} \cdots \right) = \frac{\pi^2}{8}$$

Hint. $x(t)$ is the convolution of a certain simple function with itself.

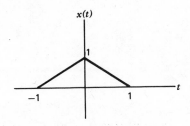

Fig. P3.   14

3.15 A periodic square wave (Fig. P3.15), is the input to a LPF with impulse response $h(t)$. Find, by Fourier transform methods, the Fourier transform of the output.

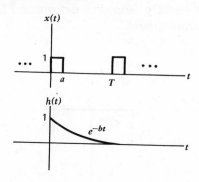

Fig. P3.   15

REFERENCES

Carlson, A. B., *Communication Systems: An Introduction to Signals and Noise in Electrical Communication*, Chapter 2, McGraw-Hill, New York, 1968.

Kuo, F. F., *Network Analysis and Synthesis*, Chapter 3, Wiley, New York, 1966.

Schwartz, M., *Information Transmission, Modulation and Noise*, Chapter 2, McGraw-Hill, New York, 1959.

CHAPTER 4

Passive Devices, Two-Port Networks, and Input Properties of Amplifiers

4.1. Introduction The first three chapters have provided a mathematical framework with which to handle an important class of system — those that possess the properties of linearity and time invariance. We next take a more detailed look at one particular kind of system, electrical networks made out of linear, passive circuit elements. These networks can be characterized by impulse responses and system functions; however, because there are usually two variables of interest at an input or output terminal — the voltage and the current — the analysis of such networks involves its own special problems.

4.2. Circuit elements and sources of excitation There are three basic circuit elements and two kinds of sources of excitation. The circuit elements are the resistor, the capacitor, and the inductor, while the sources of excitation are the ideal voltage generator and the ideal current generator. Each of these five devices is a two-terminal device, which can be characterized by the relation between the current (flow of positive charge) through it and the voltage across it. For example, the current-voltage relation for a resistor is the well-known Ohm's law:

$$V = iR \qquad (4.1)$$

Here V is the potential difference, or voltage, across the resistor with the end that the current enters being positive relative to the other end. This sign convention is illustrated in Fig. 4.1. The unit of resistance is the ohm (Ω) in the SI system of units, but often it is convenient to measure resistances in kilohms (1 kilohm = 1 K = 10^3 Ω) abbreviated kΩ, or in megohms (1 megohm = 1 M = 10^6 Ω), abbreviated MΩ.

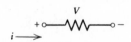

Fig. 4. 1. Resistor: symbol and sign convention.

The second circuit element is the capacitor, which is any pair of conductors separated by a nonconducting material (vacuum or dielectric). The current-voltage relation for a capacitor may be derived from the fact that the charge on either "plate" of the capacitor is proportional to the instantaneous value of the voltage across the capacitor:

$$Q = CV \qquad (4.2)$$

The constant of proportionality C is called the capacitance of that pair of conductors and is measured in farads (F) in the SI system. However, a 1-F capacitor would be quite large physically, and more useful units for capacitance are the microfarad (1 microfarad = 10^{-6} F), abbreviated μF, and the picofarad (1 picofarad = 10^{-12} F), abbreviated pF. The sign convention for charge and voltage in a capacitor is that the voltage is more positive on the plate with the positive charge. To obtain the current-voltage relation for a capacitor, we differentiate Eq. 4.2 and use the fact that the current flowing into the left end of the capacitor in Fig. 4.2 is equal to the time rate of change of the charge on the left plate:

$$\frac{dQ}{dt} = i = C \frac{dV}{dt} \qquad (4.3)$$

Thus, if the current is positive, so that the left plate is becoming more positively charged, the voltage of the left plate relative to the right plate is increasing.

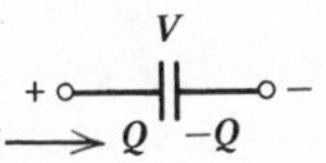

Fig. 4.   2. Capacitor: symbol and sign convention.

The third fundamental circuit element is the inductor, which has the physical form of a coil of wire, with or without a central core of iron or other magnetic material. A current through the coil establishes a magnetic field that links the turns of the coil and, if the current is changed, a voltage or electromotive force (emf) is induced in the coil according to Faraday's law of induction in such a direction as to oppose the change in current. The induced voltage is proportional to the time rate of change of the current so the relation between the current through the inductor and the voltage across the inductor is

$$V = L \frac{di}{dt} \qquad (4.4)$$

The constant L is called the inductance. The SI unit of inductance is the henry (H), however, a 1-H inductance is rather large, and it is common to

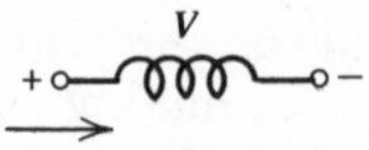

$$V$$

$$+\circ\!-\!\mathsf{000}\!-\!\circ\,-$$

$$i \longrightarrow$$

Fig. 4.   3. Inductor: symbol and sign convention.

express inductances in millihenries, mH or microhenries, μH. The sign convention for voltage and current in an inductor is shown in Fig. 4.3. The voltage V measured at the left end relative to the right will be positive if the current into the left end is increasing.

The current-voltage relations in Eqs. 4.1, 4.3, and 4.4 define ideal resistors, capacitors, and inductors. Real world resistors, capacitors, and inductors behave in a more complicated fashion for several reasons. For one thing, there is always some stray capacitance between the ends of a resistor and inductor or even between the individual turns of wire in an inductor. This stray capacitance, illustrated by dotted lines in Fig. 4.4, may amount to anywhere from a picofarad to 10^2 pF or more, depending on the size and geometry of the structures involved. Whether or not the stray capacitance needs to be included in any circuit analysis depends on the impedance (defined in the next section) of the stray capacitance relative to the impedance of the idealized device, which in turn depends on the frequencies of the signals of interest that are being applied to the circuit element.

Another stray effect that may be important is what is known as "loss" in a capacitor or inductor. In an ideal capacitor or inductor, energy is stored (in the form of electrostatic potential energy, $\frac{1}{2}CV^2$, in a capacitor and in the form of magnetic potential energy, $\frac{1}{2}Li^2$, in an inductor) and may be completely recovered. In the nonideal capacitor or inductor, some of the stored energy is lost or dissipated. Physically, there are two reasons why energy can be dissipated: one is because of radiative effects in which the device acts like an antenna for the production of electromagnetic waves that carry away energy; the other is the actual dissipation of heat in the device. In the nonideal capacitor some energy is given to the atoms of the dielectric material as they are shaken back and forth by a changing voltage across the capacitor, while in the nonideal inductor there is energy loss to the nonzero resistance of the wire used in the construction of the coil and possibly an energy loss in the form of induced currents flowing in

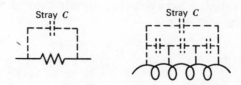

Fig. 4.   4. Models of a real resistor and inductor.

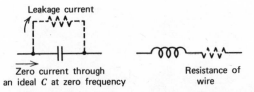

Fig. 4.   5. Models including energy dissipation in a capacitor and inductor.

nearby conductors as a result of the changing magnetic field produced by the current in the inductor itself. The energy loss per cycle of a sinusoidal current may depend on the frequency of the current in a fairly complicated way so that it is difficult to completely characterize a real circuit by a single set of idealized elements. However, at any one frequency, it usually suffices to represent the loss in a capacitor in terms of a parallel stray resistance and the loss in an inductor by a series stray resistance. Note that these models, drawn in Fig. 4.5, can represent losses even at zero frequency, where the current through the ideal capacitor is zero and the voltage across the ideal inductor is zero. The losses in the zero-frequency case would be caused solely by leakage through or around the dielectric in the case of the capacitor and by the resistance of the wire in the case of the inductor.

The other two fundamental two-terminal devices are the sources of excitation, the ideal voltage source and the ideal current source. As in the case of the three passive elements these devices are defined by their current-voltage relations, but these relations take on a particularly simple form. For the ideal voltage source, the voltage across its terminals is a definite value, independent of the current supplied by the voltage source. A lead storage battery is good condition is an approximation to an ideal voltage source at zero frequency, since it can maintain a voltage over even a quite low resistance external load. However, it cannot supply more than a couple of hundred amperes because of its internal resistance. A real voltage generator will always have an internal resistance, and the real generator may be modeled as in Fig. 4.6. Another nearly ideal voltage

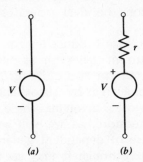

(a)   (b)

Fig. 4.   6. Symbols for (a) ideal and (b) real voltage generators.

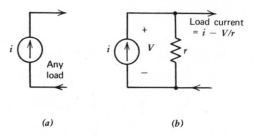

Fig. 4.   7. Symbols for (*a*) ideal and (*b*) real current generators.

source would be the 115-V, 60-Hz power line. In fact, the generators at a power generating station are capable of supplying so much current that it is imperative to introduce a current-limiting fuse or circuit breaker in any 60-Hz power circuit. These fuses or circuit breakers normally are placed in a junction box near where the power lines enter a building but, in addition, fuses that limit the current to a smaller value than the main fuses are always placed in the input power line in electronic instruments to prevent destructively large currents from flowing in case of an accidental short circuit.

The other type of excitation in a circuit is the ideal current generator (Fig. 4.7a). No commonly known device has the ideal characteristic of a current generator, which is that the current supplied from its terminals is the same independent of the circuit to which it is connected. However, it is possible to design relatively simple electronic circuits that approximate this behavior. Of course, if the terminals of a current generator were open-circuited (infinite resistance), an infinite voltage would be developed by any finite current. Thus real current generators often supply current that is relatively constant up to a certain limiting voltage across the terminals of the generator. Another way in which the behavior of a real current source may differ from the ideal is that the current it supplies may fall gradually as the voltage across its terminals rises. This nonideal behavior usually may be adequately modeled by a resistor placed across the terminals of the generator as in Fig. 4.7b, so that, as the voltage rises, more and more current from the ideal generator flows across the internal parallel resistance.

In some circuits the voltage produced by a voltage generator or the current produced by a current generator at one place in a circuit may depend on the voltage or current at a quite different place in the circuit. In particular, active devices such as transistors and vacuum tubes are often modeled by simple networks involving these *dependent* voltage and current generators. Furthermore, as we will see in a later section on two-port networks, the use of dependent generators is not restricted to models of active devices. Although it may seem strange that a current through a device may depend on the voltage across another device elsewhere in a circuit, the analysis of circuits containing these dependent generators is quite straightforward and poses no special problems. Later examples will clarify the use of dependent generators.

4.3. Time-varying voltage and current: impedance and admittance
So far we have used none of the techniques developed in Chapters 1 to 3. Those techniques are useful whenever the problem is to compute the current through or voltage across any branch of a circuit in response to a voltage or current generator in another part of the circuit. Although such a problem can be solved by applying Kirchhoff's laws (see the following section) to obtain a set of differential equations for the entire network, an easier approach is to set up and solve a set of *algebraic* equations obtained by applying the same Kirchhoff's laws. The idea is to analyze the voltage or current source waveform into a sum (or integral) of exponential functions of the form e^{st} and then to use the transform philosophy. That is, the response, the desired current or voltage elsewhere in the circuit, is computed for each separate e^{st} excitation, and the responses to all the e^{st} excitations are added to give the actual response (see the next section for examples). Now we must find the relation between current and voltage in the three basic circuit elements — the resistor, capacitor, and inductor — in order to pave the way for the application of Kirchhoff's laws.

To find the current-voltage relations for the circuit elements when the voltages and currents are exponential in form, we simply use the basic equations for the resistor, capacitor, and inductor (Eqs. 4.1, 4.3, and 4.4.). For example, if we let the current through a resistor be Ie^{st}, where I and s are constants in time, the voltage across the resistor is IRe^{st} from Eq. 4.1. Or, if we let the voltage across a capacitor be Ve^{st}, the current is, from Eq. 4.3, simply $C(d/dt)Ve^{st}$ or $sCVe^{st}$. Finally, if we let the current through an inductor be Ie^{st}, Eq. 4.4 tells us that the voltage across the inductor is $L(d/dt)Ie^{st}$ or $sLIe^{st}$. These results have a common property that a voltage or current of the form e^{st} in any one of the basic circuit elements generates a current or voltage of exactly the same form in time — namely, e^{st} In other words, e^{st} is an eigenfunction of these basic circuit elements; either the voltage or curent can be considered to be the input to the system and the other variable can then be considered to be the output. Because ideal resistors, capacitors, and inductors are linear and time invariant, we should not be surprised at this result.

Of particular use in the analysis of circuits is the concept of impedance along with its reciprocal, admittance. The impedance of any two-terminal circuit, for example, a resistor, capacitor, or inductor, is defined to be the ratio of the exponential voltage across the terminals to the exponential current into one terminal. Since the symbol for impedance is usually Z, for a resistor,

$$Z = \frac{IRe^{st}}{Ie^{st}} = R \tag{4.5}$$

Similarly, for a capacitor, the impedance is

$$Z = \frac{Ve^{st}}{sCVe^{st}} = \frac{1}{sC} \tag{4.6}$$

and for an inductor the impedance is

$$Z = \frac{sLIe^{st}}{Ie^{st}} = sL \tag{4.7}$$

Notice in particular that each impedance is independent of time but that in the case of the capacitor and inductor, the impedance does depend on the value of s. Thus for dc signals, for which $s = 0$, the impedance of an ideal capacitor is infinite while the impedance of an ideal inductor is zero.

The admittance of a two-terminal device is the reciprocal of the impedance and is usually denoted by the symbol Y. For the three ideal circuit elements, the admittances are

$$\text{resistor } Y = 1/R$$
$$\text{capacitor } Y = sC \tag{4.8}$$
$$\text{inductor } Y = 1/sL$$

Any two-terminal network made up of resistors, capacitors, and inductors has an impedance defined as the ratio of the voltage to the current for exponential signals and an admittance defined as the inverse of that ratio. For example, look at the parallel combination of a resistor and a capacitor shown in Fig. 4.8. If the voltage across the combination is Ve^{st}, the current in the upper branch is $(V/R)e^{st}$, and the current in the lower branch is $VsCe^{st}$. The admittance of the combination, the ratio of the current to the voltage, is just

$$Y = \frac{1}{R} + sC \tag{4.9}$$

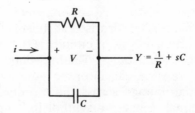

Fig. 4.   8. A simple two-terminal network.

which is the sum of the individual admittances. Of course the impedance is the reciprocal of the admittance, or:

$$Z = \frac{1}{1/R + sC} = \frac{R}{1 + sRC} \qquad (4.10)$$

The generalization of Eq. 4.9 to any number of admittances in parallel should be obvious. It is

$$Y = Y_1 + Y_2 + Y_3 + \ldots \qquad (4.11)$$

A similar rule holds for the impedance of a network made up of several impedances in series. Since $V = iZ$ for each impedance, and since voltages in series add, the total voltage across the series combination is

$$V = iZ_1 + iZ_2 + iZ_3 + \ldots = i(Z_1 + Z_2 + Z_3 + \ldots) \qquad (4.12)$$

and the impedance of the entire series network is

$$Z = Z_1 + Z_2 + \ldots \qquad (4.13)$$

While in modern books and articles on circuits it is common to express impedances and admittances in terms of s, we must not forget that, for most signals (and for all signals that are man-made and therefore begin and end at finite times) a Fourier analysis can be made. In other words, the signals of interest can always be analyzed into sums (for periodic signals) or integrals (for nonperiodic signals) of exponential functions e^{st} involving purely *imaginary* values of s. This is the frequency domain analysis that we developed in Chapter 3. Notice that even the idea of a truly periodic signal is a fiction because of the requirement that man-made signals must have a beginning and an end. However, it is a useful approximation to imagine that certain signals that repeat many times are truly periodic and can be represented by a Fourier series rather than by a Fourier integral. In any case, whether we are dealing with a Fourier series or a Fourier integral, a single Fourier component can be written in the form of a constant times $e^{j2\pi ft}$, where f is the frequency of that particular Fourier component of the total signal. As shown explicitly by theorem 1, Table 3.2, in the Fourier representation in which individual Fourier components are complex exponentials, to represent a real voltage or current we must always add a negative frequency term with a coefficient that is the complex conjugate of the coefficient of the positive frequency term. That is, the Fourier transform of a real signal has "conjugate symmetry."

Because of the importance of Fourier transformable signals and especially of sinusoidal signals, which consist of positive and negative frequency components of equal magnitude (theorems 3 and 4, Table 3.1), we next investigate the impedance and admittance of a few simple circuits for pure imaginary values of s.

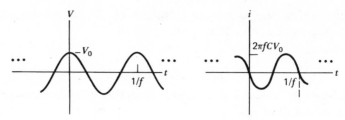

Fig. 4.   9. Relative phase of sinusoidal excitations in an ideal capacitor.

Example.   Find the current through a capacitor if the voltage across it is $V_0 \cos 2\pi ft$.

The easiest and most straightforward way of doing this problem is to return to Eq. 4.3. Thus

$$i = C \frac{dV}{dt} = - 2\pi fCV_0 \sin 2\pi ft \qquad (4.14)$$

The voltage and current are shown as time functions in Fig. 4.9. However, to illustrate the use of impedance and admittance concepts, we redo the problem by analyzing the voltage into exponential functions, finding the current that flows for each Fourier component of voltage, and adding the Fourier components of the current to get the total current. The input voltage may be analyzed into Fourier components by means of the identity:

$$V_0 \cos 2\pi ft = \frac{V_0}{2} e^{j2\pi ft} + \frac{V_0}{2} e^{-j2\pi ft} \qquad (4.15)$$

Now the admittance of a capacitor is sC; therefore, for each Fourier component of the voltage, we need only multiply by the admittance to obtain the corresponding Fourier component of the current. For one of the Fourier components, $s = j2\pi f$, and for the other Fourier component $s = -j2\pi f$. Also we note that the amplitude of each of the two Fourier components of the voltage is $V_0/2$. Thus the amplitudes of the two Fourier components of the current are $(V_0/2) \cdot j2\pi fC$ and $(-V_0/2) \cdot j2\pi fC$. If we write out the complete expression for the current through the capacitor, we get

$$i = 2\pi fCV_0 \left(\frac{j}{2} \cdot e^{j2\pi ft} - \frac{j}{2} \cdot e^{-j2\pi ft} \right) \qquad (4.16)$$

The expression in parenthesis may be recognized as $-\sin 2\pi ft$ so that the total current is $-2\pi fCV_0 \sin 2\pi ft$, which agrees with Eq. 4.14.

Because we know in advance that a real voltage across a device will generate a real current, it is not necessary to perform a detailed calculation for the coefficient of the negative-frequency Fourier component of the current, as we did in the example. It should be clear by now that because the current must be real, this coefficient must be the complex conjugate of the coefficient of the positive-frequency Fourier component. Thus in practice, one may think only in terms of the positive frequency component and say, for example, that the admittance of the capacitor is sC with $s = j2\pi f$, or:

$$Y = j2\pi fC \qquad (4.17)$$

However, it is important to remember that there is always a negative-frequency Fourier component associated with each positive frequency component.

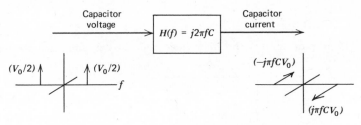

Fig. 4.   10. A systems model of an ideal capacitor.

Finally, consider the previous example from a "systems" point of view (see Fig. 4.10). The Fourier transform of the input signal, which in this case is the voltage across the capacitor, is a pair of impulses in frequency (Fig. 4.10). These impulses each have an area of $V_0/2$ and are purely real because the input voltage is a cosine function. To find the output (the current) produced by the positive-frequency voltage impulse, multiply by the admittance $sC = j2\pi fC$. Multiplication by a purely imaginary number has the effect of rotating a complex number through $90°$ in the complex plane, so the positive frequency current appears in the purely imaginary direction with magnitude $(V_0/2) \cdot 2\pi fC$. The negative frequency current must have a coefficient that is the complex conjugate so that its impulse lies in the negative imaginary direction. The combination of the pair of impulses of Fig. 4.10b is a negative sine function of amplitude $2\pi fCV_0$.

4.4. Bode plots and pole-zero plots   Impedances and admittances are often shown graphically by means of logarithmic plots in which the *magnitude* of the impedance is plotted as a function of the frequency for positive frequencies *with both variables on logarithmic scales*. Such graphs are known as Bode plots. Bode plots for the impedance and admittance of a capacitor are drawn in Fig. 4.11. Only the magnitude of an impedance is shown on a Bode plot; sometimes, however, an auxiliary plot is made on which the phase angle of the impedance is plotted as a function of f, with

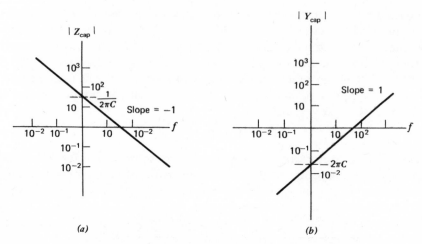

Fig. 4.   11. Bode plot for (*a*) the impedance and (*b*) the admittance of an ideal capacitor.

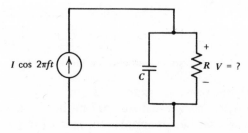

Fig. 4.   12. Example of a two-terminal network driven by a current source.

f on a logarithmic scale. The phase angles for the impedance and admittance of the capacitor are $-90°$ and $+90°$, respectively. (The $+90°$-phase angle for the admittance means that any positive frequency voltage generates a current that is rotated by $+90°$ in the complex plane as in Fig. 4.10.)

The student may wonder why we take the trouble to use frequency-domain concepts to describe the basic circuit elements when the time domain relations of Eqs. 4.1, 4.3, and 4.4 are so simple. The reason is that, as with most powerful techniques, their advantages become more apparent as the problems become more difficult. In another example (Fig. 4.12), a sinusoidal current generator is connected to a two-terminal network consisting of a parallel combination of a capacitor and resistor.

Example.   Find the voltage developed across the RC network as a function of the frequency of the current generator.

The voltage desired is just the impedance of the parallel RC network times the current $I \cos 2\pi ft$ flowing into it. We have already expressed the impedance of such a network in terms of s in Eq. 4.10. A Fourier analysis of the sinusoidal current source tells us that there are the usual positive and negative frequency components. It is, therefore, useful to make a Bode plot of the impedance for positive frequencies, obtained by substituting $j2\pi f$ for s in Eq. 4.10.

$$Z = \frac{R}{1 + j2\pi fRC} \tag{4.18}$$

For frequencies much less than $1/2\pi RC$, the imaginary term in the denominator is negligible, and the magnitude of the impedance is closely equal to R. Correspondingly, for frequencies much higher than $1/2\pi RC$, the magnitude of the impedance is closely equal to $1/2\pi fC$. Physically, this means that a low-frequency current flows primarily through R and a high-frequency current flows primarily through C. The low-frequency impedance R and the high-frequency impedance $1/2\pi fC$ are straight lines on a Bode plot because of the logarithmic scales. As shown in Fig. 4.13a, these straight lines intersect at a frequency given by $f = 1/2\pi RC$, called the cutoff frequency. The straight line Bode plot is accurate enough for many purposes. However, if an exact Bode plot is desired, one must determine the magnitude of the impedance exactly. To do this we use the mathematical identity:

$$|Z| = \sqrt{Z \cdot Z^*} \tag{4.19}$$

So that, for the parallel RC network;

$$|Z| = \sqrt{\frac{R}{1 + j2\pi fRC} \cdot \frac{R}{1 - j2\pi fRC}} = \frac{R}{\sqrt{1 + 4\pi^2 f^2 R^2 C^2}} \tag{4.20}$$

A more careful Bode plot, Fig. 4.13b, shows that the impedance actually varies smoothly in the vicinity of the cutoff frequency.

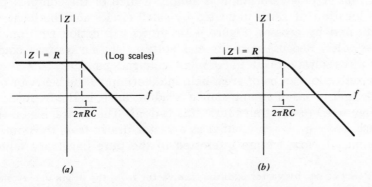

(a)   (b)

Fig. 4.   13. Bode plots of the impedance of the RC network of Fig. 4.12: (a) straight line approximations (b) actual.

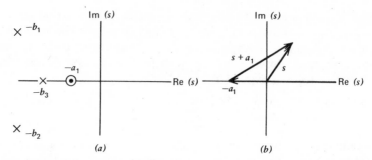

Fig. 4.   14. Pole-zero plots: (a) for a function with one zero and three poles; (b) geometrical interpretation of the factor $s + a$.

Impedances and admittances of circuits made up of interconnected ideal resistors, capacitors, and inductors will always turn out to be a ratio of polynomials in s. The parallel RC circuit of Fig. 4.8 was an especially simple example; the admittance had a constant, 1, in the denominator, and the impedance had a constant, R in the numerator. In general, if we imagine the numerator and denominator polynomials to be in factored form, an impedance (or admittance) may always be written:

$$Z(s) = K \cdot \frac{(s + a_1)(s + a_2)(s + a_3) \ldots}{(s + b_1)(s + b_2)(s + b_3) \ldots} \tag{4.21}$$

where the constants K, a_i, and b_i may be complex.

Now, even though we would ordinarily be interested in doing a Fourier analysis in which $s = j2\pi f$, let us treat s as a complex variable. Then $Z(s)$ will be zero whenever s takes on one of the values $-a_1$, $-a_2$, etc., and $Z(s)$ will be infinite whenever s takes on one of the values $-b_1$, $-b_2$, etc. The special values of s for which the impedance (admittance) function becomes zero or infinity are called, respectively, zeros and poles. The zeros and poles are often illustrated graphically on a pole-zero plot such as in Fig. 4.14a. A pole-zero graph is simply a map of the complex plane, with the location of zeros indicated by small circles and the location of poles indicated by crosses. Figure 4.14 shows a function with one zero and three poles. Because currents and voltages are always real, complex poles (or zeros) always occur in complex-conjugate pairs.

A pole-zero plot can be of great help in determining the behavior of the magnitude and phase of an impedance or admittance, as the frequency f is varied. Figure 4.14b illustrates how this is done. The central idea is that a factor such as $s + a_1$ is represented as a vector drawn from the point $-a_1$ to the point s.† Now we are interested in the pure imaginary values of

---

†This is easy to see by vector addition; the vector from the origin to $-a_1$ plus the vector $s + a_1$ from a_1 to s is equal to the vector from the origin to s. That is,

$$-a_1 + (s + a_1) = s$$

$s : s = j2\pi f$. In other words, the component of a sine or cosine signal with positive frequency is represented by a vector from the origin to a point on the positive imaginary s axis. The individual factors $(s + a_i)$, $(s + b_i)$ are then represented as vectors from the points $-a_i$, $-b_i$ to the point $j2\pi f$ on the positive imaginary axis. To find the magnitude of $Z(s)$, we make use of the fact that the magnitude of a complex number is equal to the product or quotient of the magnitudes of all of its factors.† Thus $| Z(s) |$ is the product of the lengths of all the vectors from the zeros divided by the lengths of all the vectors from the poles. Similarly the phase angle of $Z(s)$ is given by the sum of the phase angles of the numerator factors minus the sum of the phase angles of all the denominator factors.

Example.   Sketch the pole-zero plot for the impedance of the parallel RC network of Fig. 4.8.

We calculated this impedance as a function of s in Eq. 4.10, which we rewrite as:

$$Z(s) = \frac{1}{C} \cdot \frac{1}{(s + 1/RC)} \tag{4.22}$$

There is a pole at $s = -1/RC$, on the negative real axis, and there are no zeros. Thus the pole-zero plot is as shown in Fig. 4.15. To find the

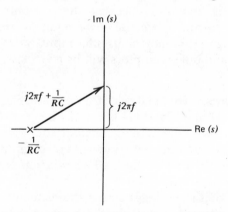

Fig. 4.   15. Pole-zero plot for the impedance of a parallel RC network.

† This is proved by writing each factor of $Z(s)$ in polar form:

$$Z(s) = \frac{A_1 e^{j\alpha_1} \cdot A_2 e^{j\alpha_2} \ldots}{B_1 e^{j\beta_1} \cdot B_2 e^{j\beta_2} \ldots}$$

The coefficient:

$$\frac{A_1 \cdot A_2 \ldots}{B_1 \cdot B_2 \ldots}$$

is the magnitude, and $(\alpha_1 + \alpha_2 + \ldots) - (\beta_1 + \beta_2 \ldots)$ is the phase of the product.

magnitude of the impedance we simply divide $1/C$ by the magnitude of the vector from the pole to the appropriate s value on the imaginary axis. The result is

$$| Z(s) | = \frac{1/C}{| s + (1/RC) |} = \frac{1/C}{\sqrt{(2\pi f)^2 + (1/RC)^2}} = \frac{R}{\sqrt{4\pi^2 f^2 R^2 C^2 + 1}}$$

(4.23)

where we have used the Pythagorean theorem to find the length of $s + 1/RC$. The phase angle θ of the impedance is given by:

$$\theta = -\tan^{-1} 2\pi f RC \tag{4.24}$$

A negative phase angle for the impedance implies that for a sinusoidal input current, the output voltage *lags* the input current by that angle. Note that the straight line portions of the Bode plot for this example (Fig. 4.13) correspond to points very close to the origin on the pole-zero plot (for low frequencies) and to points very far up on the imaginary axis (for high frequencies). The cutoff frequency $f = 1/2\pi RC$ is a special point; the phase angle is seen to be exactly $45°$ at this frequency.

The example we have just completed is an example of a first-order, low-pass filter: it is first order because there is only one factor in the denominator, and low-pass because low-frequency sinusoidal inputs are "passed," while high-frequency sinusoidal inputs are "attenuated." More examples of Bode and pole-zero plots will be given later.

4.5. Kirchhoff's laws   The analysis of a complex circuit is based on two laws known as Kirchhoff's voltage law — KVL — and Kirchhoff's current law — KCL. These laws allow one to calculate the "responses" of a circuit; that is, the voltages across and the currents through the various elements, when the "excitations" (the voltage and current sources) are given.

KVL.   The sum of the voltages across the elements and sources as one traverses a closed loop is zero. (Proper attention must be paid to the algebraic signs of the various voltages.)

KCL.   The sum of the currents into any junction point of a circuit is zero. (Again, proper attention must be paid to the algebraic signs.)

Given these two laws, there are two basic procedures by which one analyzes a complicated circuit, the loop method and the nodal method. Sometimes one of these is simpler to carry out than the other. The following example will illustrate both methods.

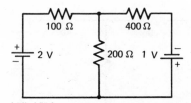

Fig. 4.   16. A two-loop circuit driven by dc voltage sources.

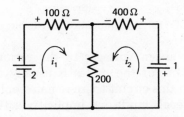

Fig. 4.   17. Assignment of loop currents.

Example. Find the currents flowing through each resistor in Fig. 4.16.

Loop method: First assign unknown "loop" currents i_1 and i_2 to the right and left loops in the circuit as in Fig. 4.17. By assigning loop currents we automatically satisfy KCL because any loop current that flows into a junction also flows out. Now go around the left loop in the direction assumed for the current, writing down the voltage across each element or source in turn, with the sign of the term being the sign of the voltage at the end of the element that we enter. Thus, starting at the lower left-hand corner we have three terms:

$$-2 + 100 i_1 + 200(i_1 + i_2) = 0 \qquad (4.25)$$

(Remember that the current into the top of the 200-Ω resistor is $i_1 + i_2$.) Do the same around the right-hand loop starting, say, at the lower right-hand corner.

$$1 + 400 i_2 + 200(i_1 + i_2) = 0 \qquad (4.26)$$

Now solve this set of algebraic equations by any method you know (i.e., brute force, determinants, or matrix methods) to find:

$$i_1 = 0.01 \text{ A}$$
$$\qquad\qquad\qquad\qquad (4.27)$$
$$i_2 = -0.005 \text{ A}$$

The minus sign on i_2 simply means that the actual i_2 is flowing around the right-hand loop in the opposite direction to the counterclockwise direction we originally assumed. The current in the 200-Ω resistor is, of course, $i_1 + i_2 = 0.005$ A in the downward direction.

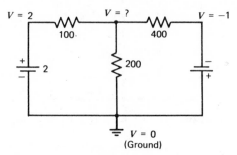

Fig. 4. 18. Assignment of nodal voltages.

Nodal method: First assign voltages to the various nodes or junction points in the circuit. One of the nodes is usually chosen to be at zero voltage or "ground." In this circuit there remains only one other node. By assigning a voltage to each mode we implicitly satisfy KVL. The nodal voltages are shown in Fig. 4.18. Now we use the KCL and equate the currents flowing into each node to the currents flowing out. These currents are written in terms of the node voltages. Thus, since the voltage at the left end of the 100-Ω resistor is 2 V, and $V = iR$ for a resistor, the current that is flowing to the right through this resistor is $(2 - V)/100$. Similarly, the current flowing to the right through the 400-Ω resistor is $[V - (-1)]/400$ or $(V + 1)/400$. The current flowing down through the 200-Ω resistor is just $V/200$. KCL tells us that

$$\frac{2 - V}{100} = \frac{V + 1}{400} + \frac{V}{200} \qquad \text{or} \qquad V = 1 \qquad (4.28)$$

Once we have calculated V, we can go back and compute the various currents, which, of course, turn out to be equal to the ones we computed by the loop method. Notice that in this particular example, which is a fairly typical one, the nodal method is simpler, involving the solution of only one algebraic equation instead of two.

In the example we have just worked out, the voltage sources were zero-frequency sources or batteries. However, Kirchhoff's laws are applicable to circuits in which the sources are sinusoidal or analyzable into any number of e^{st} time functions. The next example illustrates the use of Kirchhoff's laws for a circuit with a sinusoidal excitation.

Example. *Twin-tee network.* Find the current in the extreme right-hand branch of the circuit in Fig. 4.19 when the applied voltage is a unit sine wave $V(t) = \sin 2\pi ft$.

The first step is to assign voltages V_1 and V_2 at the two nodes where the voltages are not already known. (At the left end of the twin-tee network, the voltage is $\sin 2\pi ft$ and at the right end the voltage is zero.)

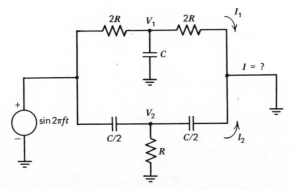

Fig. 4.   19. Twin-tee network driven by a sinusoidal voltage source.

Next we recognize that the sinusoidal input voltage can be written as the sum of two complex exponentials, so we need to find only the response to a general e^{st} excitation voltage. The actual time function e^{st} appears in every term of the Kirchhoff's current law equations for the circuit and need not be written down explicitly. That is, when we write that the current through the lower left capacitor is $(sC/2)(1 - V_2)$, we really mean that this expression is the magnitude of a current $(sC/2)$ $(1 - V_2)e^{st}$, flowing in response to exponential voltages $1 \cdot e^{st}$ and $V_2 e^{st}$ at the left and right ends of that capacitor.

The KCL law gives us two equations, one involving V_1, the other involving V_2. These may be written down by inspection:

$$\frac{(1 - V_1)}{2R} = V_1 sC + \frac{V_1}{2R} \tag{4.29}$$

$$(1 - V_2)\frac{sC}{2} = \frac{V_2}{R} + V_2 \frac{sC}{2}$$

and solved for V_1 and V_2 to give:

$$V_1 = \frac{1}{2(1 + sCR)} \tag{4.30}$$

$$V_2 = \frac{sCR}{2(1 + sCR)}$$

The currents I_1 and I_2 are found by multiplying these voltages by the appropriate admittances, $1/2R$ and $sC/2$, respectively.

$$I_1 = \frac{1}{4R(1 + sCR)} \tag{4.31}$$

$$I_2 = \frac{s^2 RC^2}{4(1 + sCR)}$$

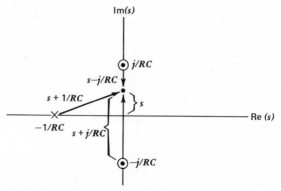

Fig. 4.   20. Pole-zero plot for the total current I in the twin-tee network.

The total current is therefore

$$I = \frac{1}{4(1 + sCR)} \cdot \left(\frac{1}{R} + s^2 RC^2\right) = \frac{C}{4} \cdot \frac{(s^2 + 1/R^2 C^2)}{s + 1/RC} \tag{4.32}$$

This result may be interpreted with the help of a pole-zero plot in Fig. 4.20. There are zeros at the complex conjugate points $\pm j/RC$ and a pole on the negative real axis at $-1/RC$. Imagine the frequency to vary, beginning at $f = 0$. As long as the frequency is near zero, the current is roughly independent of frequency, and the phase angle is zero. As s approaches j/RC (or as f approaches $1/2\pi RC$), the current goes to zero. When f passes this critical value the current increases again and, for very large frequencies, where all three factors are represented by vertical or nearly vertical vectors, the current increases in proportion to f. Just below the critical frequency the phase angle is $-45°$, $(-90° + 90° - 45°)$; just above the critical frequency the phase angle is $135°$, $(90° + 90° - 45°)$; and at very high frequencies the phase angle is $90°$, $(90° + 90° - 90°)$.

4.6. Two-port networks   Frequently electronic circuits are connected together in cascade as illustrated in Fig. 4.21. Although we might be led to believe from the discussion of cascaded LTI systems in Chapter 2 that we need only multiply the system functions of the various circuits to obtain the system function for the entire network, such is not quite the case. Additional complications arise because of what are known as loading effects. If the output of an LTI network is a voltage, this output voltage

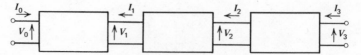

Fig. 4.   21. Cascaded two-port networks.

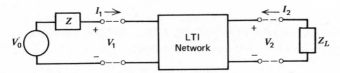

Fig. 4.   22. Two-port network connected to voltage source and load.

will depend not only on the network and the input to that network but also on the circuitry connected to the output. A typical situation is diagrammed in Fig. 4.22. Here the input is represented as a voltage source V_0 in series with an impedance Z. The voltage V_1 developed across the output terminals of the source, and the current I_1, which flows out of these terminals, depend on the properties of the LTI network to which the input is connected. If a load impedance Z_L is connected to the output terminals of the LTI network, our problem, typically, is to calculate the four quantities I_1, V_1, I_2 and V_2 in terms of the source voltage V_0, the source and load impedances Z and Z_L, and the parameters of the LTI network.

A frequency domain analysis of the system in Fig. 4.22 is considerably simpler than a time domain analysis because, of course, we can imagine that the source voltage can be expressed as a sum (or perhaps integral) of exponential time functions of the form $V_0 e^{st}$ with various values of s and the corresponding amplitude factors V_0. Each individual exponential source voltage will cause exponential input and output voltages and currents of the form $I_1 e^{st}$, $V_1 e^{st}$, etc. If we can calculate the four amplitude factors $I_1(s)$, $V_1(s)$, $I_2(s)$, and $V_2(s)$ as functions of s, a complete solution for the voltages and currents is then obtained by adding the responses to the various exponential inputs with the proper weighting factors. For example, if the source voltage, expressed as a time function, is

$$v_0(t) = \sum_n V_{0n} e^{s_n t} \qquad (4.33)$$

then the solution for, say, the time function $v_2(t)$ would be

$$v_2(t) = \sum_n V_2(s_n) V_{0n} e^{s_n t} \qquad (4.34)$$

with similar expressions for the time functions $i_1(t)$, $v_1(t)$, and $i_2(t)$.

With the basic frequency domain approach in mind, let us see how we would characterize the LTI network and compute the amplitudes of the exponential voltages and currents. The network is described completely by two linear algebraic equations in which two of the four variables are the dependent variables and the other two are the independent variables. Since there are six ways of choosing the dependent variables, we see that there is quite a bit of latitude in our exact means of describing the network. Some of these six descriptions are used more often than others; for example, in one popular representation, the dependent variables are the input voltage and output current. In this representation, the coefficients in the characteristic equations for the network are called the h parameters, and the two equations are written:

$$V_1 = h_i I_1 + h_r V_2 \tag{4.35}$$

$$I_2 = h_f I_1 + h_o V_2$$

The subscripts on the h parameters serve as aids in remembering where they occur in the equations. Thus, h_i is the input impedance (ratio of input voltage to input current), with the output short-circuited ($V_2 = 0$). Similarly, h_0 is the |output| admittance (ratio of |output| current to output voltage) with the input open-circuited, h_f is the forward current transfer ratio (ratio of output current to input current) with output short-circuited, and h_r is the reverse voltage transfer ratio (ratio of input voltage to output voltage) with the input open-circuited. It is important to remember that Eqs. 4.35 describe the network for any signal of the form e^{st}. I_1 and V_2 are the amplitudes of the exponentially varying input current and output voltage, and V_1 and I_2 are the resulting amplitudes of the exponentially varying input voltage and output current. The four h parameters usually have values that depend on s. In this respect, the h parameters behave like generalized system functions.

Equations 4.35 can be visualized with the aid of the circuit model in Fig. 4.23. Here h_i is an impedance in series with a *dependent* voltage generator $h_r V_2$, and h_o is an admittance in parallel with a *dependent* current generator $h_f I_1$.

Since there are four unknowns to be found, we need four equations involving these unknowns. Equations 4.35 provide two of these; the other two describe what we might call the current-voltage relations for the input and output circuits. Thus, an application of Kirchhoff's voltage law to the input circuit in Fig. 4.22 shows that

$$V_0 = I_1 Z + V_1 \tag{4.36}$$

and to the output circuit shows that

$$V_2 = -I_2 Z_L \tag{4.37}$$

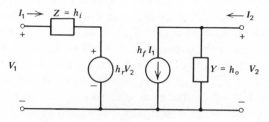

Fig. 4.   23. An h-parameter model of a two-port network.

It is a straightforward, though somewhat tedious, task to solve Eqs. 4.35, 4.36, and 4.37 for the four unknowns I_1, V_1, I_2, and V_2.

4.7. An example of the *h*-parameter representation of an LTI network

There are two ways of determining the four complex coefficients (the *h* parameters) in Eqs. 4.35. The most direct method is to measure them experimentally with sinusoidal inputs, that is, to take data from which the magnitude and phase of each of the *h* parameters can be calculated as a function of *s*. The other method of determining these parameters is to make some kind of a model, based on the actual physics of the network, that contains elements such as resistors, capacitors, inductors, and voltage and current generators. Later we will give an example of how this model is constructed for a real device such as a transistor, but now we want to show how, given the circuit model of an LTI electrical network, we can compute the magnitudes and phases of the *h* parameters as functions of *s*.

The example we will discuss is drawn in Fig. 4.24. Notice that it contains a dependent current generator, $g_m V$. This circuit happens to be one possible model (known as a hybrid-π model) of a transistor. The full hybrid-π model would contain some additional elements in the form of dc voltage generators (batteries) and dc current generators to assure that the actual dc input and output voltages and currents were correctly described. However, these additional dc generators may be viewed simply as additional inputs of the form Ve^{st} or Ie^{st}, with *s* equal to zero and, since we want to solve ultimately for the amplitudes of exponential voltages and currents with *s* generally *not* equal to zero, we can ignore the contributions to the input and output voltages and currents that are added by these dc generators. The model of Fig. 4.24 breaks down not only for dc inputs but also for ac signals that are too large in amplitude and drive the transistor into nonlinear regions of operation. Later we will examine the limitations of this model of a transistor but presently we simply want to show how the *h* parameters are computed as functions of *s*.

The simplest way of calculating the *h* parameters is to be a little clever in the application of Eqs. 4.35, which define these parameters. For

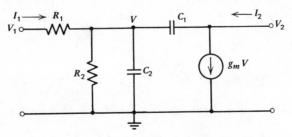

Fig. 4.   24. Example of a two-port network.

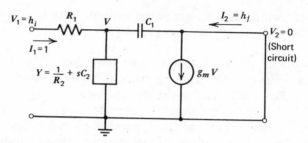

Fig. 4.   25. Short-circuiting the output to obtain h_i and h_f.

example, from the first of Eqs. 4.35, we see that the input voltage is equal to h_i if the input current is 1 and the output voltage is 0. Furthermore, from the second of Eqs. 4.35 with the same conditions of $I_1 = 1$ and $V_2 = 0$, the output current is equal to h_f. In Fig. 4.25 we have redrawn the circuit of Fig. 4.24, with the input and output voltages and currents set equal to the values described above. Also in Fig. 4.25 we have lumped the parallel combination of R_2 and C_2 into a single admittance, $Y = (1/R_2) + sC_2$.

To calculate the input voltage h_i in Fig. 4.25 in terms of the circuit parameters, note first, that from Kirchhoff's law,

$$h_i = 1 \cdot R_1 + V \tag{4.38}$$

We must eliminate V because its value depends on the input voltage, and we can do this by writing a Kirchhoff's current law equation. The sum of the downward current VY through Y and the rightward current VsC_1 through C_1 must equal the input current:

$$1 = VY + VsC_1 \tag{4.39}$$

From Eqs. 4.38 and 4.39, we solve for h_i:

$$h_i = R_1 + 1/(Y + sC_1) = \frac{R_1 + R_2 + sR_1R_2(C_1 + C_2)}{1 + sR_2(C_1 + C_2)} \tag{4.40}$$

It is even easier to solve for h_f, the output current in Fig. 4.25. The current law shows that h_f plus the rightward current VsC_1 through C_1 must equal the current $g_m V$ in the dependent generator. Thus, with the aid of Eq. 4.39, we find

$$h_f = g_m V - VsC_1 = \frac{g_m - sC_1}{(1/R_2) + s(C_1 + C_2)} \tag{4.41}$$

Note that both h_i and h_f have the general form of a ratio in polynomials in s.

It is now apparent that we can compute h_r and h_o by setting $I_1 = 0$ and

$V_2 = 1$ in the original circuit of Fig. 4.24 and finding the input voltage and output current.

4.8. Amplifier input characteristics   As its name implies, an amplifier is an instrument used to increase the level of a signal, either a voltage or a current, so that some output device can respond to the signal. Amplifiers are used extensively in electronic instrumentation. Some typical applications are: in oscilloscopes to raise the input voltage level to where the electron beam can be deflected; in radiation survey instruments to convert the tiny (e.g., 10^{-13} A) current from an ionization chamber into a current that can deflect a meter; in a pH meter to convert a small voltage (e.g., 10^{-2} V) into a current that can deflect a meter while drawing as little current as possible from the pH electrode itself; in bridge (e.g., Wheatstone bridge) amplifiers to detect very small departures from the balance condition in these sensitive comparison instruments; in communications equipment (radio, television) to raise the level of a radio-frequency signal so that other circuits can selectively extract information from them; and in sound reproduction equipment (stereos, public address systems) to convert a few millivolts of signal from a phonograph stylus or microphone into many watts of power in a loudspeaker.

While an amplifier can usually be viewed as a two-port network, it has enough special properties, particularly at its input port, to warrant separate consideration. For example, in a typical well-designed amplifier for any purpose, the load connected to the output port has essentially no effect on the input characteristics. This means that $h_r = 0$ in Eq. 4.35, and the input circuit (Fig. 4.23) reduces to the input impedance h_i. Usually, in the set of specifications of an amplifier, there are two quantities that relate to the input characteristics of the amplifier. One of these is the input impedance, and the other is the bias current. Although the input impedance is a function of s, or f, the input bias current is strictly a dc quantity. The input bias current is necessary for the proper operation of the input circuitry of the amplifier, whether that input circuitry involves a vacuum tube, a bipolar transistor, or a field-effect transistor (see Chapter 6). The input bias current, which may be anywhere from a few microamperes down to 10^{-15} A, must be supplied through some circuit element; sometimes this element is a "bias" resistor connected within the instrument between its input terminal and ground, and sometimes the external circuit must supply the entire bias current.

Three typical input configurations of amplifiers are illustrated in Fig. 4.26. In each case we model the external transducer as an ideal voltage source V in series with a resistance, R_s, called the source resistance. Figure 4.26a shows a direct connection, in which the bias current must be supplied from the source itself. This is the connection that would be used, for example, in a pH meter. For this application the

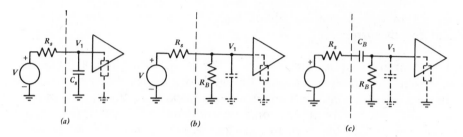

Fig. 4.   26. Typical input connections of amplifiers.

amplifier must have a very high input impedance because R_s is typically several hundred megohms; if the input impedance h_i is not much larger than R_s the amplifier input voltage V_1 will be considerably less than the voltage V to be measured. While the error, $V - V_1$, can be calculated if R_s and h_i are known, the point is that since the amplifier may have to be used with several different input devices with different (and perhaps time-varying) values of R_s, it is best to have $h_i \gg R_s$ to minimize the need for a correction. For a pH electrode application the amplifier must also have a very low bias current (e.g., $<10^{-12}$ A) because the electrode can supply very little current.

Figure 4.26b shows an input connection that can be used when the source resistance R_s is much smaller than the bias resistor R_B. Some of the bias current flows through R_B and some through R_s. This is the typical input connection in an oscilloscope when the input switch is set to "dc" or "dc coupled." The only advantage it has over the circuit in Fig. 4.26a is that when the source is disconnected entirely, there is still a path through which the bias current can flow. The amplifier input impedance in Fig. 4.26b is reduced from that in Fig. 4.26a because the bias resistor is in parallel with the amplifier input. In an oscilloscope a common value for the input resistance (the real part of the input impedance) is 10^6 Ω. If we saw a bias resistor of that magnitude in the circuit diagram and if the input resistance was specified to be 10^6 Ω, we could conclude that the real part of h_i for the remainder of the amplifier was much larger than 10^6 Ω.

The circuit in Fig. 4.26c includes a blocking capacitor C_B between the source resistance and the actual amplifier input terminal. This blocking capacitor is necessary whenever it is desired to amplify only small changes in voltage, which are superposed on a large dc or very slowly varying voltage. This might occur when testing a dc power supply for ripple. The ripple amplitude, because of feedthrough from the 60-Hz power lines, might be on the order of a millivolt, while the dc voltage might be many volts. The blocking capacitor charges to this dc voltage so that the source output potential can be many volts while the amplifier input voltage

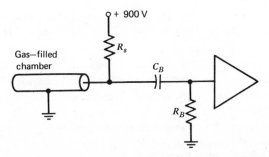

Fig. 4.   27. Blocking the bias voltage of a Geiger-Mueller tube.

remains very close to zero. This is the circuit that is connected in an oscilloscope when the input switch is on ac. Another application of the capacitatively coupled circuit in Fig. 4.26c is in a Geiger counter (Fig. 4.27). Here the detector is a Geiger-Mueller tube, which behaves somewhat as a variable resistance. When there is no ionizing radiation present the tube is nearly an open circuit. When an ionizing particle enters the cylindrical region between the electrodes and forms free electrons and positive ions in the enclosed gas, these charges carry a burst of current. This sudden current produces a small drop in voltage across the series resistor R_s. The blocking capacitor isolates the amplifier input terminal from the high dc voltage present at the anode of the G-M tube, while passing the rapid change in voltage that signals the presence of an ionizing particle.

Figures 4.26a to 4.26c all show an undesirable but unavoidable property of amplifier input connections known as stray capacitance, C_s. This capacitance exists because of the proximity to ground of the wire carrying the input signal and also perhaps because of the details of construction of the input vacuum tube or transistor. In some applications this stray capacitance, usually on the order of 1 to 100 pF (1 pF = 10^{-12} F) is negligible, but often it can pose a problem. For example, suppose we want to measure changes in the light output of a laser over times as short as a nanosecond, using a photomultiplier tube as the transducer. We must charge and discharge the stray capacitance in times of the order of 10^{-9} sec or less. For a stray capacitance of 10 pF this requires an effective source resistance R_s of less than $R_s = 10^{-9}$ sec/10^{-11} F = 10^2 Ω, which may be difficult to achieve. Nor is this problem limited to signals that change over such extremely short times. In the study (neurophysiology) of the potentials developed in nerve cells, the electrodes used are fine glass capillaries, pulled to a tip diameter of less than 10^{-6} m and filled with a conducting salt solution. Because of their small diameter, required to penetrate cell membranes, the source resistance of these electrodes is often about 100 MΩ. In order to measure cell potential changes that occur over times of the order of a millisecond, the stray capacitance must be held to less than $C_s = 10^{-3}$ sec/$10^8\ \Omega = 10^{-11}$ F.

The stray capacitance and source resistance can be viewed as an LTI system with input V, the transducer voltage, and output V_1, the amplifier input voltage. From the frequency domain point of view this system is a first-order, low-pass filter of cutoff frequency $1/2\pi R_s C_s$. From the time domain point of view it is a smoothing filter whose impulse response is an exponential pulse of time-constant $R_s C_s$.[†]

Before leaving the subject of amplifier input circuitry we will mention two applications in which an amplifier must have a *low* input impedance. One of these occurs when very high-frequency signals are to be amplified, that is, frequencies such that the wavelength of the associated electromagnetic radiation is of the same order of magnitude as the circuit dimensions, for example, length of cables. Since a wavelength of 1 m corresponds to a frequency of $f = c/\lambda = (3 \times 10^8$ m/sec)/1 m $= 300$ MHz, we should expect to have to deal with the wave aspect of signals at frequencies of this order or higher. Although the special problems that occur with these very high frequencies are beyond the scope of this book, we will mention that one of the problems that occurs is reflection of the wave travelling down a cable from the end of the cable. Reflections are undesirable because they can produce time-delayed signals that add to and interfere with the signal of interest. In order to eliminate reflections, a signal-carrying cable must be "terminated" with the proper resistance as it enters an amplifier. Since typical values of termination resistance for commercially produced cables are in the range of 50 to 100 Ω, amplifiers for these high-frequency signals have input resistances of this order of magnitude.

The other application in which an amplifier must have a low input impedance is when it is intended to amplify a current, without producing an appreciable voltage drop at its input terminal. Examples of this application are in electronic ammeters and in ionization-chamber amplifiers in radiation survey instruments. In the straightforward approach, we might use a small resistor in place of R_B in the circuit of Fig. 4.26b, but then much of the current flows through this small resistor and is, in effect, wasted. It is possible to design amplifiers with very low input impedance that nevertheless use nearly all of the current supplied by the transducer. These amplifiers employ feedback, which is the subject of the next chapter.

[†] These statements are true provided the amplifier input resistance is much greater than the source resistance.

Problems for Chapter 4

4.1. Find h parameters for the following T circuit (Fig. P4.1).

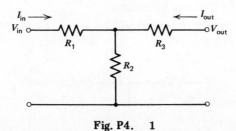

Fig. P4.   1

4.2. Find h parameters for the following π circuit (Fig. P4.2).

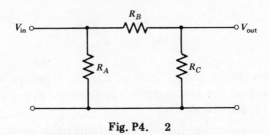

Fig. P4.   2

4.3. (a) Find the current amplitude I_2 in the output branch of the T circuit (in Fig. P4.3), if the input is a voltage source $1 \cdot e^{st}$ and the output is short-circuited to ground.

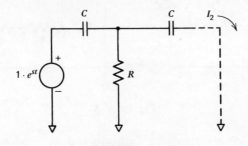

Fig. P4.   3

(b) Find the output voltage under the same input conditions if the output terminal is open-circuited.

(c) Plot the ratio $Z_T(s)$ — sometimes called the Thevenin equivalent output impedance — of open-circuit output voltage to short-circuit output current on a Bode plot ($\log |Z_T(j2\pi f)|$ versus $\log f$ and $\sphericalangle Z_T(j2\pi f)$ versus $\log f$).

(d) Show that the same ratio is obtained if we calculate the impedance looking back into the output terminal with the input voltage source replaced by a short circuit. This impedance is the ratio of a voltage applied to the output divided by the current drawn from the voltage source.

4.4. (a) Find the four h parameters, in terms of R, C, and s, for the T circuit in Fig. P4.4.

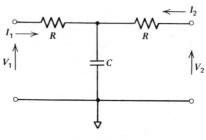

Fig. P4.   4

(b) Construct Bode plots, that is, $\log |h_o|$ versus $\log \omega$ and $\angle h_o$ versus $\log \omega$ for the output admittance, where $j\omega = s$, for the case $RC = 10^{-4}$ sec.

4.5. A photomultiplier tube, for detecting faint light levels, and its biasing network might be modeled as in Fig. P4.5 for *signal* currents.

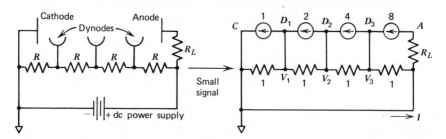

Fig. P4.   5

The emission of electrons by cathode and dynodes is represented by ideal current sources, of values 1, 2, 4, and 8, and all bias resistors are assumed to be equal to 1.

Set up and solve a set of four equations for the power supply current I and the three nodal voltages V_1, V_2, and V_3. What is the ratio of current through the power supply to anode current? Note the pattern of nodal voltages, particularly their polarities.

4.6. An important transistor circuit known as an emitter follower (Fig. P4.6a) has the simplified equivalent circuit in Fig. P4.6b.

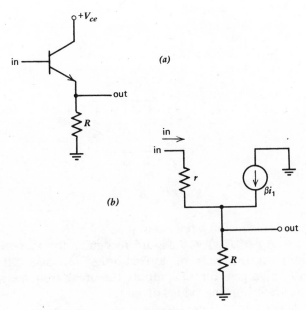

Fig. P4.   6

(a) Find the input impedance of the circuit with the output open-circuited.

(b) Show that with the output open-circuited, the output voltage is $[(1+\beta)R]/[r+(1+\beta)R]$ if the input voltage is 1.

(c) Find all four h parameters of the circuit.

4.7. The circuit (Fig. P4.7) consists of two *dependent* current generators, the second of which is connected across a capacitor C. The portion of the network between the dotted lines is known as a *gyrator*. Show, by finding the input impedance V_1/I_1, that this circuit behaves like an inductor, and find the effective inductance in henries if $g = 2$ mA/V and $C = 1\ \mu$F.

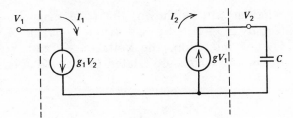

Fig. P4.   7

4.8. (a) Find, by standard circuit analysis, the system function $H(s) \equiv Y(s)/X(s)$ for the circuit shown (Fig. P4.8). Substitute numerical values.

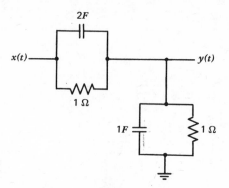

Fig. P4.   8

(b) Sketch and dimension a pole-zero graph for $H(s)$.
(c) From a study of $H(s)$ you should recognize the system as being equivalent to the sum of a first-order low-pass filter and a first-order high-pass filter. Compute the break frequency for each filter and their limiting values of gain.
(d) Sketch and roughly dimension a Bode plot for $\log |H(s)|$ versus $\log \omega$.

4.9. An ideal transformer (see Fig. P4.9) is a four-terminal device in which all of the magnetic flux ϕ, produced by the primary winding of n_1 turns, links the secondary winding of n_2 turns, and vice versa.

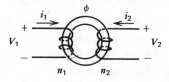

Fig. P4.   9

When the windings are as shown, the total magnetic flux in the magnetic toroid is given by $\phi = k(n_1 i_1 - n_2 i_2)$. According to Faraday's law of induction, the voltages V_1 and V_2 across the primary and secondary coils are related to the rate of change of flux by:

$$V_1 = n_1 \frac{d\phi}{dt}$$

and

$$V_2 = -n_2 \frac{d\phi}{dt}$$

(a) Show that the h parameters are $h_i = 0$, $h_r = -(n_1/n_2)$, $h_f = n_1/n_2$, $h_o = 1/skn_2{}^2$.

(b) Show that if a resistor R is connected across the output terminals, then the input circuit looks like an inductor $L_1 = kn_1{}^2$ in parallel with a resistor $n_1{}^2 R/n_2{}^2$. (Often the inductive reactance is much greater than $n_1{}^2 R/n_2{}^2$ so that the input circuit can be simplified to just the reflected resistance $n_1{}^2 R/n_2{}^2$.) Also show by eliminating I_1 that the resistor in the output is driven by an effective voltage source $(n_2/n_1) \cdot V_1$ with zero series resistance.

REFERENCES

Bleuler, E., and R. O. Haxby, Eds. *Methods of Experimental Physics, Vol. 2, Electronic Methods*, Part A, Chapters 1 and 6, Academic Press, New York, 1975.

Brophy, J. J., *Basic Electronics for Scientists*, Third Edition, Chapter 3, McGraw-Hill, New York, 1977.

Jones, B., *Circuit Electronics for Scientists* Chapters 1, 2, and 3, Addison-Wesley, Reading, Mass., 1974.

CHAPTER 5

Feedback

5.1. Introduction Feedback is a process by which the output of a system affects the input of the system. A familiar physical feedback system is the thermostatically controlled furnace, in which the input (the signal that turns the furnace on) is determined by the difference between the system output (the room temperature) and a preset temperature. Biological feedback occurs in ecological systems, where the system input (say the food supply) is affected by the output (the number of members of the species). Also feedback occurs in the chemical control systems in our bodies and even in the complex chemical reactions within single cells. In electronics, feedback is used in oscillators and signal generators and also in electronic control systems where the deviation of the output of a system from a desired value is amplified and the result used to reduce the deviation. A typical control system is illustrated in Fig. 5.1. There is a goal, or intended result, a power amplifier that works toward the goal, and a sensor that measures the deviation between the actual result and the intended result.

An example of the control system type of feedback is people picking up objects. To begin the process, the intended result is somehow generated and stored within the brain. Their motor nerves and muscles form the power amplifier and their eyes act as the sensor, telling the brain how far their hand is from the object. The brain subconsciously does the calculations necessary to correct any errors in the path of the hand. Of course, when we get pretty good at picking up objects, we can look at, say, a pencil lying on a table and then close our eyes and pick it up. But we are still using feedback. Now the sensor is the network of proprioceptive nerves that report back to the brain the positions of all the joints involved: shoulder, elbow, wrist, and fingers.

5.2. Stability and instability Whenever feedback exists there is the possibility that the overall system can be unstable, that is, that the output can either oscillate or run off to some extreme value that may cause a catastrophic failure. Although this type of feedback is often useful, as in trigger circuits and oscillators, one is generally faced with the task of eliminating an unwanted oscillation. The problem of assuring that a feedback system is stable is not quite as trivial as simply making the sign

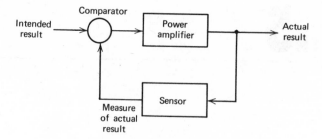

Fig. 5.   1. A control system with feedback.

of the feedback negative or, in other words, adjusting things so that deviations from the desired result that are in the positive direction, are met by a power amplifier response that drives the output in the negative direction, and vice versa. An example will show what can happen.

We are all familiar with the problem of trying to adjust the temperature of a shower so that it neither scalds nor chills us. This is a feedback problem, with the desired temperature as the goal, the combination of our muscles and the water valve as the power amplifier, and temperature-sensitive nerve cells in our skin as the sensor. Everyone has the common sense to apply negative feedback; when the water is too hot we turn the faucet toward "cold" and when the water is too cold we turn it toward "hot." However, especially if the shower is a strange one, you experience an initial difficulty in obtaining a steady, desirable temperature. Small children in particular seem to have an unusually difficult time in learning how to avoid the instability, which is accomplished by making small, rather than large, adjustments in faucet position in response to an error in temperature. The basic reason why a shower control system may become unstable is the delay in various parts of the system. There is delay both in the response of the power amplifier (because of the time required for the water to travel from the valve to our body) and in our sensing and responding to changes in temperature. We will return to the shower problem as an example later, but first we wish to develop a mathematical technique for analyzing feedback systems.

First, for nonlinear or nontime-invariant blocks within a feedback system, there is no general technique, except for trial and error or experiment, by which we can find the output of the overall system for a given input. By restricting ourselves to linear, time-invariant subsystems, however, we can hope to develop some insights and even some approximate solutions to systems that are almost LTI or are LTI for small excursions of the variables. Thus in Fig. 5.1, we will henceforth assume that both the power amplifier subsystem and the sensor subsystem are LTI and, furthermore, that the comparator element simply takes the

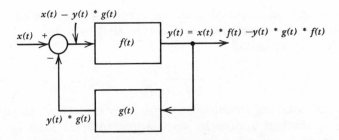

Fig. 5.   2. Time-domain description of a feedback system.

algebraic difference between the intended result and the measurement of the actual result.

Figure 5.2 shows a diagram identical to Fig. 5.1 except that the variables are labeled according to systems notation, with input $x(t)$ and output $y(t)$. The subsystems, assumed to be LTI, are described by their impulse responses $f(t)$ and $g(t)$. The input and output time functions are related by the equation:

$$y(t) = [x(t) - y(t)*g(t)] *f(t)$$

or

$$y(t) + y(t)*g(t)*f(t) = x(t)*f(t) \tag{5.1}$$

where, as in Chapter 1, the asterisk denotes convolution. If the problem is to find $y(t)$, given the input $x(t)$ and the impulse responses, then we face the formidable task of unraveling the convolutions in Eq. 5.1. The problem is similar to that of doing integration. There we work with quite definite rules for finding the derivative of a function, but we are asked to guess what function has the specified derivative. We have the same sort of *implicit* problem here. We are forced to guess at the solution and then substitute it into Eq. 5.1 to see if it works. Fortunately, it is not too hard to be a good guesser in the LTI feedback problem, provided that the input has the form e^{st} or some linear combination of e^{st}'s. In that case, we have learned to expect the output to contain a term of the form $H(s)e^{st}$ for each one of the input exponentials. Here $H(s)$ is the overall system function for the feedback system.

If we substitute $x(t) = e^{st}$ and $y(t) = H(s)e^{st}$ into the second of Eqs. 5.1 and recall that each subsystem simply multiplies its exponential input by its system function, then we arrive at:

$$H(s)e^{st} + H(s)e^{st} \cdot G(s)F(s) = e^{st} \cdot F(s)$$

or

$$H(s) \cdot [1 + G(s)F(s)] = F(s) \tag{5.2}$$

Since this is simply an algebraic equation involving functions of s, we can solve directly for the overall system function $H(s)$ in terms of the system functions of the subsystems:

$$H(s) = \frac{F(s)}{1 + G(s)F(s)} \tag{5.3}$$

Once again, the complex frequency approach proves to have an advantage over the time domain approach, since apparently, as long as we can express our input function by a weighting factor $X(s)$ times e^{st}, we will be able to compute the output as the weighting factor $H(s)X(s)$ times e^{st}. However, we must face up to one embarrassing possibility: what if, for one or more values of s, the factor $1 + G(s)F(s)$ is exactly zero? In terms of Eq. 5.3, this means that the overall system function, $H(s)$, is infinite for these values of s, and even with no input, we can have an output of the form of a sum of e^{st}'s with these critical values of s. If these critical values of the complex frequency all have negative real parts, then these "spontaneous" outputs will die away; but what if there is a value of s that has a positive or zero real part? Then the corresponding exponential time function will grow, or at least not decay, with time, and we have an unstable situation.

Another way to view the problem of instability is as follows. Let the input to the system be zero, and assume an output of the form e^{st}. If this assumption is consistent with the system equations for any s with positive (or zero) real part, then such a signal can exist and, unless the real part of s is exactly zero, will grow larger. This situation is shown in Fig. 5.3a.

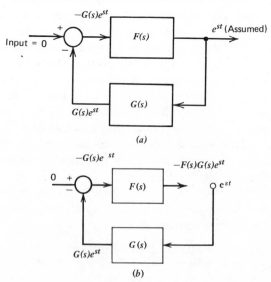

Fig. 5. 3. Determining whether a feedback system is stable.

Under the assumption that the output is e^{st}, the output of the sensor subsystem is $G(s)e^{st}$, which is inverted by the comparator to give $-G(s)e^{st}$. Finally, this signal is multiplied by $F(s)$ as it passes through the power amplifier, and the result must be e^{st}. That is,

$$-F(s)G(s)e^{st} = e^{st}$$

or

$$-F(s)G(s) = \text{OLG}(s) = 1 \qquad (5.4)$$

This equation is known as the *characteristic equation*, and the quantity $-F(s)G(s)$ is often called the open-loop gain (OLG).[†] This is the gain that results if one imagines breaking the feedback loop at any point and applying e^{st} to the free input as illustrated in Fig. 5.3b.[‡] The open-loop gain can be 1 for no s, one s, or more than one s. If any s that satisfies OLG = 1 has a positive or zero real part, then the feedback system is said to be unstable. Otherwise, the feedback system is stable. Sometimes, especially in certain electronic circuits that have circuit elements that are common between the "sensor" and the power amplifier, it is difficult to separate $F(s)$ and $G(s)$. However, it is practially always a straightforward process to determine the open-loop gain around a feedback loop and then test to see if the open-loop gain can be 1 for any s with positive or zero real part.

5.3. Tests for stability   First, it is important to recognize that it is extremely useful to have a mathematical test for stability. Although it is always possible to build a system, turn it on, and watch to see if it is unstable, these tests can be expensive, with the resulting unstable oscillation possibly driving the system to self-destruction. Furthermore, without a mathematical model, it is difficult to determine just what should be changed in order to stabilize a system, because the self-destruction may occur so rapidly that one does not have enough time to find safe values of the operating parameters.

There are several formal tests for stability. One type, known as Routh-Hurwitz tests, is applicable when the open-loop gain is a polynomial in s. These tests involve calculations based on the coefficients of the polynomial. In another type of test, known as the root-locus method, the roots or solutions of the characteristic equation are plotted on a

---

[†] Some authors call $F(s)G(s)$ the open-loop gain, in which case the analog of Eq. 5.4 is that the open-loop gain is equal to -1.

[‡] If one were to try to determine the open-loop gain in the manner suggested in Fig. 5.3b one would have to make the source impedance equal to the output impedance of $F(s)$ and terminate $F(s)$ with an impedance equal to the input impedance of $G(s)$.

complex-number diagram as some parameter, such as the magnitude of $F(s)$, is varied. There exist relatively quick methods of computing such root-loci, and for someone who is going to make a career out of designing feedback control systems they are important to learn. (See, for example, the text by Wilts listed in the references at the end of the chapter.) For our purposes we deal with just two methods: (1) direct algebraic solution for the roots of the characteristic equation and (2) a graphical method, due to Nyquist, for determining whether any roots lie in the right half-plane.

In some situations it is a simple matter to solve Eq. 5.4 algebraically. For example, when the open-loop gain is a simple linear or quadratic polynomial in s, elementary algebra gives the solutions.

Example. Investigate as a function of A the stability of the feedback system of Fig. 5.4 in which the power amplifier has gain A, and the sensor system has impulse response $y(t) = e^{-at}$; $a > 0$, $t > 0$. From number 8, Table 2.1, the system function for the sensor is $1/(s + a)$, the Laplace transform of the exponential pulse. The characteristic equation becomes

$$\text{OLG}(s) = -\frac{A}{s + a} = 1$$

which has a solution $s = -(A + a)$. If A is greater than $-a$, this solution has a negative real part and, therefore, for all positive A and small enough negative A, the system is stable. But if $A = -a$ or is more negative than $-a$, the solution s will have zero or positive real part and the system will be unstable.

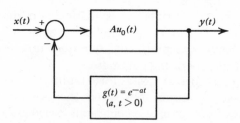

Fig. 5.   4. Example of a feedback system: for what values of A is it stable?

Example. Investigate, as a function of A, the stability of the feedback system of Fig. 5.5, which differs from that of Fig. 5.4 in that the sensor impulse response is te^{-t}; $t > 0$. From number 8, Table 2.1 and number 6, Table 2.2, the Laplace transform of the sensor impulse response is $1/(s + 1)^2$. The characteristic equation becomes

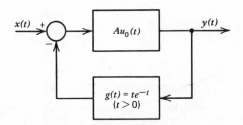

Fig. 5.   5. Another example of a feedback system.

$$\text{OLG}(s) = -\frac{A}{(s + 1)^2} = 1$$

or

$$s^2 + 2s + 1 + A = 0$$

The roots of this equation may be determined from the quadratic formula. They are

$$s = -1 \pm j\sqrt{A}$$

For all positive (or zero) values of A, the real part of the solution is -1 and the system is stable. But for values of A that are more negative than or equal to -1, since one of the solutions has a positive real part, the system is unstable.

Note that there are two steps in the algebraic test for stability:

1.   Solve the equation in which the open-loop gain is set equal to 1.

2.   Investigate the solutions to see if any have a positive or zero real part; if so, the system is unstable.

The examples in the preceding paragraphs involved first-order and second-order, low-pass filters in the sensor system and a wide-band amplifier of gain A in the power amplifier system. In each case the system was stable for positive and small negative values of A but unstable for more negative values of A. For a sensor system that is third order, the open-loop equation becomes a cubic equation. Rather than use an algebraic method of solution, we will introduce next a graphical test for stability, known as the Nyquist plot, which can be useful not only for cubic and higher-order polynomial equations but for other algebraic equations as well.

5.4. The Nyquist plot: a graphical stability test   The method we are about to introduce is analogous to the method of solving transcendental equations by graphs. Thus, to solve $e^x \sin x = 4$, one might simply plot the left side as a function of x and see where it equals 4. However, there is a

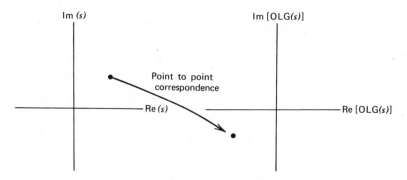

Fig. 5.   6. Mapping a function of a complex variable.

difficulty in determining the values of s for which the open-loop gain of a feedback system is unity. The difficulty is that s is a complex number so that we must investigate the value of the open-loop gain for s values in the entire right-half plane. Because both s and the open-loop gain are in general complex, we need four axes for the graph, rather than two, as in the case of a real transcendental equation. Consider Fig. 5.6. For each complex value of s, represented by a point in the complex s plane, the function OLG(s) has a value, which can be represented by a point in the complex OLG(s) plane. That is, each point in the s plane maps into a point in the OLG(s) plane. The question of stability now boils down to: Is there any point in the right-half s plane (including the imaginary axis) that maps into the point 1 in the OLG(s) plane? If the answer is "yes", then the system is unstable.

At this point readers might object and say that to map all points in the right-half s plane would be very time-consuming. They would be quite right. Fortunately, however, we usually do not need to know the guilty values of s, but only whether there are any! To find out, we use what is called the conformal property† of the complex-plane maps. The idea of a conformal map is shown in Fig. 5.7. A small corner in the s plane maps into a small corner with the same angle between its sides in the OLG(s) plane, though the position and orientation of the two corners may be very

----

†The requirement for a conformal mapping at a point is that the function OLG(s) has a unique nonzero derivative at the point and in the neighborhood of the point. That is,

$$\lim_{\Delta s \to 0} \frac{\text{OLG}(s + \Delta s) - \text{OLG}(s)}{\Delta s}$$

must be independent of the direction of the increment Δs. Such a function is said to be an *analytic* function of the complex variable s. All of the standard functions such as polynomials and exponentials are analytic over the entire s plane except, perhaps, at isolated points.

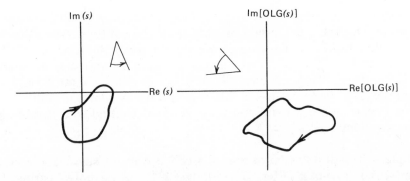

Fig. 5.   7. Properties of a conformal mapping.

different. Furthermore, a closed contour in the s plane traversed in, for example, the clockwise sense, maps into a closed contour traversed in the same sense in the OLG(s) plane. The only restriction on this second property is that the function OLG(s) must be analytic (see footnote) at every point within the s-plane contour.

In order to find whether any right-half plane s values satisfy the equation OLG(s) = 1, simply map the *boundary* of the entire right-half s plane into the F(s)G(s) plane and look to see whether the point OLG(s) = 1 lies "inside" of the resulting contour. If so, the system is unstable. The map in the OLG(s) plane of the boundary of the right-half s plane is called a Nyquist plot. Figure 5.8a shows the D-shaped contour that (when extended to infinity) encloses the entire right-half s plane, and Figs. 5.8b and 5.8c show representative Nyquist diagrams for two different open-loop gain functions. There is no question that in Fig. 5.8b the point OLG(s) = 1 lies *outside* the Nyquist contour (for $A > 0$) and therefore that the system is stable. (Earlier, we showed that a similar second-order, low-pass system is stable for $A > 0$ by brute force solution

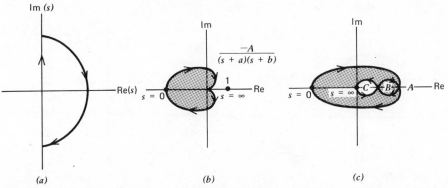

Fig. 5.   8. Nyquist diagrams: (a) mapping contour, (b) and (c) representative diagrams illustrating the meaning of "inside" and "outside."

of a quadratic equation.) However, in Fig. 5.8c it may not be so easy to determine whether 1 lies inside or outside of the contour, especially if 1 lies in either region B or region C. We should really think of "inside" as meaning "to the right of," in which case it should be clear that the shaded area is the inside. Thus, in Fig. 5.8c, if 1 lies in either region A or C, the system is stable, while if 1 lies within region B the system is unstable.

As a first example we will apply the Nyquist method to a third-order, low-pass system.

Example. Test the system of Figs. 5.4 or 5.5 with the sensor replaced by a third-order, low-pass filter, $G(s) = 1/(s + 1)^3$. As we noted earlier, the characteristic equation becomes a cubic, which we will write in the form:

$$1 = \frac{-A}{(s + 1)^3} \tag{5.5}$$

The right-hand side of Eq. 5.5 is an analytic function in the entire right-half s plane. It is nonanalytic only at the point $s = -1$, which lies in the left-half plane. The Nyquist plot $-A/(s + 1)^3$ appears in Fig. 5.9a. To obtain the Nyquist plot as s goes from zero to infinity along the imaginary axis, one can read off the magnitude and phase angle of the vector $s + 1$ from Fig. 5.9b. To find whether 1 lies inside the shaded area of the Nyquist plot, for a given value of A, we note that the phase angle of $s + 1$ must be +60° as the Nyquist plot crosses the positive real axis so that the denominator will have a phase angle of 180°, as the numerator does. Simple trigonometry shows that the magnitude of $(s + 1)$ is 2 at that point. Thus, if $A/2^3 = A/8$ is greater than 1, the Nyquist plot will enclose the point 1, and the system will be unstable.

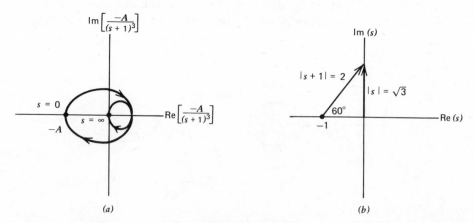

(a)   (b)

Fig. 5.   9. Nyquist diagram for a third-order LPF and the imaginary s at which it crosses the real axis.

Notice that in the Nyquist plot of $-A/(s+1)^3$, the entire curved portion of the D-shaped contour enclosing the right-half s plane maps into the origin in the $-A/(s+1)^3$ plane. Furthermore, the map of the negative imaginary axis is the complex conjugate of the map of the positive imaginary axis.

If the open-loop gain has factors that depend on s in both the numerator and the denominator, the resulting Nyquist plot can have some interesting shapes.

Example. The open-loop gain of a feedback system is given by $-A(s+100)^2/(s+1)^3$, where A is real and positive. Find the values of A, if any, for which this system is stable.

The Nyquist method applies because the function is analytic in the entire right-half s plane. We can make a fast guess at the behavior of a Nyquist plot by looking at the vectors $s+100$ and $s+1$ for imaginary $s = j\omega$ (see Fig. 5.10). For small ω, the phase angles of these vectors are near zero. As ω increases, first the phase of $s+1$, and then, much later, the phase of $s+100$, rise toward $90°$. But since the factor $s+1$ is cubed and in the denominator, its contribution to the overall phase of the OLG rapidly approaches $-270°$ as ω increases. Eventually, however, as ω becomes still larger, the $(s+100)^2$ in the numerator contributes a phase that approaches $+180°$ and therefore cancels much of the phase contribution of the denominator. When the $180°$ phase, due to the factor $-A$, is included, we might expect a Nyquist plot similar to that of Fig. 5.8c. A table of phase angles and magnitudes, Table 5.1, aids in the construction of a Nyquist plot, which appears, only approximately to scale, in Fig. 5.11. A slightly more detailed calculation shows that the Nyquist plot crosses $0°$ first at $\omega \approx 1.8$, at which point the magnitude is $1140\,A$, and then again at $\omega \approx 97$, at which point the magnitude is $0.021\,A$. Hence there is a range of values of A, between about $1/1140 = 8.8 \times 10^{-4}$ and about $1/.021 = 48$ for which the Nyquist plot

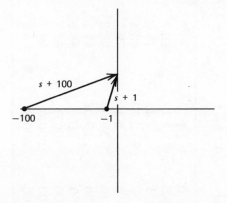

Fig. 5. 10. Illustrating the magnitude and phase of the factors $s+1$ and $s+100$.

TABLE 5.1. Computations for Nyquist diagram of Fig. 5.11

ω	$\lvert s+1 \rvert$	$\angle(s+1)$	$\lvert s+100 \rvert$	$\angle(s+100)$	$\left\lvert \dfrac{-(s+100)^2}{(s+1)^3} \right\rvert$	$\angle\left[\dfrac{-(s+100)^2}{(s+1)^3} \right]$
0	1	$0°$	100	0	10^4	$180°$
0.5	1.12	$26.6°$	100	$0.3°$	7.1×10^3	$100.8°$
1	1.41	$45.0°$	100	$0.6°$	3.6×10^3	$46.2°$
2	2.24	$63.4°$	100	$1.1°$	890	$-8.0°$
5	5.10	$78.7°$	100	$2.9°$	75.4	$-50.3°$
10	10.05	$84.3°$	100	$5.7°$	9.9	$-84.3°$
20	20.02	$87.1°$	102	$11.3°$	1.3	$-58.7°$
50	50.0	$88.9°$	112	$26.6°$	.10	$-33.5°$
100	100.0	$89.4°$	141	$45.0°$	.020	$1.8°$
200	200.0	$89.7°$	224	$63.4°$	6.3×10^{-3}	$37.7°$
500	500.0	$89.9°$	510	$78.7°$	2.1×10^{-3}	$67.7°$

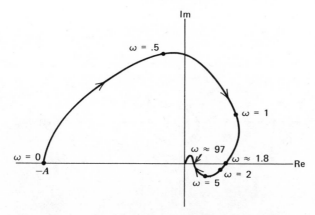

Fig. 5.   11. Example of a Nyquist diagram.

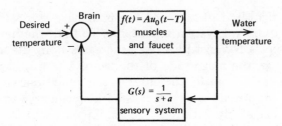

Fig. 5.   12. Model of the shower control problem.

encloses the point 1 and the system is unstable. For $A < 8.8 \times 10^{-4}$ and $A > 48$, the system is stable.

As a third example of the Nyquist method, we investigate a simple mathematical model of the shower control problem of section 5.2. We assume that the power amplifier subsystem is characterized by a gain A and a delay, T. That is, the water temperature changes exactly in proportion to the error or difference signal sent out by the brain, except that there is a delay of T seconds in the response. Furthermore, as shown in Fig. 5.12, we assume that the sensor system, consisting of temperature-sensitive nerves in our skin that communicate with our brain, is a first-order, low-pass system with system function $G(s) = 1/(s + a)$. Since the impulse response of the power amplifier is $f(t) = Au_o(t - T)$, we can find the system function $F(s)$ from number 1, Table 2.1.

$$F(s) = Ae^{-sT} \tag{5.6}$$

The characteristic equation for our model of the shower control problem is therefore

$$\frac{-Ae^{-sT}}{s + a} = 1 \tag{5.7}$$

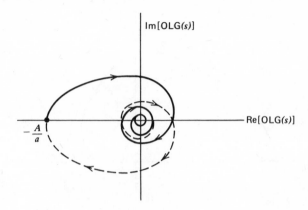

Fig. 5.   13. Nyquist diagram for the shower control problem.

If any s in the right-half plane or on the imaginary axis satisfies Eq. 5.7, then the shower control system is unstable. To find whether there is such an s, we construct a Nyquist plot. As s increases along the imaginary axis from zero to $j \cdot \infty$, the magnitude of the factor $1/(s + a)$ goes from $1/a$ to 0, and its phase goes from $0°$ to $-90°$. At the same time, the factor e^{-sT} has a constant magnitude of 1 (because s is imaginary) while the phase angle becomes negatively infinite. The combined result is a spiral in the OLG(s) plane, beginning on the real axis at $-A/a$, circling the origin an infinite number of times, and ending at the origin. This much of the Nyquist plot is shown by the solid curve in Fig. 5.13. Next for the entire curved portion of the "D," the map remains at the origin, since e^{-sT} is finite or zero and $| s + a |$ is infinite. Then as s travels up the negative imaginary axis, the remainder of the Nyquist plot is the complex conjugate (shown by the broken curve in Fig. 5.13) of the map of the positive imaginary axis. Which is the "inside," or better, the right side, of the spiral is unambiguous and, in practice, one would have to plot only the first half turn of the spiral to find whether the shower control system is stable. It is easy to see from the Nyquist plot why large values of the gain A of the "power amplifier" (the muscle-faucet system) yield instability. The primary difficulty is the delay factor; with no delay we would not have the phase-shifting factor e^{-sT}, and with a small enough delay, the increase in the magnitude of $s + a$ might bring the first turn of the spiral inside 1 before the total phase shift reaches $0°$.

Sometimes it is important to know how much a system parameter can change (because of aging, temperature changes, humidity, and so on) before the system becomes unstable. Two parameters, in particular, are commonly used figures of merit for describing how close a system is to instability. These are known as the *gain margin* and the *phase margin*. The gain margin is the factor by which the magnitude of the open-loop gain

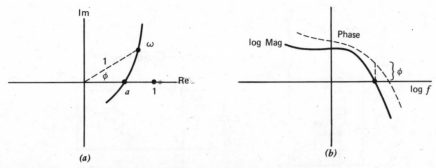

Fig. 5. 14. Two views of gain margin and phase margin

can be increased before the system becomes unstable, while the phase margin is the additional phase shift in degrees required (at the frequency at which the magnitude of the open-loop gain is one) in order for the system to become unstable.

The gain and phase margin concepts are illustrated in Fig. 5.14. In Fig. 5.14a, a segment of a typical Nyquist plot for a stable system is shown. The plot crosses the real axis at a (<1), so the gain could be increased by the factor $1/a$ before the system becomes unstable. Also, at a lower frequency, where the magnitude of the gain is one, an additional ϕ degrees of phase shift could be added before the system becomes unstable. The same concepts are shown in Bode plots in Fig. 5.14b. As the magnitude of the open-loop gain passes through unity, its logarithm passes through zero, but the phase angle ϕ is still positive. Then, at a somewhat higher frequency, the phase angle becomes zero but the magnitude of the open-loop gain has become less than one.

5.5. Design of a transistor oscillator   Our next example of a feedback system is an electronic oscillator, built from a transistor and some passive circuit elements. The circuit is shown in Fig. 5.15a. Our problem is to discover the relations among the circuit parameters that are necessary for sinusoidal oscillation to occur.

The device connected between the supply voltage and the collector of the transistor could be a resistor, an inductor (with high impedance at the frequency at which oscillations occur), or a current generator. Its purpose is to allow some dc "bias" current to flow through the transistor but to allow negligible signal currents to flow through it.

In order to analyze the circuit, to find whether oscillations can occur in the circuit and, if so, at what frequencies, we model the transistor itself by a simplified hybrid-π model (see Chapter 6) consisting of a resistor r between the base and emitter terminals and a dependent current generator $g_m V_{BE}$ between the collector and emitter terminals. The current produced by this dependent generator is a constant, g_m, times the voltage

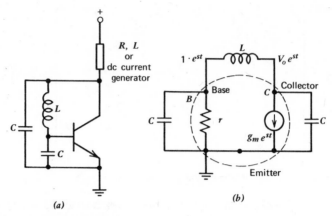

Fig. 5.   15. Circuit diagram and small-signal model of a transistor oscillator.

from base to emitter. The circuit model of the transistor itself appears within the dotted lines in Fig. 5.15b.

The capacitors and the inductor are connected in the equivalent circuit in the same way as in the actual circuit. For example, the inductor is between the collector and base of the transistor. The dc supply voltage and the current biasing element are not drawn in Fig. 5.15b because they are concerned with the dc bias current and not signal currents. Although some signal current would flow through the biasing element, its inclusion in the equivalent circuit would add complexity to the subsequent analysis without giving additional insight into the circuit behavior.

To begin the analysis we assume that there are signals of the form e^{st} in the circuit, in particular a voltage $1 \cdot e^{st}$ at the base terminal, and try to write an open-loop gain equation. The magnitude of the exponential voltage at the collector node can be written as V_o, though we do not yet know how large it is or for what values of s, if any, it can be nonzero.

Next we write the KCL equations for the two nodes, B and C. Note that the magnitude of the leftward exponential current through L is $(V_o - 1)/sL$. Equating the inductor current to the downward currents through the left-hand capacitor and r, we find

$$\frac{V_o - 1}{sL} = sC + \frac{1}{r} \tag{5.8}$$

and equating the inductor current to the upward currents through the right-hand capacitor and the current generator we have

$$\frac{V_o - 1}{sL} = -g_m - V_o sC \tag{5.9}$$

These equations are to be treated as two equations for the unknowns V_o and s. Since we are interested only in the solution for s, we eliminate V_o.

One way is to solve Eq. 5.8 for V_o:

$$V_o = sL\left(sC + \frac{1}{r}\right) + 1 \tag{5.10}$$

and then to equate the right sides of Eqs. 5.8 and 5.9 to arrive at:

$$-V_o sC = g_m + \left(sC + \frac{1}{r}\right) \tag{5.11}$$

Finally we substitute for V_o from (5.10) into (5.11) and rearrange to get:

$$s^3 + \frac{s^2}{rC} + \frac{2s}{LC} + \frac{\frac{1}{r} + g_m}{LC^2} = 0 \tag{5.12}$$

This is the equation that must be satisfied by s for an e^{st} signal to circulate in the system.

If Eq. 5.12 is satisfied by any s with positive or zero real part, the circuit will be unstable. But this is what we want in an oscillator! More precisely, we want the solutions to lie exactly on the imaginary axis. There are, in general, three solutions to a polynomial equation of third degree such as Eq. 5.12. We want two of these to be $j\omega$ and $-j\omega$ so that $e^{j\omega t} + e^{-j\omega t}$, a sinusoid at angular frequency ω, can exist.

Thus $j\omega$ and $-j\omega$ must each satisfy Eq. 5.12. If we substitute $s = j\omega$ into Eq. 5.12 we find

$$-j\omega^3 - \frac{\omega^2}{rC} + \frac{2j\omega}{LC} + \frac{\frac{1}{r} + g_m}{LC^2} = 0 \tag{5.13}$$

Now the real and imaginary terms must separately add to zero. Thus

$$\omega^2 = \frac{r}{LC}\left(\frac{1}{r} + g_m\right) \quad \text{and} \quad \omega^2 = \frac{2}{LC} \tag{5.14}$$

and we see that we must have $g_m = 1/r$ in order to obtain sinusoidal oscillations. These oscillations occur at a frequency given by:

$$f = \frac{\omega}{2\pi} = \frac{1}{2\pi}\sqrt{\frac{2}{LC}} \tag{5.15}$$

In practice, it is not possible to obtain a transistor with $g_m = 1/r$ *exactly*. However, a more advanced analysis would show that the solutions to Eq. 5.12 have negative real part (thus no oscillations) if $g_m < 1/r$ and positive real part (thus growing oscillations) if $g_m > 1/r$. In an actual circuit the growing oscillations would be limited by nonlinear behavior in the transistor, so that the circuit of Fig. 5.15 is indeed practical.

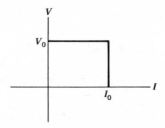

Fig. 5.   Current-voltage characteristic of a current-limited power supply.

5.6. Applications of feedback in electronic systems In conclusion we mention several additional uses of feedback in electronic systems in order to give an idea of the diverse nature of the problems it can solve. One widespread application is in dc power supplies. Often a desirable current-voltage characteristic for a power supply is that shown in Fig. 5.16. The output voltage is to remain constant, independent of current, up to some limiting current I_o, and then fall abruptly to zero. Usually a combination of a linear feedback system to maintain the voltage constant plus a nonlinear current sensor to shut down the supply when I reaches I_o can give a close approximation to this ideal behavior.

Feedback is often used to minimize variations in gain in an amplifier because of aging of components or changes in ambient temperature. In this application, illustrated in Fig. 5.17, the gain of the power amplifier might vary. But the overall system gain, from Eq. 5.3, is

$$H(s) = \frac{A(s)}{1 + A(s) \cdot \beta} \qquad (5.16)$$

If $\beta A(s) \geqslant 1$, then $H(s) \approx 1/\beta$, and since a β less than 1 can be obtained with a simple, time-and-temperature-independent, resistor divider network, the overall gain can be greater than 1 and also independent of time and temperature.

Another application of feedback is to modify input impedances. Either higher or lower than normal input impedances can be obtained with the

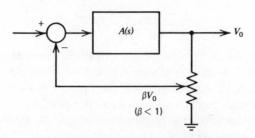

Fig. 5.   17. Using feedback to obtain a gain independent of A(s)

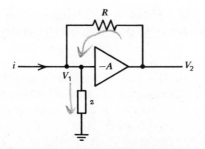

Fig. 5.   18. Using feedback to reduce input impedance.

proper use of feedback. For example, in an electronic ammeter, a circuit is required that maintains nearly zero voltage across the input terminals but gives an output voltage proportional to the input current. This is achieved by using an inverting amplifier of input impedance z, and a direct feedback connection, through a resistor, to the input terminal (Fig. 5.18). If we call the input voltage V_1 and the output voltage V_2 then we can write

$$V_2 = -A V_1 \tag{5.17}$$

and, from a nodal equation at the input,

$$i + \frac{V_2 - V_1}{R} = \frac{V_1}{z} \tag{5.18}$$

Substituting for V_2 from (5.17) in (5.18), we find

$$i = V_1 \left(\frac{1}{z} + \frac{1}{R} + \frac{A}{R} \right) \tag{5.19}$$

The input impedance is given by

$$\frac{V_1}{i} = \frac{1}{\dfrac{1}{z} + \dfrac{1}{R} + \dfrac{A}{R}} \tag{5.20}$$

which can be made very small by making A large. Since the input voltage is nearly zero, practically no current flows through z. It is then easy to see that, with these approximations, $V_2 = -iR$. That is, the output voltage is proportional to i and, as long as A is large, the constant of proportionality is virtually independent of A and z.

Another use of feedback is to remove nonlinearities in a system. For example, in an amplifier one usually desires an output that is strictly proportional to the input, at least for signals confined to a range of frequencies. Often, however, the transistors or other devices used for amplification have an input—output relationship that is nonlinear, but when a fraction of the output is fed back to the input, for example, through a resistor divider network, the overall system can become quite linear.

In summary, feedback has many uses in electronics and in control systems of various types. The use of feedback usually results in a system that does a job better or cheaper than could be done with nonfeedback systems. We will see in Chapter 8 how a class of devices known as operational amplifiers (or op-amps for short) can indeed do some wonderful things at a miniscule cost when used with feedback.

Problems for Chapter 5

5.1. Consider the function $F(s) = 1/[(s + 2)^2 (s + 1)]$.
  (a) Find the pure imaginary values of s at which the phase of $F(s)$ is 180°.
  (b) Find the map in the $F(s)$ plane of the point $s = j$. Give either magnitude and phase or the real and imaginary parts.
  (c) Sketch and dimension a Nyquist graph of $F(s)$.

5.2. In the system shown (Fig. P5.2), the amplifier system function is $H(s) = A/(s + 1)$ where A is a positive, real constant and x, y are the input and output voltages. The input impedance of $H(s)$ is large, and the output impedance is small.
  (a) Find the range of values of A for which the system is stable.
  (b) If A is chosen to give sinusoidal oscillations, find their frequency in Hz.

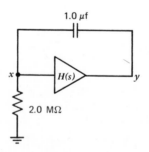

Fig. P5.   2

5.3. (a) Construct the appropriate Nyquist diagram and determine whether the following feedback system is stable (Fig. P5.3).
  (b) Find the s-plane solutions of $F(s)G(s) = -1$ by direct calculation.

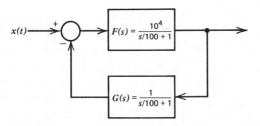

Fig. P5.   3

5.4. (a) Find, by any method, the range (+ and −) of the values of A for which this feedback circuit is stable (Fig. P5.4).

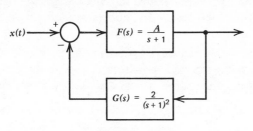

Fig. P5.   4

(b) If A is chosen so that $F(s)G(s) = -1$ for pure imaginary s, calculate the frequency of the resulting sinusoidal oscillation.

5.5 The characteristic equation for a system is $[-A(s + 10)^2]/(s + 1)^3 = 1$ where A is positive. Investigate the stability of this system by making a Nyquist plot. Are there any positive values of A for which the system is unstable? If so, find them. *Hint*. Find the pure imaginary values of s for which the phase of $(s + 10)^2/(s + 1)^3$ is $180°$.

5.6. A "phase-shift" oscillator can be constructed from three followers and an inverting amplifier (see Fig. P5.6). (A is real and positive and all four amplifiers have very high input impedance and low output

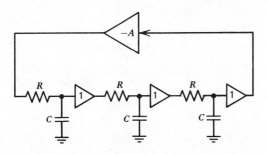

Fig. P5.   6

impedance.) If $RC = 0.5$ sec, make a Nyquist plot of the open-loop gain and determine the frequency of the sinusoidal oscillations. Find the value of $A(> 0)$ for which sinusoidal oscillations occur.

5.7. Sometimes it is possible to tell that there is a right-half plane solution to the characteristic equation by inspection of the coefficients, even if the expression, $F(s)G(s) + 1$, is a polynomial of degree greater than 2. For example, show that two *necessary* conditions (but not sufficient ones) for stability are: (1) all coefficients in the polynomial must have the same sign, and (2) no coefficient of any power of s can be zero. (Thus, if any coefficients are missing or if any two have opposite signs, the system is *unstable.*) *Hint.* Write the product of factors that represent solutions with negative real part, such as $(s + a)$; $a > 0$. Note that all complex solutions must occur in conjugate pairs.

5.8. In Fig. P5.8, the triangle represents a high input-impedance voltage amplifier with gain A, where A is positive and independent of frequency. Find the range of values of A for which the system is stable.

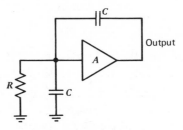

Fig. P5.   8

5.9. One of the most popular and useful integrated circuits, the 555 timer, is used to control the feedback circuit in Fig. P5.9. The 555 senses the output voltage and then opens the switch if $V_{out} \leq (1/3) V_{cc}$ and closes the switch if $V_{out} \geq (2/3) V_{cc}$.

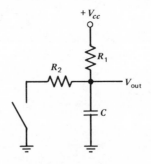

Fig. P5.   9

(a) Suppose at $t = 0$ the capacitor is uncharged, so that $V_{out} = 0$. Solve the differential equation (with switch open) that governs the output voltage and find the time at which $V_{out} = (2/3)V_{cc}$.

(b) Now with the switch closed and V_{out} starting at $(2/3)V_{cc}$, solve the new differential equation to find the time required for V_{out} to fall to $(1/3)V_{cc}$. (What condition on R_2 is necessary in order that V_{out} can equal $(1/3)V_{cc}$?)

(c) Finally, find the time required for V_{out} to rise from $(1/3)V_{cc}$ to $(2/3)V_{cc}$ when the switch opens.

(d) What must be the ratio R_1/R_2 in order that the times between switchings are equal?

REFERENCES

Barbe, E. C., *Linear Control Systems*, Chapters 1, 8, 10, 11, and 12, Scranton International Textbook Co., 1963.

Wilts, C. H., *Principles of Feedback Control*, Chapters 1, 4, 5, and 6, Addison-Wesley, Reading, Mass., 1960.

CHAPTER 6

Circuit Models of
Transistors

6.1. Introduction Here we discuss the physics of bipolar and field-effect transistors and develop circuit models for them. These extremely important devices cannot be represented by entirely linear elements within the model because the currents within them are very nonlinear functions of the terminal voltages. However, for small enough signals, we can approximate a nonlinear system by a linear one. Our basic program for bipolar transistors is (1) to develop a large-signal, nonlinear model from which we can find the dc "operating point" of a transistor when it is connected within a larger circuit, and (2) to develop a linear model for small signals. The small-signal model is used as a two-port network in the amplification and filtering of signals and can be analyzed by the basic frequency domain techniques developed in Chapters 2, 3, and 4. To understand the circuit models of a transistor and their ranges of applicability, one must have some knowledge of the physics underlying the flow of charges within it. We devote the next three sections to a discussion of this physics.

6.2. Electron and hole conduction in semiconductors A bipolar transistor is a three-terminal device constructed from a single crystal of a semiconductor, usually either germanium (Ge) or silicon (Si). A field-effect transistor may contain insulating material in addition to the semiconductor crystal. In the pure crystalline form, semiconductors have resistivities that lie between the very low values for metals (e.g. 1.6×10^{-6} Ω-cm for silver) and the very high values for insulators (e.g., 5×10^{14} Ω-cm for fused quartz). The resistivities for pure or "intrinsic" germanium and silicon at room temperature are about 50 and 200,000 Ω-cm, respectively. The reason for the use of semiconducting materials for transistors lies, however, not in their intermediate values of resistivity, but in the fact that an electric current in a semiconductor is carried by two distinguishable types of charge carriers — electrons and holes. To understand the mechanisms for these two types of conduction, one must look at the atomic structure of semiconductor atoms and crystals.

Neutral atoms of germanium and silicon have 32 and 14 electrons, respectively, surrounding the nucleus. In each case the outer shell contains four electrons, which are known as valence electrons. When atoms of germanium or silicon condense to become a crystal of germanium or silicon, the four valence electrons of each atom form covalent bonds with those of the four nearest neighbor atoms. Energy is required to remove an electron from a covalent bond, and the thermal energy of vibration within a crystal causes a fraction of the valence electrons to be shaken loose from the covalent bonds and wander through the crystal.

If an electric field is applied to the crystal, these free or "conduction" electrons move opposite to the field and carry current. The motion of electrons thus provides part of the conduction mechanism in semiconductors. However, it is also possible for the vacancy in a covalent bond to move through the crystal. We may imagine an electron from a nearby covalent bond "jumping" into the vacancy and causing a movement of electric charge. The vacancy or "hole" behaves as if it had a positive charge, and it moves under the influence of an electric field and carries current independently of the electron that originally filled it.

In an intrinsic (very pure) semiconductor crystal, the numbers of conduction electrons and holes must be equal because each hole arises as a result of an electron being excited from a covalent-bond state into a conduction state. The number of electrons or holes per unit volume is called the intrinsic concentration, n_i, and since this concentration represents an equilibrium between the competing processes of thermal generation of hole-electron pairs and the spontaneous recombination of electrons with holes, its value is a strong function of temperature. The variation of the intrinsic concentration with absolute temperature T is expressed by the approximate formulas:

$$\text{for Ge: } A = 3.40 \times 10^{32} \text{ cm}^{-6} \quad E_g = 0.78 \text{ eV}$$

$$n_i{}^2(T) = AT^3 e^{-E_g/kT} \quad \text{for Si: } A = 1.50 \times 10^{33} \text{ cm}^{-6} \quad E_g = 1.2 \text{ eV} \qquad (6.1)$$

$$k = 8.617 \times 10^{-5} \text{ eV/}^\circ\text{K}$$

where A is a parameter depending primarily on the density (number of atoms per unit volume) of the crystal, k is Boltzmann's constant, and E_g is a parameter with the units of energy. The significance of E_g is that minimum energy required to free an electron from a covalent bond. The intrinsic concentration can be determined experimentally by combining measurements of the conductivity of a crystal with measurements of the mobilities of holes and electrons. (The mobility is the ratio of the average drift velocity of holes or electrons to the electric field that propels them; it can be measured directly as the time to travel a certain distance or indirectly by means of the Hall effect.[†])

[†] See, for example, E. M. Conwell, *Proc. IRE*, **40**, 1327—1337 (1952) and *Proc. IRE*, **46**, 1281—1300 (1958) for discussions of the experimental determination of mobility, conductivity, and hole-electron concentrations.

So far we have been describing very pure samples of a semiconductor crystal. However, the crystals used in transistors are intentionally contaminated with minute amounts (a few parts per million) of either of two types of impurity atoms. The addition of impurity atoms into pure germanium or silicon can destroy the exact equality of the concentrations of holes and electrons and leads to the construction of a transistor.

The two impurity types are known as n type and p type. Examples of n-type impurities are atoms of arsenic (As), antimony, (Sb), and phosphorus (P), which have outer shells of five instead of four electrons. These impurity atoms can be added as the semiconductor crystal is grown or introduced later, for example, by heating a pure crystal in a gaseous atmosphere of the desired impurity atoms. In either method, impurity atoms can substitute for some of the host atoms of the crystal; each impurity atom provides an extra electron over and above the four required to share in the four covalent bonds with nearest-neighbor Ge or Si atoms. The extra electron turns out to be much more loosely bound to the impurity atom than an ordinary covalent bond electron and, at room temperature, practically all of the extra electrons of the impurity atoms are shaken loose by thermal vibrations and become conduction electrons. The impurity atoms, which contribute or donate their extra electrons to the pool of conduction electrons already present because of thermal production of hole-electron pairs, are called donor atoms. After the extra electron has wandered away from its original site, that site is left with a net positive charge equal in magnitude to the electronic charge, and we can think of it as a fixed, positively charged, donor ion.

The additional conduction electrons contributed by the donor impurity atoms provide more opportunities for holes to recombine with electrons and, as a result of the addition of donor impurities, there is a suppression of the concentration, p, of holes as well as an enhancement in the concentration, n, of conduction electrons. The recombination rate, which is proportional to the product of n and p, must, under thermal equilibrium conditions, be equal to the rate of thermal production of hole-electron pairs from ordinary covalent bonds. Since the production rate is independent of the impurity concentration, we must have the same rate of recombination as in an intrinsic crystal, or:

$$np = n_i^2(T) = AT^3 e^{-E_g/kT} \qquad (6.2)$$

Equation 6.2 says that the product of the hole- and conduction-electron concentrations is independent of the concentration of impurity atoms in the crystal and is equal to the product of the (equal) intrinsic concentrations.

Another important relation exists between the electron and hole concentrations. This relation is based on the fact that any macroscopic region of the crystal with added impurities must be electrically neutral, since we have added only neutral impurity atoms. This gross neutrality can be expressed by the equation:

$$n = N_D^+ + p \tag{6.3}$$

where n is the electron concentration (negative charge), N_D^+ is the donor *ion* concentration (positive charge) and p is the hole concentration (also positive charge). As we suggested earlier, practically all of the donor atoms are ionized at room temperature, so if the donor *atom* concentration can somehow be determined, along with $n_i{}^2(T)$, then Eqs. 6.2 and 6.3 can be solved to give both n and p. Typical numerical values of N_D^+ and n_i might be, for the germanium used in a transistor, $10^{16}/\mathrm{cm}^3$ and $2.5 \times 10^{13}/\mathrm{cm}^3$. These numbers and Eqs. 6.2 and 6.3 yield $n = 10^{16}/\mathrm{cm}^3$ and $p = 6 \times 10^{10}/\mathrm{cm}^3$. In this case the electron concentration is almost exactly equal to the donor atom concentration and is more than five orders of magnitude greater than the hole concentration! Thus, in an n-type semiconductor crystal, the electrons are called *majority* carriers and the holes are called *minority* carriers.

The graph of Fig. 6.1 is a log—log plot of the ratios n/n_i and p/n_i as a function of N_D^+/n_i. It is obtained by solving Eqs. 6.2 and 6.3 for these ratios. The graph shows that if the impurity ion concentration N_D^+ is more than about an order of magnitude smaller than the intrinsic concentration

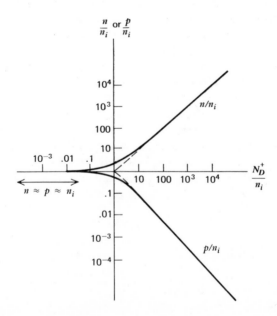

Fig. 6.   1. Majority and minority carrier concentrations as a function of impurity atom concentration.

n_i, then the hole and electron concentrations are practically equal, and the crystal is essentially intrinsic despite the impurities. On the other hand, if N_D^+ is more than an order of magnitude larger than n_i, the electron concentration is very nearly equal to the impurity concentration, and the hole concentration is two or more orders of magnitude smaller. Thus, only if N_D^+ is within a factor of 10 in either direction of n_i, will a detailed algebraic solution of Eqs. 6.2 and 6.3 have to be carried out to compute n and p.

It is also possible to produce crystals in which holes are the majority carriers and electrons are the minority carriers. Instead of adding impurities with five valence electrons, one adds impurities such as boron (B), aluminum (Al), gallium (Ga), or indium (In), which contain only three electrons in the outer shell. When such an atom substitutes for a Ge or Si atom in a semiconductor crystal, a vacancy occurs in one of the covalent bonds joining the impurity atom with its nearest neighbors. Thermal energy can excite an electron from a nearby semiconductor–semiconductor covalent bond into the "acceptor" vacancy near one of these acceptor atoms. Because the excitation energy required to do this is quite small and because of the enormous number of ordinary covalent-bond electrons available in comparison to the number of acceptor vacancies, nearly all of the acceptor vacancies are filled in a crystal at room temperature. The electron attached to an acceptor vacancy produces a stationary negative acceptor ion and, since that electron left behind a hole in a normal covalent bond, we have succeeded in producing a hole without simultaneously producing a conduction electron.

As in the case of the donor impurity discussion the recombination rate for holes and electrons (still proportional to the product of the concentrations n and p) must be equal to the rate of thermal production of hole-electron pairs, which is not changed by the addition of acceptor impurity atoms. Therefore, Eq. 6.2 applies again, but we must write a new charge neutrality equation to replace Eq. 6.3. Since acceptor ions are negative, we have, writing N_A for the acceptor ion concentration,

$$n + N_A^- = p \qquad\qquad (6.4)$$

Given the acceptor *atom* concentration, which is essentially equal to the acceptor *ion* concentration, the calculation of n and p follows a familiar line. In fact, the graph of Fig. 6.1 can be used if we simply replace N_D^+ there by N_A^- and interchange n and p.

The discussion in the preceding paragraphs has shown that we can alter the balance between holes and conduction electrons in a drastic way by the judicious introduction of either donor-impurity atoms to produce a crystal in which electrons are the majority carriers, or acceptor-impurity atoms to produce a crystal in which holes are the majority carriers.

6.3. The semiconductor junction diode   A semiconductor diode is a single crystal of semiconducting material that has been "doped" with impurity atoms to produce a boundary or junction between an n-type region and a p-type region. Wire leads are connected to the n and p regions. It is found experimentally that current flows far more easily in one direction (from p to n inside the diode) than in the other, thus giving a "rectifying" characteristic. To explain this rectifying properly requires some understanding of the *minority* carrier flows within the diode. Since bipolar transistors also owe their unique properties to the dynamics of minority carrier flow, we are laying the groundwork for an understanding of transistors by beginning with the simpler diode.

Our goal is to understand the way in which the current through a semiconductor junction diode (hereafter simply called a diode) depends on the voltage across it. Suppose first that no external device is connected across the leads of the diode. Because of the impurity atoms there are many more holes per unit volume than electrons on the p side of the junction and many more electrons than holes on the n side.

As a result of the enormous concentration gradients of both holes and electrons across the junction, there is a strong tendency for electrons to diffuse from the n type side of the boundary to the p type side and for holes to diffuse in the opposite direction. This diffusion cannot proceed indefinitely, however, because the n-type side acquires a positive charge and the p-type side a negative charge. These charges produce an electric field across the junction that is directed from the n side toward the p side. An equilibrium condition results, in which the diffusional flows are balanced by the tendency of this field to push electrons toward the n side and holes toward the p side. Since the holes that diffuse from the p side to the n side tend to recombine quickly with majority electrons on the n side (and vice versa), the positive charge on the n side of the junction resides not in positive holes, but in positive donor ions that no longer are neutralized by negative electrons. There exists then a thin ($\sim 10^{-7}$ m) layer of charge on either side of the junction, consisting of positive donor ions on the n side and negative acceptor ions on the p side. This double charge layer, called a depletion layer because it is depleted of majority carriers by recombination, is the source of the electric field that prevents further diffusion.

The equilibrium situation we have just described for an open-circuited diode must also exist for a short-circuited diode since, if a current could be maintained in an external wire connected to a diode, we would have the makings of a perpetual motion machine![†]

[†] However, if photons are absorbed at the junction (or if the junction is hotter than the ends where the leads are connected), there is a flow of current in a short-circuited diode. This is how solar batteries and other photodiodes work.

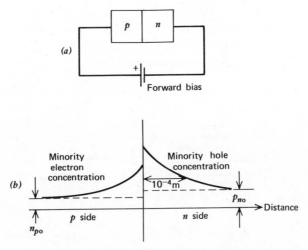

Fig. 6.   2. A forward-biased p-n junction diode and the minority carrier distribution.

Next suppose that a voltage source is impressed across the diode so that the p side is made positive with respect to the n side, as in Fig. 6.2a. This is called forward biasing. It can be shown that for relatively small voltages (a few tenths of a volt), nearly all of the voltage drop occurs at the depletion layer; the applied forward bias actually reduces the electric field that opposes diffusion and permits electrons to again diffuse across from n side to p side and holes to diffuse from p side to n side. This flow of holes into the n side and electrons into the p side is called minority-carrier injection; very *large* currents result from relatively *small* forward bias voltages.

Figure 6.2b illustrates the way the concentration of injected minority carriers depends on distance from the junction. In this figure, the width of the depletion layer is entirely negligible. The holes injected into the n side gradually disappear by recombining with majority electrons until the hole concentration is reduced to the thermal equilibrium value, p_{no}. The excess hole concentration $p'(x)$ can be shown to fall off exponentially with a characteristic length L_h, called the diffusion length, of about 0.1 mm.

The current carried across the junction by the diffusing holes is proportional to the excess concentration $p'_n(0)$ at the junction, which in turn is a very nonlinear function of the voltage V across the diode. In fact, a theory based on statistical mechanics suggests that the injected hole current should be proportional to $(e^{qV/kT} - 1)$ where q is the electronic charge, k is Boltzmann's constant, and T is the absolute temperature of the junction. Since a similar argument can be made for the current carried by electrons injected into the p side and since the currents

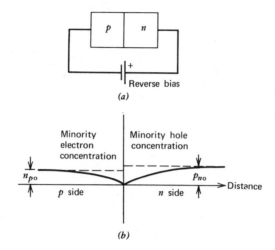

Fig. 6.   3. A reverse-biased *p*-*n* junction diode and the minority carrier distribution.

add, we may write for the total diode current:

$$I = I_o(e^{q\,V/kT} - 1) \qquad\qquad (6.5)$$

Notice that this relation agrees with the fact that the diode current is zero if the diode is short-circuited ($V = 0$).

The factor I_o in Eq. 6.5 is a constant of proportionality having dimensions of current. Its magnitude is the value of the diode current when V is made negative enough (a few times kT/q will do) so that the exponential term is small compared with 1. A negative diode voltage is called reverse bias; the minority carrier distributions under reverse bias are shown in Fig. 6.3*b*. Under reverse bias the electric field at the junction is *increased* to the point where a minority carrier that wanders into the depletion layer is immediately pulled across to the other side of the junction. Only about 100 mV of reverse bias is needed to maintain nearly zero minority carrier concentration at the junction.

The current-voltage characteristic expressed by Eq. 6.5 is plotted in Fig. 6.4*a* for voltages of the order of kT/q ($kT/q = 26$ mV at room temperature). This theoretical curve agrees with measurements made on real diodes fairly well, except that there is some difficulty in equating the voltage across the junction with the actual voltage measured at the diode terminals. Ohmic voltage drops, especially for large forward currents, occur in the bulk *n*- and *p*-type regions and cause the junction voltage to be less than the terminal voltage. Typical values of I_o are a few microamperes in germanium diodes and a few nanoamperes in silicon diodes. On a greatly compressed scale, the characteristic of a typical silicon diode is shown in Fig. 6.4*b*. The "knee" of the characteristic occurs at about 0.6 V of forward bias.

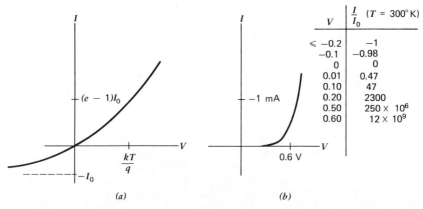

V	$\dfrac{I}{I_0}$ $(T = 300^\circ\text{K})$
$\leqslant -0.2$	-1
-0.1	-0.98
0	0
0.01	0.47
0.10	47
0.20	2300
0.50	250×10^6
0.60	12×10^9

(a)   (b)

Fig. 6.   4. Current—voltage characteristics of a p-n junction diode.

6.4. The bipolar transistor   There are many ways of fabricating a bipolar transistor, but the basic idea is to create in a single crystal a narrow region of one type of impurity sandwiched between more heavily doped regions of the opposite impurity type. The two possibilities are known as n-p-n and p-n-p transistors. Wires are connected to each of the three regions of the transistor, which are known respectively as the emitter, base, and collector. In this section we discuss an oversimplified, one-dimensional, physical model of a p-n-p transistor. In a transistor with planar symmetry, as in our model, the emitter and collector are interchangeable; in an actual transistor the electrical characteristics would depend on which side of the sandwich was considered to be the emitter and which the collector.

When a transistor is used as an amplifier, the emitter-base junction is forward biased and the collector-base junction is reverse biased. The reason for this choice of biasing is, as we will see, that small changes in emitter-base voltage can then cause large changes in collector current. The operation of a transistor can best be visualized in terms of minority carrier flow. Consequently, a graph of the minority carrier concentrations in a transistor biased for amplification is drawn in Fig. 6.5.

The forward bias at the emitter junction causes a positive excess minority carrier concentration at the emitter junction, while the reverse bias at the collector junction causes a negative excess carrier concentration at the collector junction. The base region is always made much thinner than the characteristic diffusion length for holes. As a result there is very little recombination of the holes injected from the emitter into base, and practically all of the injected holes diffuse to the collector junction, where they are pulled into the collector by the electric field at the collector depletion layer.

The current of holes from emitter through the base into the collector is

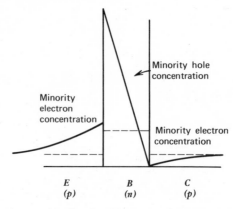

Fig. 6.   5. Minority carrier distribution in a transistor biased for amplification.

the dominant current in a *p-n-p* transistor biased for amplification, but it is not the only current. There is also injection of electrons from the base into the emitter and extraction (because of reverse bias) of electrons from collector into base. The *difference* between these electron currents is supplied as an outward base current. Also there must be an outward component of base current to replace the few electrons that recombine with the minority holes diffusing through the base.

In addition to the biasing conditions for amplification, discussed above, there are two important biasing situations that occur when a transistor is used as an off-on switch, for example, to control a lamp or a relay, or in an automobile electronic ignition system. The current to be switched is the large current of holes (in a *p-n-p* transistor) that flow from emitter to collector as described above. In the "ON" state in a switching application the voltage V_{CE} between collector and emitter is very close to zero. Since the base voltage must be negative with respect to the emitter in order to forward bias the emitter junction, the collector junction is also forward biased. Only a slight difference in the two forward-bias voltages is sufficient to produce a concentration gradient of minority carriers in the base region and a large current flow. The current flow can be in either direction depending on the polarity of V_{CE}; Fig. 6.6 shows the minority carrier concentrations in the ON state of a transistor switch when the current is from emitter to collector (*a*) and when the current is from collector to emitter (*b*).

In order to turn the transistor OFF, both junctions must be reverse biased. This is accomplished in our *p-n-p* transistor by making the base more positive than either collector or emitter. The minority carrier concentrations for the OFF state are shown in Fig. 6.6c. Note particularly that the minority-hole concentration in the base is depressed below its thermal equilibrium value. Therefore, holes are being generated thermally

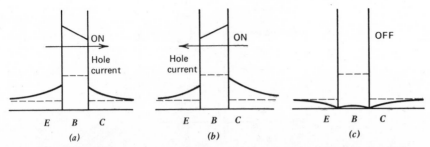

Fig. 6.   6. Minority carrier distributions in a transistor used as a switch.

faster than they are recombining, which requires a continual small flow of current into the base. Also, electrons are being extracted from both collector and emitter into the base by the reverse-bias field, thus adding to the small base current required. Nevertheless, the transistor switch permits a large emitter-collector current to be turned on and off by relatively small base currents.

6.5. The Ebers-Moll large-signal model

The occurrence of two p-n junctions in a bipolar transistor suggests that we might try to understand the current flows in a transistor in terms of the diode behavior discussed in Section 3. In particular, if the voltage across one of the junctions is held at zero by a short circuit, the current across the other junction obeys a diode characteristic, that is, it varies with the junction voltage according to Eq. 6.5. Nearly all of this "diode" current flows across the short-circuited junction, though recombination limits the fraction to a little less than 1.

Figure 6.7a shows a model of a p-n-p transistor with base-collector junction shorted ($V_{BC} = 0$). The emitter-base junction is modeled by a diode, whose reverse saturation current is I_{ES}, shown in parentheses. The current through the diode is I_E', and a fraction α_F of this current continues on to the collector. This current across the collector junction, which flows as a result of a nonzero voltage across the emitter junction, is modeled by a dependent current generator $\alpha_F I_E'$. Here F stands for forward, since the current is flowing from the emitter, *forward* toward the

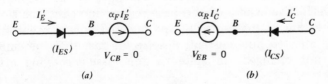

Fig. 6.   7. Ebers-Moll model of a p-n-p transistor with (a) base-collector junction shorted and (b) emitter-base junction shorted.

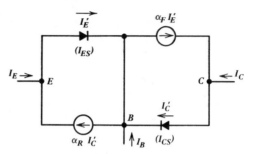

Fig. 6.   8. Full Ebers-Moll model of a *p-n-p* transistor.

collector. The fraction α_F depends on construction details such as the width of the base region.

An exactly similar situation occurs if the emitter junction is shorted ($V_{EB} = 0$), except that the model diode carrying current $I_C{}'$ between collector and base has a reverse saturation current I_{CS}, and the dependent current generator between base and emitter is $\alpha_R I_C{}'$ where R stands for "reverse."

Now what if neither junction is shorted? Can we somehow combine the models of Fig. 6.7a and 6.7b into a complete composite model of the transistor? The answer is yes; we can simply connect the two "partial" models in parallel to get the complete model in Fig. 6.8a. This is called the Ebers-Moll model.[†] The two partial models can be combined in this way because the currents are linear functions of the excess-minority carrier concentrations at the junctions. In effect, the currents caused by the bias voltage across one junction are independent of the currents caused by the bias voltage across the other junction. Having made this statement, we should point out that it is only an approximation. For example, suppose that the collector junction is reverse biased and the emitter junction is forward biased. The Ebers-Moll model suggests that a reverse current I_{CS} flows from base to collector through the collector diode, and that a current $\alpha_R I_{CS}$ flows from emitter to base through the emitter current generator. So far so good. But the model also predicts that the currents $I_E{}'$ and $\alpha_F I_E{}'$ depend only on the (forward) emitter-base voltage and not on the (reverse) collector-base voltage. In the actual transistor, however, a larger reverse voltage at the collector junction causes the depletion layer at the collector junction to widen. This widening of the depletion layer reduces the effective width of the base region, increases the slope (and, therefore, the rate of diffusion) of the injected hole concentration, and increases the injected hole current, despite the fact that the emitter-base voltage has not changed. In spite of this base-narrowing effect, the

[†] J. J. Ebers and J. L. Moll, "Large Signal Behavior of Junction Transistors." *Proc. IRE*, **42**, 1761–1772 (December 1954).

Ebers-Moll model is useful in that it provides a first approximation to the large-signal behavior of a bipolar transistor. Notice that only four numbers, I_{ES}, I_{CS}, α_F, and α_R, are necessary to characterize a transistor by the Ebers-Moll model.

To show how the Ebers-Moll model can be used to find the currents, we do the following example.

Example. Given a *p-n-p* transistor with the following parameters at $T = 300°\text{K}$,

$$I_{ES} = I_{CS} = 10^{-9} \text{ A}$$

$$\alpha_F = \alpha_R = 0.95$$

Sketch graphs of I_C versus V_{CB} for two values of emitter-base current: $I_E = 0$ and $I_E = 1$ mA (inward emitter current). The collector-base voltage determines the current $I_C{}'$ by means of Eq. 6.5.

$$I_C{}' = 10^{-9} \cdot (e^{V_{CB}/0.026} - 1) \tag{6.6}$$

For values of V_{CB} more negative than about 0.10 V, $I_C{}' = -1$ nA, so that 1 nA flows to the *right* through the collector diode and 0.95 nA flows to the right through the $\alpha_R I_C{}'$ current generator. If $I_E = 0$, the 0.95-nA current comes entirely by way of reverse current through the emitter diode, and only 0.90 nA flows to the left through $\alpha_F I_E{}'$. In this case, a net current of about 0.1 nA must flow *out* of the collector, so that $I_C = -0.1$ nA. On the other hand, since an emitter current of 1 mA would completely swamp the 0.95 nA from $\alpha_R I_C{}'$, the rightward current through $\alpha_F I_E{}'$ would be 0.95 mA and $I_C \approx -0.95$ mA. We conclude that for a reverse-biased collector, the outward collector current is independent of V_{CB} and nearly equal to the inward emitter current.

For a forward-biased collector junction ($V_{CB} > 0$), $I_C{}'$ increases rapidly with V_{CB}, and 0.95 $I_C{}'$ flows to the left through $\alpha_R I_C{}'$. If $I_E = 0$, then $I_E{}' = \alpha_R I_C{}'$, and we can deduce that the *inward* collector current is $(1 - \alpha_R \alpha_F) I_C{}'$. Since an inward emitter current I_E just adds to the rightward current through $I_E{}'$ and $\alpha_F I_E{}'$, its effect on I_C at a fixed V_{CB} is to shift I_C toward more negative (outward) values by an amount 0.95 I_E.

Figure 6.9 summarizes these results. Notice that $-I_C$ is plotted versus $-V_{CB}$ so that the first quadrant represents the amplifying conditions.

A set of curves as in Fig. 6.9 for different emitter currents is called the grounded-base (or common-base) collector characteristics.

Another set of characteristic curves, called the grounded-emitter collector characteristics is plotted in Fig. 6.10a for the transistor of the previous example. Note that the horizontal axis is $-V_{CE}$ (which equals $V_{EB} - V_{CB}$) and the parameter is the base current. Since neither junction voltage is determined directly by the external circuitry, the transistor

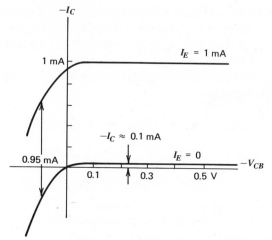

Fig. 6.   9. Transistor common-base characteristics.

currents are best found by trial and error. Calculation shows that for V_{CE} more negative than about 100 mV and provided that $-I_B$ is much greater than 1 nA, the emitter junction is sufficiently forward biased and the collector junction sufficiently reverse biased that the current $I_C{}'$ can be ignored. Then $-I_C = \alpha_F I_E$ and $-I_B = (1 - \alpha_F)I_E$, so that $-I_C = -[\alpha_F/(1 - \alpha_F)]I_B = -19\,I_B$. A similar analysis applies when V_{CE} is more positive than 100 mV except that now $-I_E = -19\,I_B$.

If $-I_B$ is comparable to the diode saturation currents, a more careful analysis must be made. For example, if $I_B = 0$ and V_{CB} is different from zero, one of the diodes, say $I_C{}'$, must be reverse biased and therefore passes a small ($\sim$1 nA) reverse current. The other diode will then be forward biased only slightly, so that 0.05 nA can flow through the collector diode; the resulting I_C will be about -1.95 nA. The grounded-

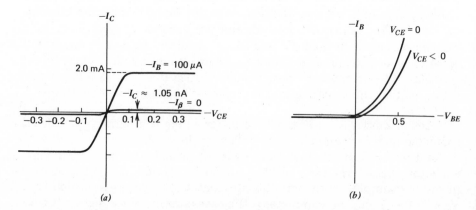

Fig. 6.   10. Transistor common-emitter characteristics.

emitter collector characteristics are shown in Fig. 6.10a. The symmetry in the first and third quadrants occurs because $\alpha_F = \alpha_R$; in a real p-n-p transistor, $\alpha_F > \alpha_R$ and the magnitude of the collector current is larger for negative collector-emitter voltages.

Another important set of grounded-emitter characteristics called the base characteristics is sketched in Fig. 6.10b. Here $-I_B$ is plotted versus $-V_{BE}$ for different values of V_{CE}. The steep rise in base current occurs for $-V_{BE} > 0$ (or $V_{EB} > 0$), that is, a forward-biased emitter junction. A typical voltage drop from emitter to base in a silicon p-n-p transistor biased for amplification is about 0.6 V.

6.6. A hybrid-pi, small-signal model

The large-signal, Ebers-Moll model developed in the previous section has the great disadvantage of being nonlinear in the junction voltages. Only if a system is LTI can we use the powerful frequency-domain methods for synthesis and analysis. However, if we consider only small changes v_{eb} in V_{EB} while maintaining a reverse voltage across the collector junction, the changes in the various currents will be proportional to v_{eb}, so that the system is LTI for small signals.

Next we develop a small-signal model for a transistor biased for amplification, that is, with collector junction reverse biased and emitter junction forward biased.

We begin by returning to the physical model for diffusional flow of minority carriers presented in Fig. 6.5. We ask what circuit elements must be connected between emitter, base, and collector terminals inside the model in order to model adequately the linear, small-signal behavior. Suppose, for example, that we hold the collector-base voltage fixed and increase the base-emitter voltage a little bit. Figure 6.11a shows the situation before and after (dotted) the change in base-emitter voltage. Three important effects occur. These are:

1. The current of holes that diffuse across from emitter to collector *decreases*, thus *increasing* the inward collector current, I_C .

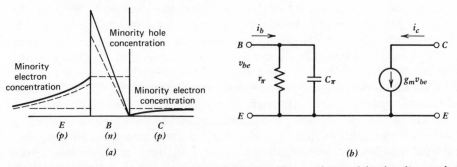

Fig. 6.   11. Effect of a small change in base-emitter voltage: (*a*) minority carrier concentrations and (*b*) small-signal equivalent circuit.

2. The rate of recombination of holes in the base *decreases*, because of the smaller excess concentration, thus *decreasing* the inward flow of electrons on the base lead to support the recombination and *increasing* the inward base current, I_B .

3. The quantity of charge stored in the base region (which consists of positive excess minority holes and an equal number of negative majority electrons required to give zero net charge density) decreases.

In the equivalent circuit, Fig. 6.11*b*, process 1 is modeled by a current generator $g_m v_{b\,e}$ between collector and emitter, process 2 is modeled by a resistor r_π between base and emitter, and process 3 is modeled by a capacitor C_π between base and emitter.

The constant g_m that describes the current generator is called the transconductance of the transistor. Since it is simply the rate of change of collector current with base-emitter voltage, we can evaluate it by referring to the Ebers-Moll model. Furthermore, since the collector junction has a reverse voltage across it, we can neglect the reverse current through the I_{CS} diode and use the partial circuit of Fig. 6.7*a*. We find

$$g_m = \frac{\partial I_C}{\partial V_{BE}} = -\frac{\partial I_C}{\partial V_{EB}}$$

$$\approx \frac{\partial}{\partial V_{EB}} [\alpha_F I_{ES}(e^{qV_{EB}/kT} - 1)] = \frac{q}{kT}\alpha_F I_{ES}\, e^{qV_{EB}/kT} \approx \frac{q}{kT} |\, I_C\,| \quad (6.7)$$

This rather surprising result suggests that the transconductance is proportional to I_C but that it is independent of the details of construction of the transistor! The parameters C_π and r_π, however, do depend on the details of construction.

Next we must introduce elements that represent the effect of a change in collector-base junction voltage. With the emitter-base voltage held constant, the primary effect of making the collector-base voltage more negative (in a *p-n-p* transistor) is to widen the depletion layer at the collector junction and to reduce the effective base width. Again, we can observe three distinct effects on the terminal currents. First, there is an increase in the magnitude of the collector current that originates at the emitter, because of the increased slope of the minority-carrier concentration in the base (Fig. 6.12*a*). In the circuit model of Fig. 6.12*b*, this additional current flows through the resistor r_o. Second, there is a decrease in the excess hole concentration in the base, which leads to a smaller recombination rate and less electron flow into the base from the external base lead. In the circuit model the additional, inward base current caused by the decreased (more negative) collector-base voltage flows through the resistor r_μ, connected between base and collector. Finally, a

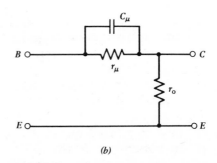

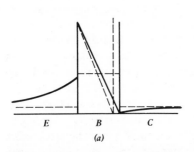

(a)	(b)

Fig. 6.   12. Effect of a small change in base-collector voltage: (a) minority carrier concentrations and (b) small-signal equivalent circuit.

change in charge must accompany the change in collector-base voltage. This charge resides mainly in the depletion layer at the collector junction, but there is an additional component of this charge, which is the decrease in minority- and majority-carrier concentrations in the base region. These two mechanisms for the change in charge produced by a change in junction voltage are both represented in the circuit model by the capacitor C_μ, connected between collector and base.

The complete hybrid-π, small-signal model is presented in Fig. 6.13. Here we have combined the elements r_π, C_π, and the dependent voltage generator $g_m v_{be}$, which describe the effects of changing the emitter-junction voltage and the elements r_π, r_o, and C_μ, which describe the effects of changing the collector-junction voltage. There is also one new element, the resistor r_x. This resistor has been added because the actual voltage within the base, or what might more accurately be called the spatial average of the voltage within the base region, may be slightly different than the external voltage on the base lead because of ohmic voltage drops caused by a flow of base current parallel to the junctions. Since the thickness of the base region is usually much less than diffusion length, there is considerable resistance to flow parallel to the junctions. The voltage that controls the dependent current generator is the internal

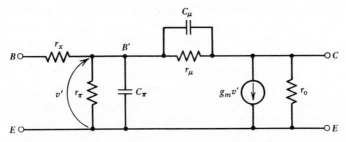

Fig. 6.   13. A hybrid-π, small-signal model of a transistor.

base-emitter voltage, designated v' in the circuit model, rather than the external base-emitter voltage, v_{be}.

It is important to understand clearly the meaning and limitations of the small-signal model we have just obtained. The input and output voltages and currents in the model, which has the form of a two-port network, are not the actual base-emitter and collector-emitter voltages and base and collector currents, but only the small deviations of these voltages and currents from their quiescent, or average values. Furthermore, the values of the parameters in the model depend very markedly on the quiescent operating point of the transistor and on its temperature. For example, we have seen in Eq. 6.8 that the parameter g_m, which is the ratio of a change in collector current to the change in internal base-emitter voltage, is proportional to the quiescent collector current I_C and inversely proportional to the absolute temperature.

The other hybrid-π parameters also are functions of the operating conditions. For example, in addition to g_m, $g_o = (1/r_o)$, $g_\pi = (1/r_\pi)$, and C_π are proportional to the collector current, while g_π and C_π decrease as the magnitude of the collector-junction reverse bias is increased.

After all is said and done, we must be critical of the entire idea of modeling a transistor by means of elements that have a more or less direct connection to the physics of the device. Such a procedure may be important to the designer of a transistor prototype, who is concerned with, for example, decreasing the value of C_μ in the hybrid-π model by appropriate physical construction. However, the user of the device can change the parameters only by changing the operating point and, as we have noted, a change in operating point may affect several of the hybrid-π parameters. In Chapter 4, we discussed how linear two-port networks can be represented by four parameters such as the h parameters that are functions of the complex frequency, s. For the user of a device, it would seem that the most important information is the variation of these four complex functions with operating point and not the variation of the elements of some particular internal model for the device. The h parameters can be determined experimentally in a fairly straightforward manner. It is a different matter to try to unravel the values of the four resistors, two capacitors, and transconductance g_m from measurements at the terminals. Furthermore, even after these seven element values are computed, considerable calculation must be done in order to obtain the h parameters from the element values. (An example of this type of calculation, using a simpler hybrid-π model in which r_o and r_μ were assumed to be open circuits, was performed in Section 4.7.) At best, the calculated values of the h parameters are only approximations to the experimentally determined "actual" values.

6.7. Field-effect transistors   Although the bipolar transistor has been more widely applied, another type of transistor, known as the field-effect transistor (FET), has electrical characteristics that make it superior to the junction transistor for certain circuit applications. There are two basic types of FET. In both types the fundamental mechanism is the modulation of the conductivity of a semiconducting *channel* by means of a voltage applied between the channel and a structure known as the *gate*. In the junction FET (JFET), the conductivity modulation is achieved by changing the effective cross-sectional area of the channel, while in the metal-oxide-semiconductor FET (MOSFET), the conductivity modulation is achieved by changing the average majority-carrier concentration within the channel.

6.8. Structure and characteristics of the JFET   A schematic diagram of a junction FET is shown in Fig. 6.14. The device consists of an *n*-type channel of moderate doping level and a pair of *p*-type gate regions that are very heavily doped. Electrical contacts are soldered or alloyed to the ends of the channel and called the source contact and drain contact, and a common electrical contact is made to the gate regions. Although the following discussion applies to the *n* channel JFET of Fig. 6.14, it should be clear that an analogous *p*-channel device can be constructed. Its terminal voltages and currents have opposite algebraic signs to those of the *n*-channel JFET.

The most important region of the JFET is the depletion layer, which straddles the junction between the *p*-type gate and the *n*-type channel. Because the gate is doped much more heavily with impurity atoms than is the channel, this depletion layer extends farther into the channel than into the gate. For small values of drain-source voltage of either polarity, the JFET functions as a resistor, the current being carried between source and drain almost entirely by majority electrons in the nondepleted portion of the channel. The depletion layer part of the channel is an

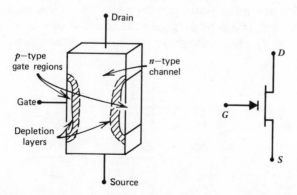

Fig. 6.   14.   An *n*-channel junction FET and its circuit symbol.

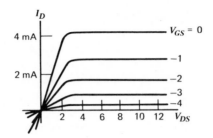

Fig. 6.   15. Characteristic curves for a junction FET.

excellent insulator because it is depleted of majority carriers. In normal operation of the n-channel JFET, the gate is biased negatively (reverse bias) with respect to the source and drain, and the more negative the gate bias the wider is the depletion layer. Eventually, as the gate is made more negative, a point is reached where the depletion layer extends all the way across the channel, thereby "pinching off" the conducting region of the channel and reducing the conductance between source and drain to zero. The value of gate-source voltage at which the source-drain conductance drops to zero is known as the pinch-off voltage. As long as the magnitude of the drain-source voltage is small compared with the difference between the actual gate voltage and the pinch-off voltage, the edges of the depletion layer will be nearly parallel, and the source-drain current will be proportional to the source-drain voltage. In this linear region of operation, the actual value of conductance depends on the gate-source voltage.

A set of characteristic curves for a JFET is drawn in Fig. 6.15. The linear region of operation is the region in which the curves are straight lines through the origin. We discuss next the reasons for the deviation from this linear behavior as the drain-source voltage, V_{DS}, is made more positive.

Figure 6.16 shows representative profiles of the depletion layer as the drain voltage is made more positive with respect to the source voltage. The gate voltage is held constant at a value below the pinch-off voltage. For positive values of V_D, the voltage across the depletion layer is greater near

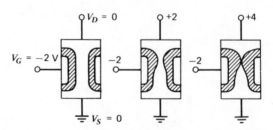

Fig. 6.   16. Depletion layer profiles for different drain-source voltages.

the drain that it is near the source, and therefore the depletion layer is wider and the conducting portion of the channel narrower nearer the drain end of the channel. If the drain voltage becomes sufficiently positive, the channel is pinched off almost completely near the drain, and the current is forced to flow in a very narrow conducting region. Further increases in drain-source voltage are taken up in the dense space-charge of the majority electrons in this narrow channel, and very little change in the depletion-layer profile occurs. In this operating region, known as the saturation region, the characteristic curves are nearly horizontal straight lines, with the current depending only on the gate voltage, V_G. A smooth transition between the linear region and the saturation region completes the curves.

One of the most important applications of the FET occurs when the current that can be supplied by an input transducer (such as an ionization chamber used to measure x-ray intensities) is very small. The transducer is simply connected to the gate of an FET, and the transducer needs to supply only the small current of a reverse-biased diode. A similar requirement occurs when a pH electrode is used, or when the voltage across the membrane of a biological cell is to be measured.

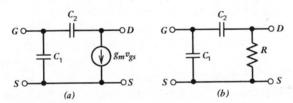

Fig. 6.   17. Small-signal models for a JFET: (a) saturation region and (b) linear region.

The small-signal model for a JFET is quite simple when the JFET is in its linear region or its saturation region. Because of the extremely high resistance of a reverse-biased diode, the gate-source circuit is modeled by a capacitance. Similarly, the gate-drain circuit is modeled by a capacitance. The source-drain equivalent circuit of a JFET depends on the operating region. For the saturation region, it is a current generator dependent on the gate-source voltage (Fig. 6.17a) and for the linear region it is a resistor, whose conductance is inversely proportional to the gate-source voltage (Fig. 6.17b). Both the transconductance, g_m, in the saturation region model and the resistance R in the linear-region model, are strongly dependent on the quiescent value of the gate-source voltage.

6.9. Structure and characteristics of the MOSFET   The MOSFET achieves terminal characteristics similar to those of the JFET depicted in

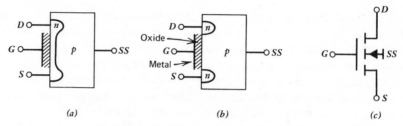

Fig. 6.   18. *n*-channel MOSFETS: (*a*) depletion type, (*b*) enhancement type, and (*c*) circuit symbol.

Fig. 6.11, but by a different mechanism for changing the effective channel conductance. There are two basic types of MOSFET, known as the depletion-mode type and the enhancement-mode type. Also, for either type, the conduction in the channel can be either by electrons or by holes. Figure 6.18 shows the basic construction of the depletion-mode and enhancement-mode MOSFETs. In the depletion-mode MOSFET of Fig. 6.18*a*, *n*-type source and drain regions and a continuous *n*-type channel are grown on a *p*-type substrate by the diffusion of *n*-type impurity atoms into the substrate. Then an oxide layer is grown over the *n*-type channel, and finally a metal gate contact is deposited on the insulating oxide layer. When the voltage between gate and source is zero, the channel provides a good conducting path between drain and source. However, as the gate voltage is made more negative with respect to the channel, positive charges must be induced within the channel. These positive charges are positive donor ions that are no longer neutralized by mobile majority electrons. Thus the channel becomes depleted of majority carriers and becomes a poor conductor.

The enhancement-mode MOSFET of Fig. 6.18*b* is of similar construction to the depletion-mode MOSFET, except that there is no *n*-type channel connecting the *n*-type source and drain regions. Thus, with 0 V between gate and source and a small voltage of either polarity between drain and source, one or the other of the *p-n* junctions interposed between drain and source is reverse biased, and the current flow between drain and source will be correspondingly small. If the gate-source voltage is now made positive, a charge of negative electrons is induced in the *p*-type substrate next to the oxide layer. If the voltage is made sufficiently positive, the region of substrate next to the oxide will actually be changed to an *n*-type semiconductor, with more electrons than holes, and there will then be a continuous *n*-type channel between drain and source, resulting in a current flow proportional to the drain-source voltage.

The electrical characteristics of a MOSFET are similar to those of a JFET with the exception that, for the enhancement-type *n*-channel MOSFET, there is no drain current for values of V_{GS} below a certain positive value. Furthermore, there is no fundamental limit on how positive the gate-source voltage can be except that the oxide layer can be broken down by an excessively high electric field. Because of the extremely high dc resistance (up to about 10^{15} Ω) between gate and source, there is a very real danger that during storage or shipment, enough static charge can build up on the gate to cause breakdown of the insulating layer. Nowadays, some manufacturers build in a Zener diode† right on the silicon chip from which the MOSFET is constructed so that excess static charges will be shunted through the Zener diode.

Problems for Chapter 6

6.1. Approximately how large a voltage swing at the base must occur to switch a bipolar transistor from ON to OFF? (i.e., is it 0.1, 1, 10, or 100 V). Explain.

6.2. The equilibrium *minority*-carrier concentration in a well-designed bipolar transistor is usually several times higher in the base region than in the collector or emitter regions. Explain why. *Hint.* Consider the case of zero base current.

6.3. Sketch the minority carrier concentrations in a *p-n-p* transistor with open-circuited base ($I_B = 0$) for the two cases: (1) collector positive with respect to emitter, and (2) emitter positive with respect to collector. Is the resulting current flow large or small in each case? Also, at which junction does most of the voltage drop occur in each case? Explain.

6.4. A *n-p-n* transistor is connected in the circuit of Fig. P6.4. The *BE* junction is forward biased and so there is approximately a 0.6 V drop across *BE*, making the emitter current I_1 about 1.40 mA. Since a collector current of this magnitude would produce a 14 V drop across the 10 K resistor, the collector junction must be forward biased. Assuming $V_{BC} \approx 0.6$ V, we must have $I_2 \approx 0.86$ mA.

(a) Assuming an Ebers-Moll transistor model, write equations that allow you to compute $I_C{}'$, $I_E{}'$, and the base current I_3 in terms of I_1, I_2, and the α_F and α_R of the model.

---

† A Zener diode is simply a junction diode, which for reverse voltages, begins to conduct above a certain voltage level because of one of two breakdown mechanisms. One is avalanche breakdown, in which extracted minority carriers receive enough energy from the strong electric field at the junction to create electron-hole pairs by collisions. The other breakdown mechanism is the quantum-mechanical tunnel effect in which majority carriers can tunnel through the narrow energy barrier at the reverse-biased junction if the electric field is large enough. Either mechanism leads to an almost vertical rise in reverse current with reverse voltage above the "Zener" voltage. Zener diodes are available commercially with a wide range of breakdown voltages and are used primarily as voltage regulating, or voltage limiting devices.

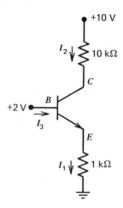

Fig. P6.   4

(b) Solve these equations for I_C', I_E', and I_3 if $\alpha_F = \alpha_R = 0.90$ and I_1, I_2 are the values assumed above.
(c) If you were given the numerical values of the diode saturation currents I_{ES} and I_{CS} and the temperature, how would you find a better approximation to V_{BE} and V_{BC}?
(d) Sketch minority-carrier concentration curves for the solution in (b).

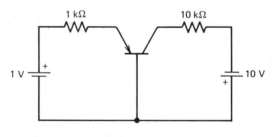

Fig. P6.   5

6.5. A *p-n-p* transistor is connected in the circuit of Fig. P6.5. The Ebers-Moll parameters are $I_{ES} = I_{CS} = 10^{-8}$ A and $\alpha_F = \alpha_R = 0.98$. Use a hand calculator to find I_C, I_E, I_B, V_{EB}, and V_{CB}.

REFERENCES

Gray, P. E., and C. L. Searle, *Electronic Principles: Physics, Models and Circuits*, Chapters 6, 7, 8, 9, and 10, Wiley, New York, 1969.

Searle, C. L., A. R. Boothroyd, E. J. Angelo, Jr., P. E. Gray, and D. O. Pederson, *Elementary Circuit Properties of Transistors*, S.E.E.C., Vol. 3, Chapters 1, 2, and 3, Wiley, New York, 1964.

CHAPTER 7

The Differential Amplifier

7.1. Introduction The differential amplifier is one of the most important building blocks of linear electronic systems. Although its original development in the late 1930s and 1940s was largely motivated by biological applications such as brain wave or electroencephalogram (EEG) recording, and measurement of the potential difference between electrodes placed on a nerve bundle, differential amplifiers are now found in nearly every modern linear integrated circuit. One important application of differential amplifiers is in the detection and measurement of the unbalance signal in various electrical bridges used in determining resistance, capacitance, and inductance, for example, the Wheatstone bridge. Often the input amplifier in a laboratory oscilloscope is a differential amplifier; such a configuration is indispensable when one wishes to know the voltage waveform across a circuit element, neither end of which is grounded.

7.2. Input—output configurations Two general circuit configurations of differential amplifiers are illustrated in Fig. 7.1. In each of these configurations there is a reference terminal, or ground, and two input terminals. In Fig. 7.1a the output consists of the voltage between two terminals 3, and 4 and in Fig. 7.1b the output consists of the voltage between one terminal, 3, and the reference terminal.

The input voltages V_1 and V_2 at any instant of time, each measured with respect to the reference terminal, may be thought of as combinations of a "differential" voltage V_d and a "common" voltage V_c, where V_d and V_c are defined† in Eq. 7.1.

$$V_d = \frac{V_1 - V_2}{2}$$

$$V_c = \frac{V_1 + V_2}{2} \tag{7.1}$$

---

† Some authors define the differential voltage to be $V_1 - V_2$. We have chosen to define the differential voltage with a factor of ½ because, if the differential output voltage is defined in the same way in Fig. 7.1a, then the differential gain is the same for Fig. 7.1a as for 7.1b.

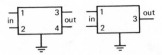

Fig. 7. 1. Differential-amplifier configurations: (*a*) differential output and (*b*) single-ended output.

With these definitions, V_1 is the sum of V_c and V_d while V_2 is the difference.

$$V_1 = V_c + V_d \qquad\qquad (7.2)$$
$$V_2 = V_c - V_d$$

The purpose of a differential amplifier is to amplify the difference signal, V_d, and to reject the common signal, V_c. For example, if the input voltages are exactly equal, then $V_d = 0$, and we want $V_3 - V_4$ to be zero in Fig. 7.1*a* or V_3 to be zero in Fig. 7.1*b*. On the other hand, if $V_1 = 1$ V and $V_2 = -1$ V, we want $V_3 - V_4$ to be large in Fig. 7.1*a* and V_3 to be large in Fig. 7.1*b*. For time-varying signals we could define a differential impulse response and a common impulse response or, alternatively, a differential system function and a common system function, from which we could compute the output time function given the input time functions. Furthermore, because there are both voltages and currents at the various input and output terminals, we really need a set of differential h parameters and a set of common h parameters for a complete description.

In the next three sections we discuss a very common form of differential amplifier, one constructed from bipolar transistors. Our goal is modest: to investigate the voltage gain and input impedance of both differential and common signals for a *symmetric* differential amplifier.[†]

7.3. Bipolar transistor configuration: dc operating point   One standard configuration for a bipolar-transistor, differential amplifier is shown in Fig. 7.2. The numbers next to the base and collector terminals of the two *n-p-n* transistors correspond to the terminals in Fig. 7.1*a*; the reference terminal, not shown, is at some constant potential between the supply voltages V^+ and V^-. The element I_0 is a dc current generator or constant current source, which can be constructed from one or more additional transistors. We consider the current generator separately. Note that a bias current must be supplied to the base of each transistor. Depending on the application, the magnitude of the bias current could range from fractions

---

[†] For a very readable discussion of differential amplifiers constructed from vacuum tubes or field-effect transistors and of the effect of dissymmetry in the various elements of a differential amplifier, refer to the excellent monograph by Giacoletto (L. J. Giacoletto, *Differential Amplifiers*, Wiley–Interscience, New York, 1970).

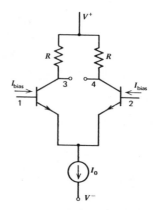

Fig. 7.   2. Differential amplifier with *n-p-n* transistors.

of a microampere, for low-level signal amplifiers, to many milliamperes for power amplifiers such as the driver stage in a sound system. The load resistors R are chosen to fit the application. These values might range from several hundred kilohms in a low-level amplifier to a few ohms in a power amplifier.

To determine the quiescent conditions in our differential amplifier, that is, the various terminal voltages and currents when the input signals are zero, we use the Ebers-Moll model for an *n-p-n* transistor. Figure 7.3 shows such a model connected in one branch of the differential amplifier of Fig. 7.2 with an emitter current of 2.5 mA, a collector load resistor of 2 kΩ, a collector supply voltage of 10 V, and an input biasing resistor of 10 kΩ.

In Fig. 7.3 we may assume that the diode I_{CS} is reverse biased, so that it makes an entirely negligible contribution of a 0.010 nA to the downward collector current. Nearly the entire emitter current of 2.5 mA must flow through the diode I_{ES}, and a fraction $\alpha_F = 0.98$ of 2.5 mA flows through the α_F current generator and therefore down through the 2-kΩ resistor. We can immediately find the collector voltage to be $10 - \alpha_F I_E R = 10 - 0.98 \times 2.5$ mA $\times 2$ kΩ $= 5.10$ V. Similarly, the base current is $(1 - \alpha_F) \times 2.5$ mA, and the base voltage is $-(1 - \alpha_F) \times 2.5$ mA $\times 10$ kΩ $= -0.50$ V. To find the base-emitter voltage drop we use the Ebers-Moll equation

$$I_E = I_{ES}(e^{q V_{BE}/kT} - 1) \tag{7.3}$$

where we have neglected the current $\alpha_R \times 0.01$ nA through the lower current generator. We solve Eq. 7.3 with $I_{ES} = 1.0 \times 10^{-11}$ A, $I_E = 2.5$ mA, $kT/q = 0.026$ V. to find $V_{BE} = 0.50$ V. Since we have already found that $V_B = -0.50$ V, we have $V_E = -1.0$ V. The quiescent conditions are now determined completely.

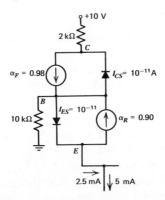

Fig. 7.   3. Using the Ebers-Moll model to find the quiescent voltages and currents.

Before turning to an analysis of the response of our differential amplifier to small signals look at one common form of current source, as illustrated in Fig. 7.4a. Recall, from the basic definition in Chapter 4, that an ideal current generator supplies a current that is independent of the voltage across it. That voltage is actually determined by the external circuit connected to the current source; in the present discussion that external circuit is a pair of transistors and the associated supply voltages.

In the circuit of Fig. 7.4a the base-emitter voltage drop will be about 0.6 V so that the voltage across emitter resistor R_E is approximately $V_B - 0.6$. The emitter current is then $(V_B - 0.6)/R_E$, and the collector current I_0 is just slightly less than this emitter current. Furthermore, the collector current depends only slightly on the actual value of the collector-emitter voltage. This is true because the small variations in base-emitter voltage caused by large changes in collector voltage cause only very small changes in the voltage across R_E, and small changes in I_B represent very small fractional changes in I_C.

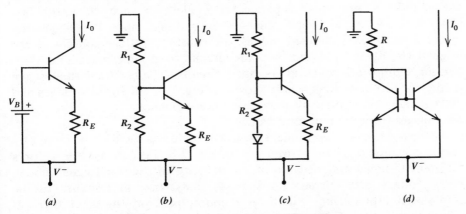

Fig. 7.   4. Current-source configurations.

Various refinements can be made in the basic current source. In Fig. 7.4b the bias voltage V_B is replaced by a resistor-divider network. The resistors R_1 and R_2 should be chosen small enough so that the base current is at least an order of magnitude less than the current through the resistors with the intent that the base voltage is approximately independent of the base current.

One undesirable feature of bipolar transistors is a rather strong dependence of their characteristics on temperature. In the simple circuits of Fig. 7.4a or 7.4b a rise in ambient temperature can be shown, from the Ebers-Moll model, to result in an increase in V_{BE} and hence a drop in I_E and I_C. This undesirable temperature dependence in a current source can be compensated by placing one or more diodes in series with one of the bias resistors (Fig. 7.4c). As the temperature increases, the voltage across the diode also increases, thereby tending to raise the voltage at the base. This counteracts the slight rise in base-emitter voltage and keeps the emitter voltage more nearly constant.

Figure 7.4d shows another temperature-compensation circuit — one that is widely used in integrated circuits. Because the temperatures of the two transistors tend to be closely equal when the transistors are formed in close proximity on a single silicon chip and because the base-emitter voltages of the two transistors are equal, their collector currents will tend to be equal. But the collector current on the left is $(-V^- - V_{BE})/R_1$ and, as long as the supply voltage is much larger than V_{BE}, this current is quite insensitive to changes in V_{BE} brought about by changes in temperature. An advantage of the circuit of Fig. 7.4d over that of 7.4c is that the actual voltage across the current source can approach zero in Fig. 7.4d while in 7.4c several volts must be maintained across R_E.

7.4. Bipolar-transistor configuration: differential small-signal analysis

In order to find the voltage gain and input impedance for a differential signal we employ a simplified h-parameter model for each transistor (see Chapter 6). As in all small-signal circuit models, since the signals are *changes* from the quiescent values, all constant voltage sources in the original circuit are replaced by short circuits to ground, and all constant current sources are replaced by open circuits. The resultant small-signal model for differential signals is shown in Fig. 7.5. All signal voltages and currents are represented by lowercase letters to emphasize that the signals are changes from the quiescent values.

Next we make an additional simplification in the small signal circuit by taking advantage of the symmetry in Fig. 7.5. Because $v_2 = -v_1$ for a differential signal and because of the linearity of the small-signal model, every voltage and current on the left that exists in response to the differential input signal will have a counterpart on the right that has opposite polarity. In particular, since $v_{e1} = -v_{e2}$ as a result of the

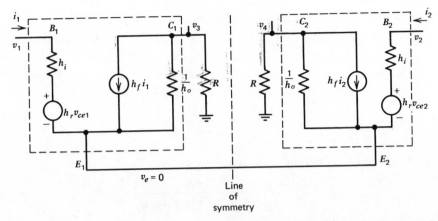

Fig. 7.   5. Transistor differential amplifier: small-signal model for differential input.

symmetry argument, and $v_{e1} = v_{e2}$ because the emitters are connected together, we must have $v_{e1} = v_{e2} = 0$. By using this "boundary condition" on v_{e2}, we need only analyze the left half of the entire circuit of Fig. 7.5; all signals in the right half will necessarily be the negative of their counterparts on the left. The simplified model is drawn in Fig. 7.6. We emphasize that this model is correct only for a differential input; the model for a common input is different and is discussed in the following section.

We begin the analysis by choosing the signal current into the base, B_1, to be unity. Then the current in the dependent current generator must be h_f. The input voltage v_1 and the output voltage v_3 are unknown; we solve for them by writing one KVL and one KCL equation as follows.

$$\text{KVL: } v_1 = h_i + h_r v_3 \tag{7.4}$$

(Here we have used the fact that $v_{e1} = 0$, so that $v_{ce} = v_3$.)

$$\text{KCL: } h_f + v_3 h_o + v_3/R = 0 \tag{7.5}$$

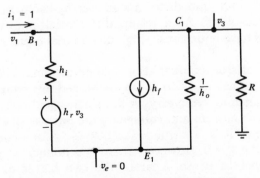

Fig. 7.   6. Simplified version of Fig. 7.5.

These equations can be solved by inspection to give:

$$v_3 = \frac{-h_f}{h_o + 1/R} \tag{7.6}$$

$$v_1 = h_i - \frac{h_r h_f}{h_o + 1/R} \tag{7.7}$$

Now the differential voltage gain is given by:

$$A_d = \frac{v_3}{v_1} = \frac{-h_f}{h_i(h_o + 1/R) - h_r h_f} \tag{7.8}$$

and the differential input impedance v_1/i_1 is given by Eq. 7.7, since $i_1 = 1$ by assumption.

These equations can be used with experimentally or theoretically determined values of the h parameters of the transistor, which are, in general, functions of the frequency parameter s. At frequencies up to several kilohertz it is usually true that $h_o \ll 1/R$, $h_r \approx 0$, and h_f and h_i are constants. Therefore we have for this low-frequency range:

$$\text{differential voltage gain} \approx \frac{-h_f R}{h_i}$$
$$\text{differential input impedance} \approx h_i \tag{7.9}$$

If we choose as typical values, $R = 1$ kΩ, $h_i = 1$ kΩ, $h_f = 100$, we see that we have achieved a differential voltage gain of approximately 100 with an input impedance of 1 kΩ.

One advantage of performing the circuit analysis in terms of the h parameters of the transistors is that we can utilize the results when the circuit is constructed from different active devices. For example, when properly biased MOSFETS are used in place of ordinary bipolar transistors, the input impedance parameter h_i has a real part of many thousands of megohms. As a result, differential amplifiers constructed from MOSFETS have enormously high differential input impedances.

7.5. Bipolar transistor configuration: common small-signal analysis
A common input signal ($v_d = 0$) is characterized by equal input voltages $v_1 = v_2$. If we assume symmetry in the small-signal equivalent circuit of Fig. 7.5, then each current and voltage on the left has an equal counterpart on the right. Since it is not true that $v_{e1} = -v_{e2}$ for a common signal, we cannot set $v_{e1} = v_{e2} = 0$. However, there is a boundary condition on the emitters that enables us to separate the two halves of the circuit and analyze only the left half. This boundary condition is a result of equality

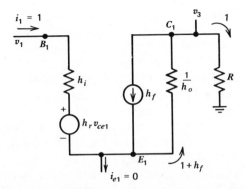

Fig. 7.   7. Transistor differential amplifier: small-signal model for common input.

of the emitter currents. Since the two equal emitter currents flow into a *constant* current generator, neither one can change; therefore, in the small-signal model we must have $i_{e1} = i_{e2} = 0$. The small-signal model for common input signals is drawn in Fig. 7.7.

Again we begin the analysis by setting $i_1 = 1$. The current generator produces current h_f and, because no current can flow out of the emitter lead, a current $1 + h_f$ must flow upward through resistance $1/h_o$. The downward current through R is just 1. Since $v_{ce1} = -(1 + h_f)/h_o$, we can write expressions for v_3 and v_1 by inspection:

$$v_3 = R \tag{7.10}$$

$$v_1 = h_i - \frac{h_r(1 + h_f)}{h_o} + \frac{1 + h_f}{h_o} + R \tag{7.11}$$

The common voltage gain is

$$A_c = \frac{v_3}{v_1} = \frac{R}{h_i + \dfrac{(1 + h_f)(1 - h_r) + R}{h_o}} \tag{7.12}$$

and the common input impedance is given by the same expression as v_1.

Again, for low to moderate frequencies, bipolar transistors have $h_o \ll 1/R$ and $h_r \approx 0$. When these approximations are substituted into Eqs. 7.10 and 7.11 we find that

$$\text{common voltage gain} \ll 1 \tag{7.13}$$

$$\text{common input impedance} \gg R$$

The analyses in this section and the last have been performed on a perfectly symmetric equivalent circuit. We cannot expect a practical differential amplifier to exhibit this precise symmetry, since there are several transistor parameters that are determined during the manufacturing process and cannot be adjusted thereafter. (Of course any dissymmetry in the values of the load resistors R could be eliminated by making one of them adjustable.) Nevertheless our analyses have shown that the differential, small-signal gain can be much larger than unity and the common small-signal gain much smaller than unity.

Note that the differential amplifier we have been discussing, not only conforms to the two-output configuration of Fig. 7.1a, but also conforms to the single-output configuration of Fig. 7.1b. The single output is taken between C_1 and ground (or between C_2 and ground).

An example may help clarify the difference between common and differential inputs and gains and between the single- and two-output configurations.

Example.   In a certain differential amplifier with A_c = 0.01, A_d = 100, the input signals are v_1 = 3, v_2 = 1. Find v_3 and v_4.

The common input is $v_c = (v_1 + v_2)/2 = 2$ and the differential input is $(v_1 - v_2)/2 = 1$. From Eqs. 7.8 and 7.12, the responses of v_3 to the differential and to the common inputs are $A_d v_d$ = 100 V and $A_c v_c$ = 0.02, respectively. The total voltage v_3 is the sum of these or 100.02 V. When computing the other output voltage, v_4, we must realize that Eq. 7.8 must be used with a negative differential input. Thus v_4 is $-A_d v_d$ = -100 V because of the differential input and $A_c v_c$ = 0.02 V because of the common inputs or -99.98 V altogether.

7.6. Gain modification in a differential amplifier   So far our discussion of differential amplifiers has assumed that the current generator in the common emitter circuit of the amplifying transistors produces a precisely constant current. As we will show, the differential voltage gain is a function of the current supplied by the current generator, and advantage can be taken of this fact in a variety of systems in which gain modification is useful.

We return to the result, Eq. 7.9, for the differential voltage gain of our bipolar-transistor, differential amplifier. Since we learned in Chapter 6 that the ratio h_f/h_i is closely equal to $(kT/q)I_c$ in a bipolar transistor that is biased for amplification, we see that the differential gain can be written as:

$$A_d = -\frac{kT}{q} |I_C| R = -\frac{kT}{2q} |I_0| R \qquad (7.14)$$

where I_o is the current supplied by the current source. That is, the differential gain is directly proportional to I_0.

One application of gain modulation is in electronic switches. Here the problem is to turn a signal on or off by means of a separate control signal.

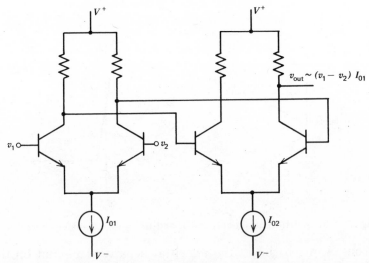

$$v_{out} \sim (v_1 - v_2)\, I_{01}$$

Fig. 7.   8. A pair of differential amplifiers used as an electronic switch.

An electronic switch can be made from a pair of differential amplifiers with the outputs of the first serving as the inputs of the second, as in Fig. 7.8. The current source of the first differential amplifier is switched between zero and some nonzero value (by changing the base drive voltage, for example, in Fig. 7.4a), while the current source of the second supplies a steady current. The reason for the second differential amplifier is that both outputs of the first amplifier switch to approximately the supply voltage when I_{01} is made zero; therefore, it is only the difference between these outputs that is zero, and it is this difference signal that is amplified by the second amplifier.

The circuit of Fig. 7.8 can also be used as an analog multiplier circuit, in which the output voltage is proportional to the instantaneous product of two other signals, in this case, $v_1 - v_2$ and I_0. This circuit, called a two-quadrant multiplier, is not the most general multiplier circuit, however; because, while $v_1 - v_2$ can have either polarity, I_0 is restricted to be nonnegative. Analog multiplier circuits have a number of uses and are discussed again in Chapter 12. For example, it will be seen there that an extension of the basic multiplier of Fig. 7.8 can be made in which the output signal is the product of two signals, either of which can be positive or negative. A multiplier with this property is called a four-quadrant multiplier.

7.7. The trigger circuit: a differential amplifier with feedback   A trigger circuit is a system that produces a well-defined step in its output signal when its input signal crosses a particular value or level. Trigger circuits are used to produce timing pulses; a familiar application is in oscilloscopes

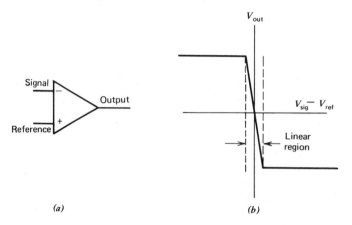

(a)   (b)

Fig. 7. 9. A differential comparator and its voltage transfer characteristic.

where one wants to begin the sweep at a particular point on the input
waveform or some other, external, waveform.

The basic operation in a trigger circuit is to compare the input signal
level with a reference level. The output is then switched to one of two
values, depending on whether the input signal is above or below the
reference level. Thus a trigger circuit is a sort of hybrid; it is not a linear
system, because the output is a nonlinear function of the input signal; nor
is it a purely digital system, because the input signal may have a
continuous range of values. A schematic representation of a trigger circuit
is shown in Fig. 7.9a and its "transfer characteristic" in Fig. 7.9b. Note
that there is a narrow range around $V_{sig} - V_{ref} = 0$ in which the system
behaves as a differential amplifier, with the output a linear function of the
input. For larger values of $V_{sig} - V_{ref}$ the output "saturates" at one of
two values. The differential voltage gain in the linear region is often quite

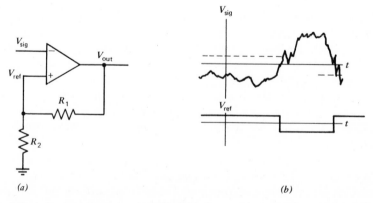

(a)   (b)

Fig. 7. 10. A comparator circuit with hysteresis; the reference voltage depends on
the history of the input voltage.

high; for example, in one popular integrated circuit trigger, the LM 319, this voltage gain is about 40,000.

Because all input signals contain a certain amount of noise, or unwanted random fluctuations, a trigger circuit is seldom used in the manner suggested in Fig. 7.9a. For suppose the input signal is a sine wave of amplitude 1 V, but with about 10 μV of noise. The reference signal might be a dc signal, say zero volts, but even it undergoes random fluctuations, for example, of the order of magnitude of 1 μV, which are usually independent of the random fluctuations of the input signal. Then as the sine wave passes through the linear region, the fluctuations are amplified so that fluctuations in the output occur, which for the LM 319 would have an amplitude on the order of 40,000 x 10 μV = 0.4 V. The output signal would then be a "ragged" rather than "clean" transition between its high state and its low state, with the duration of the raggedness depending on the frequency and amplitude of the input sine wave.

Fortunately, there is a simple way to eliminate the raggedness in the output of a trigger circuit. The idea is to introduce "hysteresis," or an intentional variation of the reference level. Hysteresis is achieved by using feedback between the output and the reference input of the trigger circuit, usually with a simple resistor-divider network. Figure 7.10a shows a trigger circuit with hysteresis. Figure 7.10b shows how the reference signal changes as the input signal crosses the level of the reference signal. If the change in reference signal is made somewhat larger than any expected random fluctuations in $V_{sig} - V_{ref}$, then the sudden change in V_{ref} will be large enough to ensure that $V_{sig} - V_{ref}$ will not change sign again immediately. The output signal for the situation of Fig. 7.10 is just an amplified version of the reference signal since, assuming a large input impedance at the reference terminal, V_{ref} is just a fraction $R_2/(R_1 + R_2)$ of V_{out}.

Note that when hysteresis is used, the actual trigger levels for positive-going and negative-going signals are different. In fact, if too much hysteresis is used, the reference level change can exceed the normal variation in signal level. For this reason the amount of hysteresis in a trigger circuit is often made adjustable to accommodate signals with different amplitudes and different amounts of noise.

Sometimes the external circuitry required to provide hysteresis and the correct reference level can be somewhat involved. For example, the LM 319 integrated-circuit trigger has, as an output stage, an open-collector transistor, whose emitter is connected to some convenient supply voltage, for example, ground. Two other supply voltages, say ±15 V, must be connected to appropriate terminals of the integrated circuit. The problem might be to design a feedback circuit in which V_{ref} switches symmetrically above and below ground by 10 mV as the output transistor switches

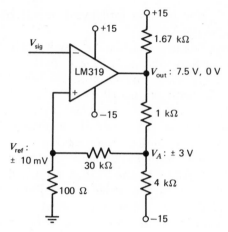

Fig. 7.   11. Biasing a comparator for upper and lower trigger levels symmetrical around zero.

between an "open circuit" (no collector current) and a "short circuit" to ground. A solution is shown in Fig. 7.11.

When the output transistor is open $(V_{sig} < V_{ref})$, a current of approximately $30/(4 + 1 + 1.67) = 4.5$ mA flows down through the 1.67-kΩ, 1-kΩ, and 4-kΩ resistors so that $V_{out} \approx 7.5$ V and $V_A \approx 3$ V. The divider action of 100 Ω and 30 kΩ then gives a reference voltage of about 10 mV. On the other hand, when the output transistor is saturated $(V_{sig} > V_{ref})$, V_{out} is held at approximately 0 V. The current down through the 1.67-kΩ resistor is about 9 mA, and the current down through the 1-kΩ and 4-kΩ resistors is about 3 mA. Thus $V_A \approx -3$ V, and the divider action makes $V_{ref} \approx -10$ mV. Note that the above analysis has been an approximate one. It ignores the current in the 30-kΩ resistor in comparison to the currents in the 1-kΩ and 4-kΩ resistors. The amount of hysteresis can easily be made adjustable by replacing the 100-Ω resistor with an adjustable resistor, for example, 0 to 500 Ω.

Another form of trigger circuit, known as a Schmitt trigger, is illustrated in Fig. 7.12. Its similarity to a differential amplifier is apparent, with the emitter current source replaced by a single resistor to ground. When $V_{sig} < V_{ref}$, transistor Q_1 is nonconducting and part of the current from the + supply flowing down through R_1, R_3, and R_4 is diverted into the base of transistor Q_2 so that it is saturated. The resultant IR drop in resistor R_2 places V_{out} in its "low" state. In this state the emitter voltage will be about 0.6 V below V_{ref}.

Next, suppose V_{in} approaches V_{ref} so that base current begins to flow in Q_1. As Q_1 turns on, it "steals" current that had been flowing through R_3, and thus tends to turn Q_2 off. If R_1 is somewhat greater than R_2 it turns out that the increase in emitter current of Q_1 is smaller than the

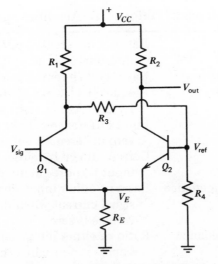

Fig. 7. 12. A Schmitt trigger circuit.

decrease in emitter current of Q_2, and V_E tends to fall, resulting in an increased base current in Q_1, a decreased and eventually zero base current in Q_2, and a "high" state at the output. It is the regenerative feedback from the collector of Q_1, through Q_2 and back to the emitter of Q_1 that provides the hysteresis in the Schmitt trigger. An analogous regenerative action occurs when V_{sig} falls toward the now lower reference voltage. An advantage of the Schmitt trigger is its obvious simplicity; a disadvantage is that it does not lend itself to variable hysteresis or variable trigger-level applications.

7.8. Differential-amplifier specifications and maximum ratings   To select or design a differential amplifier for a particular application in an instrumentation system, one must be aware of a host of specifications and maximum ratings that apply to these amplifiers. Since different manufacturers may list different specifications we can give only a general guide here.

Usually a set of *absolute maximum ratings* is supplied. These include values of supply voltages and input voltages that, if exceeded, may destroy the amplifier. Sometimes, in addition to, or in place of the absolute maximum ratings, a set of *maximum ratings* is specified. These are similar to absolute maximum ratings except that when they are exceeded, the amplifier may not burn out but will simply not meet its other specifications. Usually maximum and minimum temperatures are included in the maximum ratings.

The *specifications* of a differential amplifier consist of typical average values, and sometimes guaranteed upper or lower limits, of various operating

TABLE 7.1   Definitions of Differential—Amplifier Specifications

Differential gain	Ratio of differential output voltage to differential input voltage
Common-mode rejection	Ratio of differential gain to common gain
Frequency response	Frequency at which differential gain falls to $1/\sqrt{2}$ of its low-frequency value
Input-offset voltage	Input difference voltage required to give zero differential output voltage
Input bias current	Current drawn by either input when the input is short-circuited to ground
Differential input impedance	Ratio of either input signal voltage to its signal current when the common input voltage is zero
Common input impedance	Ratio of either input signal voltage to its signal current when the differential input voltage is zero
Output impedance	Ratio of open-circuit output signal voltage to output signal current when output is shorted to ground
Input noise voltage	Output noise voltage (with both inputs shorted to ground) divided by the differential gain. This specification may be given in terms of peak-to-peak noise voltage or as root-mean-square voltage. The range of frequencies passed by the noise measuring device must be specified.
Output slew rate	Maximum rate at which the output voltage can change when a step change in differential voltage occurs at the input

parameters. The specifications assume either a particular set of supply voltages and ambient temperature or else a range of those variables. The definitions of typical specifications are listed in Table 7.1. For some of the more important specifications such as gain, bias current, and input-offset voltage, the specification may include an indication of how that parameter varies with temperature, power supply voltage, and time.

Problems for Chapter 7

7.1. A diode ($I_s = 10^{-11}$ A) is forward biased by a 10-V battery in series with a 1 kΩ resistor. Assume that I_s is independent of temperature and solve by numerical trial and error for the voltage across the diode for the temperatures $273°K$ and $300°K$, thereby showing that the diode voltage increases with temperature.

7.2. The biasing circuit of a single-ended transistor amplifier is shown in Fig. P7.2. Use the Ebers-Moll parameters for the transistor of Fig. 7.3 and find the value of R_E such that the collector voltage is half-way between +10 V and the emitter voltage. The temperature is $300°K$. (*Hint.* As a first approximation, assume negligible base current and that $V_{BE} = 0.5$ V. Then find, in the following order: V_B, V_E, V_C, I_C, I_E, R_E, I_B. Then, as a second approximation, use the value of I_B to recalculate V_B, V_E, etc.) Answer: $R_E \approx 0.481$ kΩ.

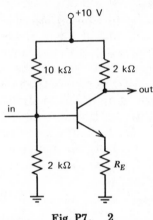

Fig. P7.   2

7.3. The simple circuit shown in Fig. P7.3 is called an emitter follower.
  (a) If $V_{in} = 5$ V, find the approximate collector current.
  (b) Use an Ebers-Moll model, with $\alpha_R = \alpha_F = 0.95$ and $I_{ES} = I_{CS} = 10^{-11}$ A, to find a better value of I_C.
  (c) Use a small-signal, h-parameter model to show that for a small change in the input voltage V_{in}, the output voltage changes by an

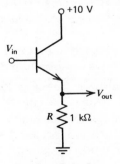

Fig. P7.   3

amount

$$\Delta V_{in} \cdot \frac{1 + h_f}{(1 + h_f)(1 + h_r) + h_i h_o + h_i/R}$$

Evaluate the voltage "gain" if $h_f = 100$, $h_r = 0.01$, $h_o = 10^{-6}\ \Omega^{-1}$, $h_i = 10^3\ \Omega$. (*Hint.* In the small-signal equivalent circuit, the collector is grounded.)

7.4. At high frequencies the gain of a differential amplifier decreases, due partly to unavoidable capacitance between the collector and the base of each transistor. This capacitance exists in the junction region between collector and base and is increased by stray wiring capacitance.

(a) Redo the derivation for the differential gain of the amplifier of Fig. 7.2, using a simplified h-parameter equivalent circuit with $h_r = 0$, $h_o = 0$, but with a capacitance C connected between collector and base. You should be able to show that the differential gain is $-[(h_f R/h_i - sRC)/(1 + sRC)]$.

(b) Make a pole-zero graph and a Bode plot to illustrate the dependence of differential gain on frequency. Use the values $h_f = 100$, $h_i = 10^3\ \Omega$, $R = 10^3\ \Omega$, $C = 10^{-11}$ F. *Warning.* This is an oversimplified analysis; in actual practice the h parameters themselves depend on frequency, so that additional poles and zeros come into play.

7.5. A commonly used method for increasing the gain of a single-ended amplifier (see Fig. P7.2) is to put a capacitor C across the emitter resistor R_E. The idea is that R_E provides negative feedback and reduces the small-signal gain; its effect can be eliminated for certain frequencies by "shorting it out" with the capacitor, which is therefore called a "bypass" capacitor.

(a) Sketch the small-signal equivalent circuit for the amplifier, letting $h_r = h_o = 0$, $h_i = 10^3\ \Omega$, $h_f = 100$, $R_E = 500\ \Omega$. Include C as an unknown parameter.

(b) Solve the circuit for the small-signal voltage gain v_{out}/v_{in}, and thereby show that there is a zero at $s = 2 \times 10^{-3}/C$ and a pole at $s = 0.103/C$.

(c) What value of capacitance results in a break frequency of 100 Hz, above which the gain is essentially independent of frequency?

7.6. A cascode emitter follower (see Fig. P7.6) can be used when the circuit must be used to drive a capacitive load, such as a coaxial cable, at high frequencies. An ordinary emitter follower can cause difficulties in this application because, when the input signal goes negative the transistor tends to cut off, and the capacitive load can discharge only slowly through the cathode resistor. There is no

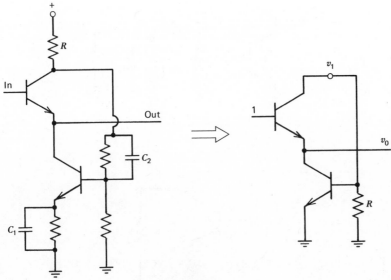

Fig. P7.   6

problem when the input signal goes positive because the load capacitance charges quickly by means of the emitter current of the transistor.

For ac signals the circuit simplifies because the bypass capacitors C_1 and C_2 act as short circuits. Use a simplified h-parameter model for the transistors ($h_r = h_o = 0$) and show through the small-signal equivalent circuit that, for a unit input voltage, $v_1 = 0$ and $v_0 = 1$.

7.7. A type of differential amplifier known as a push-pull amplifier is often used in high-power applications such as the driving of loudspeakers in sound systems (Fig. P7.7). The idea is that very little current flows through either transistor or the primary of the transformer when the signal input voltage v is zero. When $v > 0$ the upper transistor conducts and current is driven in one direction through R_L by means of the upper half of the step-down transformer. When $v < 0$ the bottom transistor conducts, and current is driven in the other direction through R_L by means of the lower half of the transformer. The ac equivalent circuit in either half cycle for frequencies such that $sL_1 \gg n^2 R_L$ (where L_1 is the inductance of either half of the primary winding and n is the ratio of the number of turns on either half of the primary to number of turns on the secondary) utilizes the reflected impedance concept (see problem 4.9) in which the equivalent input circuit of the transformer is simply n^2 times the load resistance.

Draw the small-signal equivalent circuit using a simplified h-parameter model of the transistor ($h_o = h_r = 0$) and show that the

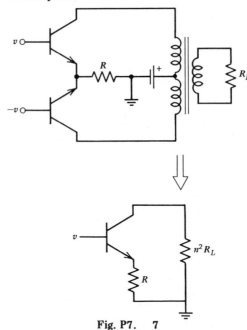

Fig. P7.   7

power $i^2 \cdot n^2 R_L$ delivered to the load is given by

$$P = \frac{v^2 n^2 R_L h_f{}^2}{[h_i + (1 + h_f) \cdot R]^2}$$

Although it seems from this expression that the power could be increased without limit if n^2 is made larger, there is always a limitation on either the power dissipated in the transistor itself or the peak voltage across the transistor during the nonconducting half of the cycle. Note that if R is small, this peak voltage could be twice the supply voltage. Why?

7.8. For a Schmitt trigger circuit (see Fig. 7.12), with $V_{cc} = 10$ V, $R_1 = 2$ kΩ, $R_2 = 1.5$ kΩ, $R_3 = R_4 = 10$ kΩ, $R_E = 0.5$ kΩ, assume that one or the other of the transistors is saturated ($V_{CE} \approx 0$) and find the voltages at both collectors, the common emitter, the base of Q_2, and the collector current in the conducting transistor for each of the two states: Q_1 on, Q_2 off and Q_1 off, Q_2 on.

7.9. In an LM319 comparator circuit with ±15-V supply voltages (see Fig. 7.11), find a set of resistor values such that the upper switching level is 3 V (for positive-going V_{sig}) and the lower switching level is 2 V (for negative-going V_{sig}). As in the text example, the actual output voltage of the comparator cannot be negative and is zero when $V_{sig} > V_{ref}$.

REFERENCES

Bleuler, E., and R. O. Haxby, Eds., *Methods of Experimental Physics, Vol. 2, Electronic Methods*, Part A, Chapter 6, Academic Press, New York, 1975.

Brophy, J. J., *Basic Electronics for Scientists*, Third Edition, Chapters 6 and 7, McGraw-Hill, New York, 1977.

Giacoletto, L. J., *Differential Amplifiers*, Wiley—Interscience, New York, 1970.

Gray, P. E., and C. L. Searle, *Electronic Principles: Physics, Models and Circuits*, Chapters, 13, 14, 15, and 16, Wiley, New York, 1969.

Jones, B., *Circuit Electronics for Scientists*, Chapters 8 and 9, Addison-Wesley, Reading, Mass., 1974.

CHAPTER 8

Operational Amplifiers

8.1. Introduction There exists a class of very high-gain, dc-coupled, differential amplifiers, which has an enormous number of applications in instrumentation systems. Developed after World War II, in the days when amplification still required bulky, expensive vacuum tubes, these operational amplifiers, or op-amps for short, were one of the first instrumentation-system building blocks to be made in integrated-circuit forms. Nowadays, integrated-circuit op-amps can be purchased for less than a dollar, despite the fact that they typically contain about 20 transistors plus their resistor biasing networks. Because of their high dc-voltage gain, sometimes as high as 10^5, op-amps always require external feedback circuitry in order to prevent saturation of the output or instability; it is the wide variety of possible feedback circuits that give these devices their versatility.

The purpose of this chapter is to introduce the operational amplifier philosophy, by means of some basic system configurations; to mention some of the important departures of op-amp parameters from the ideal and how to measure them; to take the reader "inside" the popular type-741 integrated-circuit op-amp to discover, at a transistor by transistor level, how it achieves its terminal characteristics; and finally to show a few of the many applications of op-amps to instrumentation systems. For other applications, and there are hundreds, consult any of the many excellent reference books on operational amplifiers (see the end-of-chapter references).

8.2. Basic principles of op-amps For our initial discussion, in order to get a "feel" for operational amplifier circuits, we make use of a very idealized model of the amplifier itself. The idealized model is represented in Fig. 8.1. There are two inputs, known as the inverting input (−) and the noninverting input (+). The output voltage is equal to the differential voltage gain, A, times the difference $(V_+ - V_-)$ between the input voltages. In real operational amplifiers this voltage gain may be typically between 5000 and 100,000 at low frequencies, and it falls off at high frequencies; in our idealized model we simply assume that it is large and that there is neither a falloff of gain at high frequencies nor any phase shift between an input sinusoidal voltage and the output sinusoidal

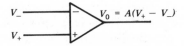

Fig. 8.   1. Idealized operational amplifier.

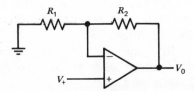

Fig. 8.   2. Noninverting amplifier configuration.

voltage. In addition, the input impedance is assumed to be infinite, so that no current flows into either input terminal, and the output impedance is assumed to be zero so that variations in the load attached to the output cause no change in either the amplitude or phase of the output signal. In addition to the signal inputs and output, an op-amp must have supply voltage inputs. In most applications, two supply voltages, symmetrical around zero, such as +15 V and −15 V are used. Usually there is no "ground" terminal on an op-amp; the reference voltage that we refer to in circuit diagrams involving op-amps is the common terminal of the power supplies.

Circuits employing the idealized operational amplifier described above are subject to an "intuitive" analysis when feedback is applied from the output terminal to the inverting input terminal. We give an example of such a negative feedback connection in Fig. 8.2. The circuit is known as the noninverting amplifier configuration. To analyze this circuit by the "intuitive" method we note that the voltage at the inverting input of the operational amplifier is determined by V_0 and the voltage-divider action of R_1 and R_2. The voltage at the inverting input is therefore

$$V_- = \frac{R_1}{R_1 + R_2} \cdot V_0 \qquad (8.1)$$

If V_- happens to be less positive than the input voltage V_+ applied to the noninverting input, then the infinite gain will force V_0 to be higher. But if V_0 is too high, then V_- becomes more positive than V_0. The actual situation must be that the difference signal $V_+ - V_-$ is always very small, and the output voltage V_0 must adjust itself so that V_- as given by Eq. 8.1 is almost exactly equal to the input signal V_+. If V_+ happens to be negative, then V_0 will be negative, but the difference signal will be only very slightly negative.

The effect, then, of the negative feedback is that the inverting and noninverting inputs must remain at very nearly equal potentials, almost as if they were short-circuited! However, there is one very important difference between this practically zero difference voltage and a true short circuit. That is, no current flows directly between the two input terminals as would perhaps flow if there were actually a zero resistance between them. To distinguish between the present situation and a true short circuit, we say that there is a "virtual short circuit" between the inputs in the feedback operational amplifier.

The result of the intuitive analysis of the noninverting amplifier circuit of Fig. 8.2 is that the voltage gain of the circuit may be found by substituting the virtual short relation:

$$V_+ = V_- \tag{8.2}$$

into Eq. 8.1 and solving to arrive at:

$$\frac{V_0}{V_+} = \frac{R_1 + R_2}{R_1} \tag{8.3}$$

We note the rather amazing fact that the gain A of the operational amplifier itself does not enter into this expression for the gain of the overall system. Thus, even if the op-amp gain drifts with time or temperature or is perhaps not as large for large signals as for small signals (actually a nonlinearity), the overall gain depends only on the ratio of resistances given by Eq. 8.3. This lack of dependence of the system gain on the actual parameters of the operational amplifier is one of the enormous advantages of the operational amplifier philosophy.

A widely used modification of the basic noninverting amplifier configuration, known as the voltage follower, is shown in Fig. 8.3. Here the output is connected directly to the inverting input. Because of the virtual short relation, the output must be at the same voltage as the input; hence the name voltage follower. The basic use of this circuit is as an impedance transformer, or isolating buffer. The input may be a high-impedance source such as a glass pH electrode or a microelectrode inserted into a nerve cell. The voltage follower transforms such a high impedance source into one with essentially zero impedance, which can drive any subsequent circuitry without that circuitry affecting the terminal voltage at the original electrode.

Both of the preceding circuits have the input applied to the

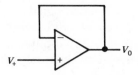

Fig. 8.   3. Op-amp voltage follower.

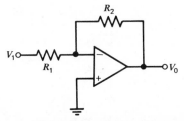

Fig. 8.   4. Inverting amplifier configuration.

noninverting input, and the resulting output has the same polarity as the input. Another class of circuit introduces the signal at the inverting input and keeps the noninverting input at a constant voltage, usually ground. An inverting amplifier circuit is shown in Fig. 8.4. Again the virtual-short concept applies. This time, since the inverting input is "virtually" shorted to ground, its potential is held very near zero. However, no current can flow into the amplifier from the inverting input because of the infinite input impedance (and because the voltage across this infinite impedance is practically zero); thus the input current V_1/R_1 drawn from the input voltage source must flow through the feedback resistor R_2. Since the voltage on the left side of R_2 is zero, the voltage on the right side of R_2 (the output voltage) must be

$$V_0 = -I_1 R_2 = -V_1 \cdot \frac{R_2}{R_1} \qquad (8.4)$$

Once again the gain $-R_2/R_1$ of the overall system is seen to be independent of the gain of the operational amplifier.

The circuit of Fig. 8.4 can be extended to have several inputs as shown in Fig. 8.5. Now the sum of the input currents flows through R_2, and the output voltage is given by:

$$V_0 = -\frac{R_2}{R_1} \cdot (V_1 + V_2 + V_3 + \dots) \qquad (8.5)$$

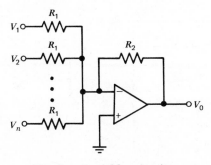

Fig. 8.   5. Adder circuit.

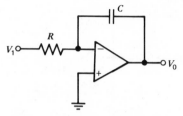

Fig. 8.   6. Op-amp integrator.

The input and feedback elements are not limited to resistors. An exceedingly useful circuit results if the feedback resistor in Fig. 8.4 is replaced by a capacitor as in Fig. 8.6. Now the input current V_1/R flows onto the left plate of the capacitor, and the capacitor sums or integrates the elements of charge flowing into it. Since the current into a capacitor is equal to C times the rate of change of capacitor voltage, we have in this case:

$$\frac{V_1}{R} = -C \cdot \frac{dV_0}{dt} \quad \text{or} \quad V_0 = \frac{-1}{RC} \cdot \int_{t_0}^{t} V_1 \, dt \qquad (8.6)$$

In words, the input is proportional to the derivative of the output or the output is proportional to the integral of the input, from some time t_0 in the past when the output voltage was zero until the present time t.

Circuits such as the integrator of Fig. 8.6 and the adder of Fig. 8.5 provide the basis for analog computation, in which the dependent variables, sometimes many of them in a physical problem, are modeled as voltages and the independent variable is modeled as time.

As an example of how integrators may be used to "solve" a differential equation, refer to Fig. 8.7. There a pair of integrators is used and feedback is employed both around the individual operational amplifiers and around the entire circuit. If we call the output of the second amplifier $V(t)$, then from Eq. 8.6 the output of the first amplifier should be labeled $-RC \cdot dV/dt$. But we also know that the sum of the currents through the input resistors to the first amplifier must equal the capacitor-charging current for the first amplifier.

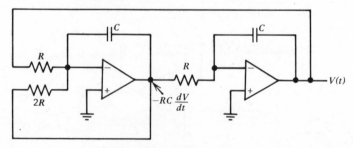

Fig. 8.   7. Example of solution of a differential equation with op-amps.

$$\frac{V(t)}{R} - \frac{dV}{dt} \cdot \frac{RC}{2R} = -C \frac{d}{dt} \left(-\frac{dV}{dt} \cdot RC \right) \tag{8.7}$$

As is often the case in commercial analog computers, if the product of R and C is chosen to be 1 second, Eq. 8.7 may be rearranged to appear as:

$$\frac{d^2 V(t)}{dt^2} + \frac{1}{2} \frac{dV(t)}{dt} - V(t) = 0 \tag{8.8}$$

In other words, the circuit voltages must obey the above differential equation at all times. Of course if the capacitors both happen to be uncharged at some instant of time, then all the input and output voltages would start and remain at zero, which is a perfectly acceptable solution of Eq. 8.8, though a trivial one. More interesting solutions can be obtained by starting with voltages on either or both capacitors. This corresponds physically to nonzero initial conditions on either the physical variable, its first derivative, or both.

Figure 8.8 shows another member of the inverting amplifier family. The input current is $C \, dV_1/dt$ and it must all flow through R, forcing the output voltage to be:

$$V_0 = -RC \frac{dV_1}{dt} \tag{8.9}$$

So we have a differentiator! Unfortunately, however, differentiators do not work well in practice, especially when built with an idealized operational amplifier; there is always unwanted noise at the input of an amplifier, and the noise has Fourier components even at extremely high frequencies. Ordinarily an amplifier, by its very nature as a band-pass device, eliminates the high-frequency components of the noise. However, our idealized operational amplifier does not eliminate the high-frequency components of the noise. To make matters worse, in the differentiator configuration of Fig. 8.8, the amplifier with feedback acts like a filter with system function $H(f)$ that increases linearly with frequency, so that the high-frequency components of the input noise are actually amplified rather than suppressed. When a real operational amplifier is used in a differentiator circuit, there is a high-frequency rolloff of gain, and it is

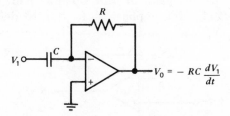

Fig. 8. 8. Op-amp differentiator.

possible to achieve a compromise between noise suppression, which requires a limited bandwidth, and faithful differentiation, which requires infinite bandwidth.

8.3. The effect of finite bandwidth of an operational amplifier   In our introductory explanation of operational amplifier circuits in section 8.2, we used an idealized model of the amplifier in which the gain remained at the same very high value even at very high frequencies. Next we examine the behavior of operational amplifier circuits in which the amplifier itself has an open-loop gain that is a function of frequency.

As a first example, we will re-examine the voltage-follower circuit of Fig. 8.3 and take into account the variation of A with frequency. (We retain, however, the idealization of infinite input impedance.) Our problem is to find the actual system function $H(s)$, treating the voltage applied to the noninverting input as the system input function and the output voltage of the follower as the system output function. A straightforward way of finding $H(s)$ is suggested in Fig. 8.9. If the input function is e^{st}, the output function, by definition, must be $H(s) \cdot e^{st}$. As usual we will drop e^{st}, which appears in every term of the equations describing the system. The difference voltage $(V_+ - V_-)$ at the input is $1 - H(s)$, and when this is multiplied by the amplifier gain $A(s)$, the result is the output voltage $H(s)$. Thus

$$[1 - H(s)] A(s) = H(s) \qquad (8.10)$$

This equation may be solved for $H(s)$ to obtain

$$H(s) = \frac{A(s)}{1 + A(s)} \qquad (8.11)$$

(Note the similarity between this equation and Eq. 5.3. Here, of course, $F(s) = A(s)$ and $G(s) = 1$, because the output is connected directly to the inverting input. The need for a comparator is eliminated because the op-amp itself is a differential amplifier.)

As always we must be aware of the possibility of instability. Because of inherent limitations on the frequency response of the transistors or other amplifying devices within the op-amp itself, for very high frequencies the magnitude $|A(s)|$ of the gain of the op-amp will fall below unity. According to Eq. 8.11, if the phase shift, due to delays or capacitive

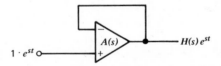

Fig. 8.   9. The voltage follower revisited; the op-amp gain is frequency dependent.

effects, is $180°$ for sinusoidal signals when $|A(s)| = 1$, the denominator will become zero for signals of that frequency, and the output will oscillate at that frequency. Somehow we must guard against not only the possibility of instability for sinusoidal signals but also the possibility that $1 + A(s) = 0$ for any s with positive real part. Usually stability can be guaranteed by an intentional and controlled spoiling of the gain of the op-amp gain at high frequencies. A typical op-amp frequency response characteristic is

$$A(s) = \frac{A_0 a}{s + a} \quad \begin{matrix} a > 0 \\ A_0 > 0 \end{matrix} \tag{8.12}$$

in which A_o may be recognized as the gain for low frequencies ($|s| \ll a$). Equation 8.12 describes a first-order, low-pass system; the gain clearly diminishes inversely with frequency at high frequencies.

We can obtain the system function of the follower circuit by substituting Eq. 8.12 into Eq. 8.11 and rearranging slightly to obtain

$$H(s) = \frac{1}{1 + \dfrac{s + a}{A_0 a}} = \frac{A_0 a}{s + a(1 + A_0)} \tag{8.13}$$

A Bode plot of the magnitude of this function for $s = j \cdot 2\pi f$ is shown in Fig. 8.10.

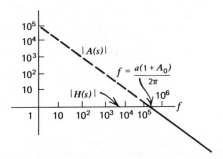

Fig. 8.   10. Bode plot of voltage gain of the follower.

For a typical op-amp, A_0 might be about 10^5 and a about 10, so that the low-frequency gain of the follower is $A_0 a/a(1 + A_0) = 10^5/(1 + 10^5) \approx 1$, and the break frequency occurs where $2\pi f = a(1 + A_0)$ or at:

$$f = \frac{a(1 + A_0)}{2\pi} \approx \frac{10^6}{2\pi} \text{ Hz} \tag{8.14}$$

Also shown, dashed, in Fig. 8.10 is the Bode plot for the magnitude of

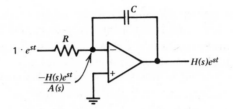

Fig. 8.   11. The integrator revisited.

$A(s)$. Note that at high frequencies the follower's gain is essentially equal to $A(s)$. The voltage follower achieves a much higher break frequency than the op-amp itself, but pays for the extended frequency response with a reduction in low-frequency gain.

Because the open-loop gain, $-A(s)$, is 1 only if $s = -a(A_0 + 1)$, which is real and negative, the follower circuit is stable.

Another example may prove illuminating. Consider, again, the integrator circuit, redrawn in Fig. 8.11. If the circuit is an ideal integrator its impulse response must be a step function, and we should find that $H(s)$ is proportional to $1/s$. However, because of the finite low-frequency open-loop gain and the falloff of open-loop gain at high frequencies (Eq. 8.12), the actual integrator does not behave in ideal fashion.

To find $H(s)$ for the real integrator we equate the currents through R and C, with the help of Fig. 8.11:

$$\frac{1 + \dfrac{H(s)}{A(s)}}{R} = \left[-\frac{H(s)}{A(s)} - H(s) \right] sC \tag{8.15}$$

Note that the inverting input voltage must be $-H(s)e^{st}/A(s)$ because the noninverting input is at 0 V. We solve Eq. 8.15 for $H(s)$.

$$H(s) = \frac{-\dfrac{1}{RC}}{s\left[1 + \dfrac{1}{A(s)} \right] + \dfrac{1}{A(s)RC}} \tag{8.16}$$

We can look at this expression in three different frequency ranges.

1. Very low frequencies $|s| \ll 1/A(s)RC$. Here the term involving s is negligible and $H(s) \approx -A(s) = -A_0$. In this frequency range the circuit behaves like an amplifier rather than as an integrator. For example, if $RC = 10^{-2}$ sec and $A_0 = 10^5$, the frequencies for which the circuit does not integrate would be less than $1/2\pi A_0 RC = 1/2\pi \times 10^3 = 1.6 \times 10^{-3}$ Hz.

2. Moderate frequencies $|s| \gg 1/A(s)RC$ and $|A(s)| \gg 1$. For these frequencies Eq. 8.16 reduces to $H(s) = -1/RCs$, and we recognize the system as an ideal integrator.

3. High frequencies, such that $|A(s)| \leqslant 1$. Here we must use a particular model for the high-frequency op-amp gain as given in Eq. 8.12. Substitution of Eq. 8.12 into 8.16 results in:

$$H(s) = \cfrac{-\cfrac{1}{RC}}{s\left(1 + \cfrac{s+a}{A_0 a}\right) + \cfrac{s+a}{A_0 aRC}} = \cfrac{-\cfrac{A_0 a}{RC}}{s^2 + s\left(A_0 a + a + \cfrac{1}{RC}\right) + \cfrac{a}{RC}} \qquad (8.17)$$

At very high frequencies, $|s| \gg A_0 a$ or, in our example, $f \gg A_0 a/2\pi \approx 160$ KHz, the overall system function is proportional to $1/s^2$, and the "integrator" really behaves like two integrators in cascade. The integrator circuit can be investigated for stability by testing the system function of Eq. 8.17. Instability will occur if any of the poles of $H(s)$ lie in the right-half s plane. Since the coefficients of s^2, s, and the constant in the denominator of $H(s)$ are all positive, there can be no right-half plane poles. Thus the integrator circuit is stable.

8.4. Input-offset voltage and bias currents   Operational amplifiers suffer numerous deviations from the pure and idealized behavior that we ascribed to them in our introductory discussion. Perhaps the most important of these are the input-offset voltage and input bias currents. An input-offset voltage occurs because of unavoidable and unpredictable asymmetries in the + and − input circuitry. The offset voltage may be modeled by introducing a battery V_{os} in one of the input leads (usually the − input lead by convention) of an otherwise ideal operational amplifier, as depicted in Fig. 8.12a. (We must emphasize that the battery V_{os} is not a *real* component, in the sense that a dc voltage V_{os} appears

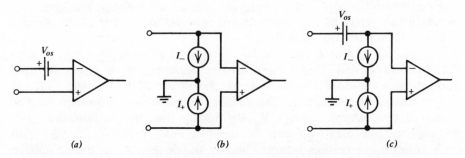

(a)   (b)   (c)

Fig. 8.   12. Input offset voltage and bias currents added to an ideal op-amp.

directly between two points in the circuit.) Typical values of the offset voltages of practical operational amplifiers are on the order of ±1 mV, but often a manufacturer provides a way of adjusting the effective offset voltage toward zero, sometimes with a built-in potentiometer (in the case of discrete component operational amplifiers) and sometimes with the addition of an external potentiometer (in the case of integrated-circuit op-amps). One of the most obvious effects of the input-offset voltage can be to drive the output to saturation when both input terminals are grounded. For example, suppose that the offset voltage is 1 mV and the dc gain of the op-amp is 10^5. Then with both input terminals grounded in Fig. 8.12a the output voltage would be 100 V, which for most op-amps is far beyond the possible range of output voltages. (Typical maximum output voltages are ±10 V.) It is primarily for this reason that in all circuits in which one wishes to keep an operational amplifier in its linear range of operation, feedback must be employed between the output and the inverting input. In fact, the difference between an ordinary differential amplifier and an op-amp is that the op-amp requires this negative feedback to avoid saturation. This negative feedback assures that any tendency for the output to saturate in either direction is countered by a compensating signal at the inverting input.

Input bias currents arise from the necessity of supplying the proper currents to the input transistors in a differential amplifier. In an amplifier in which the input transistors are ordinary bipolar transistors, these bias currents are typically in the range of 0.01 to 1 μA. In contrast, for amplifiers with MOSFETs in the input stages, the input bias currents can be as low as 100 fA (1 fA = 10^{-15} A). The effect of input bias currents may be modeled in an otherwise ideal amplifier by introducing a dc current source between each input terminal and ground (Fig. 8.12b). A composite model, including both the bias currents and the input-offset voltage is shown in Fig. 8.12c. While in an ordinary (low-gain) differential amplifier, the input-offset voltage and bias current can be measured in straightforward ways, in a high-gain op-amp, care must be taken to avoid saturation of the output.

To measure the input-offset voltages, we can use a circuit configuration in which the effect of the bias currents is negligible. Either of the circuits in Fig. 8.13 achieve this condition.

The circuit of Fig. 8.13a can be recognized as a simple follower with the input grounded. Since there are no resistors across which the bias currents can develop a voltage drop, the effect of bias current on the output voltage is negligible. The offset voltage V_{os} can be expressed in terms of the output voltage V_0 by noting that the differential input voltage to the ideal op-amp is $0 - (V_0 - V_{os})$ or $V_{os} - V_0$. This differential input voltage is multiplied by the dc gain A_0 of the op-amp to obtain V_0. That is,

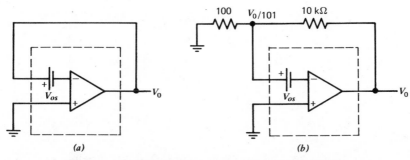

Fig. 8.   13. Circuits for measuring input offset voltage.

$$A_0(V_{os} - V_0) = V_0$$

or

$$V_{os} = \frac{V_0}{A_0} + V_0 \approx V_0 \qquad (8.18)$$

Thus, to a very good approximation, the output voltage, which can be measured directly, is equal to the offset voltage.

If the output voltage of Fig. 8.13a is too small to be measured by an available meter, the circuit of Fig. 8.13b may be useful. Here, only a fraction $100/(100 + 10,000) \approx 1/101$ of the output voltage is fed back to the actual inverting input of the op-amp. Since the differential input voltage can be seen to be $0 - [(V_0/(101) - V_{os})]$, in analogous fashion to the follower circuit we have

$$A\left(V_{os} - \frac{V_0}{101}\right) = V_0$$

or

$$V_{os} = \frac{V_0}{A} + \frac{V_0}{101} \approx \frac{V_0}{101} \qquad (8.19)$$

Now the measured output voltage is merely divided by 101 to obtain the offset voltage.

In the circuit of Fig. 8.13b the input bias current I_- flows partly through the 100-Ω resistor and partly through the 10-kΩ resistor. For typical bias currents of 100 nA or less, a detailed analysis (see one of the problems) shows that the iR drops produced by I_- have negligible effect on the output voltage, V_0. However, if a large resistor, for example, 10^6 Ω, is added in series with the inverting input lead, the bias current effect can dominate.

Thus, in the circuits of Fig. 8.14, if $I_-R \gg V_{os}$, we can replace V_{os} by

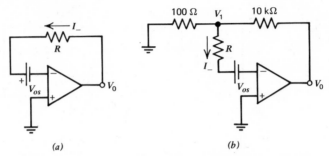

Fig. 8.   14. Circuits for measuring input bias current.

$I_- R$ in the analyses above to obtain the relations:

$$I_- R = V_0$$

and

$$I_- R = \frac{V_0}{101} \qquad (8.20)$$

for the configurations of Fig. 8.14a and 8.14b. Note that for a typical value of $I_- = 100$ nA, and an offset voltage of 1 mV, a resistor $R = 10^6$ Ω makes $I_- R = 100$ mV, which is sufficient. Note that the circuit of Fig. 8.14b would give $V_0 \approx 101 \cdot I_- R = 10.1$ V. This shows that we are close to saturation of the output and illustrates that, if the input bias current were somewhat larger, a smaller value of R might have to be used to prevent saturation.

The bias current I_+ into the noninverting op-amp input can be measured by introducing, in either of the circuits of Fig. 8.13, a resistor R of sufficient magnitude in series with the noninverting input. If $I_+ R$ is much greater than V_{os}, an analysis shows that $I_+ R = -V_0$ or $-(V_0/101)$ depending on which of the circuits of Fig. 8.13 is used.

It is not difficult to show that the effective resistance through which I_- flows is the parallel combination of the resistors from the inverting input to ground and to the output in a configuration such as in Fig. 8.13b. It is usually considered good design practice to place a resistor, equal to that parallel combination, in series with the noninverting input so that the nearly equal bias currents give iR drops whose effects on the output voltage cancel.

8.5. Measurement of common-mode and differential gain of an op-amp   We saw in Chapter 7 that the effects of unequal amplification of signals at the + and − inputs of a differential amplifier can be described in terms of a common-mode and a differential gain, each of which may be a function of frequency. An op-amp may be modeled in the same way; however, when we try to measure these gains, we must be careful to avoid

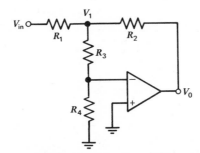

Fig. 8.   15. Circuit for measuring differential gain.

saturation of the op-amp due to its huge differential gain at low frequencies.

The circuit of Fig. 8.15 is useful for measuring the differential gain; it avoids saturation by supplying feedback to the inverting input. Also it provides built-in amplification, as in the circuit of Fig. 8.13b, by supplying this feedback through a voltage-divider network, R_3, R_4. To measure the differential gain A_D at a particular frequency, a sinusoidal signal is applied at V_{in}. The voltages V_0 and V_1 are measured (note that it is not necessary to measure V_{in}) and then, since the differential input voltage is $-V_1 \cdot [R_4/(R_3 + R_4)]$, we have

$$-V_1 \frac{R_4}{R_3 + R_4} \cdot A_D = V_0 \qquad (8.21)$$

The differential gain can be calculated from this equation if R_3 and R_4 are known. Typical values for the resistors in the circuit of Fig. 8.15 are $R_1 = R_2 = R_3 = 10 \text{ k}\Omega$, $R_4 = 100 \ \Omega$, in which case $A_D = -(101 \ V_0)/V_1$.

Note that we have not shown the offset voltage and the bias currents in the circuit of Fig. 8.15. These will be present, however, when a measurement is made and may interfere with the measurement of A_D at zero frequency. The best way of measuring the zero-frequency differential gain is to change the input voltage and to use measured *changes* in V_o and V_1 in Eq. 8.21.

The common-mode gain of an op-amp can be measured with the aid of the circuit of Fig. 8.16. If we employ the usual approximation of zero current into either input terminal of the op-amp, we readily obtain the following relations:

$$\frac{V_1 - V_-}{R_1} = \frac{V_- - V_0}{R_2}$$

$$\qquad (8.22)$$

$$\frac{V_1 - V_+}{R'_1} = \frac{V_+}{R'_2}$$

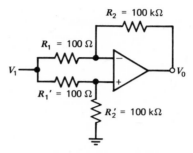

Fig. 8.   16. Circuit for measuring common-mode gain.

When these equations are solved for V_- and V_+ and used in the relation $V_0 = A_D(V_+ - V_-) + A_C V_+$ (see Chapter 7) we obtain

$$\frac{V_0}{V_1} \cdot \left(1 + \frac{A_D R_1}{R_1 + R_2}\right) = A_D \left(\frac{1}{1 + \frac{R_1'}{R_2'}} - \frac{1}{1 + \frac{R_1}{R_2}}\right) + A_C \left(\frac{1}{1 + \frac{R_1'}{R_2'}}\right) \qquad (8.23)$$

If the first term on the right is negligible in comparison with the second and provided that $A_D R_1/(R_1 + R_2) \gg 1$, then Eq. 8.23 reduces to:

$$A_C = \frac{V_0}{V_1} \cdot \frac{A_D R_1}{R_2} \qquad (8.24)$$

and a knowledge of the quantities on the right permits A_C to be calculated. This result is deceptively simple. In order to satisfy the conditions for its validity, the resistors in the circuit must be quite accurately matched. For example, suppose $A_D = 10^5$, $A_C = 10$, and 1% resistors are used. Then $R_1'/R_2' = 0.001 \pm 2\%$, $R_1/R_2 = 0.001 \pm 2\%$, and the denominators $1 + R_1'/R_2'$ and $1 + R_1/R_2$ may differ by about 0.004%. Thus the factor multiplying A_D may be as large as about 4×10^{-5}, which means that the entire A_D term can be almost as large as A_C in Eq. 8.23.

The ratio of differential gain to common-mode gain is known as the common-mode ratio or the common-mode rejection ratio (CMRR). Typical values of CMRR at low frequencies are in the range of 10^4 to 10^6. (Often the common-mode rejection ratio is expressed in dB (decibels). The number of dB is given by 20 times the log to the base 10 of the actual ratio.) In circuits such as the inverting amplifier, in which the + input stays at or near ground potential, the common-mode gain is seldom a problem. However, in noninverting circuits such as the voltage follower, in which both inputs may be far removed from ground potential, the common-mode gain can play a measurable role in the expected gain expression. For example, for the follower circuit, one can prove that the gain is given by $(A_D + A_C)/(1 + A_D)$, instead of simply by $A_D/(1 + A_D)$.

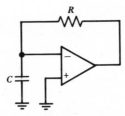

Fig. 8.   17. Is this op-amp circuit stable?

8.6. Stray capacitance at the inverting input   A potential difficulty in op-amp circuits, because of the presence of feedback, is instability. Although circuits such as the follower, the inverting amplifier, and the integrator (discussed earlier) appear upon a first analysis to be inherently stable, we must look a little closer, particularly at a possible phase shift due to stray capacitance between the inverting input and ground.

To appreciate the problem, look at the circuit of Fig. 8.17. This might be viewed as a grounded-input follower with resistive feedback (rather than a direct connection) between the output and the inverting input and with stray capacitance represented by C. We can compute the open-loop gain by postulating a voltage $1 \cdot e^{st}$ at the output. This voltage is attenuated by the factor $(1/RC)/(1/RC + s)$ (a low-pass characteristic) before being applied to the inverting input and then amplified by $-A(s)$. If the op-amp has the usual gain characteristic, $A(s) = Aa/(s + a)$, then the open-loop gain equation is

$$-\frac{Aa}{(s + a)} \cdot \frac{1/RC}{(s + 1/RC)} = 1 \qquad (8.25)$$

The condition for stability (see Chapter 5) is that all s values that satisfy this equation must lie in the left-half s plane. Now an algebraic solution for s by quadratic formula would show that the real part of the solutions is negative and therefore that the system is stable. However, to achieve a better insight into the stability problem, construct a Nyquist plot.

Figure 8.18a shows the vectors $(s + 1/RC)$ and $(s + a)$ drawn on a pole-zero plot from the poles of the open-loop gain at $-1/RC$ and $-a$ to the tip of a pure imaginary s vector, $s = j\omega$. The Nyquist plot is drawn with the help of the pole-zero plot. When $s = 0$, the open-loop gain is $-A$, so that the Nyquist plot begins at $-A$ on the negative real axis. As s climbs the imaginary axis, the vectors $s + (1/RC)$ and $s + a$ increase in magnitude and acquire positive phase angles. But the factors $s + (1/RC)$ and $s + a$ are in the denominator of the open-loop gain, so $OLG(s)$ gets smaller in magnitude and its phase angle becomes less than $180°$. At very large imaginary values of s the open-loop gain curls into the origin along a path that passes through the first quadrant. As s travels around the infinite

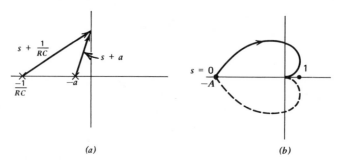

(a)   (b)

Fig. 8.   18. Pole-zero plot and Nyquist diagram for circuit of Fig. 8.17.

semicircle in the right-half s plane, the Nyquist plot remains at the origin and finally, as s goes from $-j\infty$ to 0, the Nyquist plot is the complex conjugate (dashed in Fig. 8.18b) of the first part of the plot.

The point 1 is outside the Nyquist contour. Or is it? If there is some additional phase shift in the open-loop gain not accounted for in the second-order, open-loop gain expression in Eq. 8.25 (see the shower problem in Chapter 5), then the Nyquist plot might enclose the point 1 and produce an unstable system. The margin of error decreases as RC increases. One way of understanding this is to compute the value of ω at which the *magnitude* of the open-loop gain has fallen to 1 and then to compute the phase angle of the open-loop gain at that value of ω. That phase angle is called the phase margin; if the phase margin is zero then the open-loop gain = 1 at that ω and the circuit is unstable.

Example.   Calculate the value of ω at which the open-loop gain falls to 1 and the phase margin for the three cases, $RC = 10^{-10}$ sec, 10^{-7} sec, and 10^{-4} sec, in the circuit of Fig. 8.17. Choose $A = 10^5$ and $a = 10$ sec^{-1}.

Using Eq. 8.25, we see that we have, for the magnitude of the open-loop gain,

$$| \text{OLG} | = \frac{10^6}{\sqrt{\omega^2 + 10^2}} \cdot \frac{1/RC}{\sqrt{\omega^2 + (1/RC)^2}} \qquad (8.26)$$

We can find, either by trial and error or by quadratic formula, the values of ω that make this expression equal to 1 for the three required values of RC. They are

RC	ω	*phase margin*
10^{-10}	10^6	$90°$
10^{-7}	10^6	$84°$
10^{-4}	10^5	$6°$

Also shown are the phase angles of the open-loop gain at these values of

ω, which, from Eq. 8.25, are seen to be given by:

$$\sphericalangle \text{OLG} = 180° - \tan^{-1}\left(\frac{\omega}{10}\right) - \tan^{-1}\left(\frac{\omega}{1/RC}\right) \qquad (8.27)$$

Thus, when $RC = 10^{-4}$ sec, which could be achieved by the reasonable values of $R = 1$ MΩ, $C = 100$ pF, the phase margin is only 6°. Any additional phase shift of 6° (at an angular frequency of 10^5 rad/sec) would give an oscillation at that frequency. Such phase shifts might occur in the op-amp gain expression or might even occur as a result of a small inductive contribution to the impedance of R, perhaps because of excessive length of the wires connecting R to the rest of the circuit.

The point of this section is that the conditions for oscillation may be met, inadvertently, in practical op-amp circuits. Unwanted oscillations can be a source of great frustration to the novice; perhaps the single most helpful tool in eliminating them is having a good understanding of the fundamental reason for oscillation — an open-loop gain equal to 1 at the frequency of oscillation.

8.7. Transient response of an op-amp: maximum slew rate   There are two effects that limit the response of an operational-amplifier circuit to rapidly changing signals. The most obvious is the limitation imposed by the compensated gain, which as we saw earlier, must have the form of a first-order, low-pass filter.

The second effect that limits the transient response of an op-amp circuit is a nonlinear one. Just as it is possible for the input voltage to a circuit to be of such a value that it demands an output voltage outside of the range that the op-amp is capable of supplying, so it is possible for a change in the input voltage to demand a greater rate of change of output voltage than the op-amp is capable of supplying. The maximum rate of change that an op-amp can produce at its output terminal is called its maximum slew rate and is usually specified in volts per microsecond. A typical value of maximum slew rate for a moderate-speed op-amp would be about 1 V/μsec. The maximum slew rate is limited by the speed of response of the transistors within the op-amp and, unlike the limitation imposed by the compensated gain, has nothing to do with the actual circuit in which the op-amp is placed.

It is a simple matter to measure the maximum slew rate of an op-amp. As shown in Fig. 8.19, an input step voltage is applied to, for example, a voltage follower, and the slope of the output voltage is measured on an oscilloscope. The magnitude of the input step function must be large enough that the follower does not respond with the linear step response characterized by an exponential rise. In a follower, the time constant of the linear step response may be a microsecond or smaller. For a time constant of 1 μsec and a 1-V step input, the maximum slope of the step

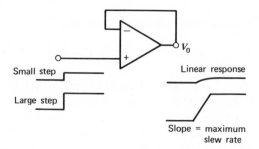

Fig. 8.   19. Measurement of maximum slew rate.

response is 1 V/μsec. Thus the input step voltage would have to be larger than 1 V if a maximum slew rate of 1 V/μsec were to manifest itself. In an amplifying circuit the effective value of the time constant of the linear step response may be much larger than for the follower circuit (see the exercises) so that the limitation of maximum slew rate is more readily reached. Whenever op-amps are used in a switching mode, the maximum slew rate may become the most important property of the op-amp.

8.8. An example of an integrated-circuit operational amplifier The complete circuit diagram of a type-741, integrated-circuit, operational amplifier is shown in Fig. 8.20. This type of op-amp was one of the first to employ internal frequency compensation, which is achieved with the aid of a 30-pF capacitor built right onto the silicon chip. In order to analyze the complicated circuit of a type 741, we must break it down into subsystems that we are able to handle with less difficulty than the entire circuit.

Of the 20 transistors in the circuit, 6 are concerned only with providing the proper biasing currents and voltages to the other 14. These six are Q_8, Q_9, Q_{10}, Q_{11}, Q_{12}, and Q_{13}. Three of these, Q_8, Q_{11}, and Q_{12}, have their bases and collectors connected together, so that they are being used as diodes rather than as transistors. The voltage across Q_{11}, which is perhaps 0.75 V, is applied to the base of Q_{10}, and Q_{10} behaves as a constant-current source. That is, the current into the collector of Q_{10} remains practically constant (at around 30 μA at supply voltages of ±15 V) despite possible changes in its collector voltage. Q_9, controlled by Q_8, is also a current source; the difference between the collector currents of Q_{10} and Q_9 is drawn as a constant current from the bases of the p-n-p transistors Q_3 and Q_4. The other current source, Q_{13}, controlled by Q_{12}, provides a constant bias current to the frequency-compensation subsystem comprising Q_{16}, Q_{17}, and Q_{18}. In addition, the collector resistance of Q_{13} provides a path to ground for signal currents in the frequency-compensation subsystem. The entire biasing system is designed to give an output voltage that tends to be near zero for equal values of + and − input

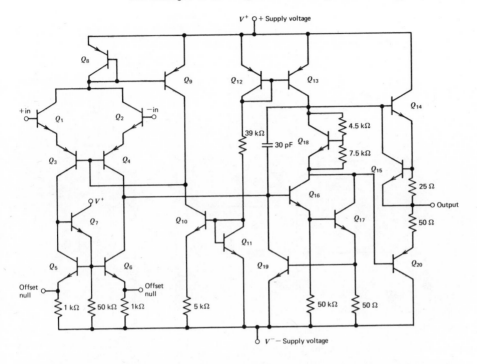

Fig. 8.   20. Circuit diagram of the type-741 integrated circuit op-amp.

voltages over a wide range of input voltage (roughly ±13 V for a ±15-V supply). Fine control over the offset voltage may be achieved by connecting an external 10-kΩ pot between the offset-null terminals, with the wiper arm of the pot connected to the negative supply voltage. This, in effect, provides a fine adjustment on the values of the 1-kΩ emitter resistors of Q_5 and Q_6.

Two of the transistors in the circuit — Q_{15} and Q_{19} — act as current limiters. Q_{15} operates as follows. As long as the current in the 25-Ω resistor in the emitter circuit of output transistor Q_{14} remains below about 25 mA, Q_{15} is cut off and has no effect in the circuit. However, when the current reaches about 25 mA, the voltage from base to emitter of Q_{15} reaches about 0.6 V and Q_{15} begins to turn on, thus diverting current from the base of Q_{14} and preventing any further increase in Q_{14}'s emitter current. Transistor Q_{19} works in a similar way, with the 50-Ω resistor in the emitter circuit of Q_{17}, to limit the collector current of Q_{17} and the current flowing out of the base of Q_{20}. The design of the output circuitry makes the type-741 operational amplifier almost foolproof, in the sense that the output terminal may be short-circuited to ground or to either supply voltage for an indefinite period without damaging the device. Earlier integrated circuit op-amps did not have this built-in,

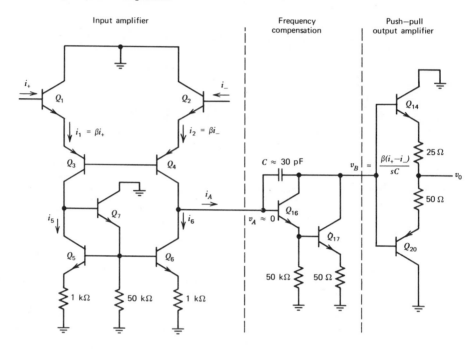

Fig. 8.   21. Simplified circuit diagram for the 741 op-amp.

short-circuit protection, and an inadvertent short could burn out the op-amp.

After this brief discussion of the biasing circuitry and quiescent state of the type-741 op-amp, we must attempt to understand the small-signal, linear behavior. For this purpose we divide the circuit into three sections. These are:

1. Input amplifier, comprising Q_1 through Q_7.
2. Frequency compensation circuit, comprising Q_{16}, Q_{17}, Q_{18}, and C.
3. Output amplifier, comprising Q_{14} and Q_{20}.

A simplified circuit diagram of the entire op-amp appears in Fig. 8.21. In drawing this diagram we have omitted all constant-current sources, since there can be no *signal* current through a *constant*-current source, and we have replaced all constant-voltage sources by short circuits to ground. Also Q_{18} is replaced by a short circuit, since it serves only to maintain the constant voltage drop necessary to supply the proper bias voltages to the bases of Q_{14} and Q_{20}. In other words, in Fig. 8.21, we have retained only those circuit elements through which a *changing* current can flow — such as resistors, transistors, and the capacitor — *and* that can support a *changing* voltage difference.

In discussing the flow of signal currents we employ the standard convention of lower-case letters. Thus the currents into the inverting and noninverting inputs are denoted by i_- and i_+, respectively. These base currents are amplified in Q_1 and Q_2 to give emitter currents $i_1 = \beta i_+$ and $i_2 = \beta i_-$, which flow through Q_3 and Q_4. The function of Q_7 is to force Q_5 and Q_6 to have equal collector currents; therefore, since $i_5 \approx i_1$ and $i_6 + i_A = i_2$, where i_A is the signal current to the frequency-compensation circuit, we have $i_A \approx i_2 - i_1 = -\beta(i_+ - i_-)$. Thus the input amplifier amplifies the *difference* between the input currents and supplies a signal output current to the frequency-compensation circuit.

Transistors Q_{16} and Q_{17} form what is called a Darlington pair. This combination is characterized by a high input impedance and a very large current amplification factor (ratio of total collector current to base current into Q_{16}). If we neglect for a moment any current supplied to the push-pull output amplifier, then practically all of the signal current i_A flows through C and then down through the collectors of Q_{16} and Q_{17} and finally through the 50-Ω resistor to ground. Since the signal voltage across this 50-Ω resistor is negligible, the emitter of Q_{17}, the base of Q_{17}, and the base of Q_{16} are all essentially at constant potential or ground, as far as signal voltages are concerned. Now the current i_A produces a voltage drop i_A/sC in the capacitor, so the signal voltage at B is $-i_A/sC$ or $\beta(i_+ - i_-)/sC$. This voltage drives the emitter follower output stage so that it is also approximately the output voltage v_0. If we replace $i_+ - i_-$ by $(v_+ - v_-)/R$, where R is the differential input impedance, we see that the overall differential voltage gain $v_0/(v_+ - v_-)$ is given by β/sCR, and we have explained the first-order falloff of gain as a function of frequency. Now the specification sheets show that the input impedance is about 1 MΩ. If we use $C = 30$ pF and a reasonable value of β, say 200, we find that the gain is $(6.7 \times 10^6)/s$, which is about right. Although this equation predicts that the dc gain is infinite, there is a low-frequency cutoff (see problem 8.14) due to the finite effective resistance of the output amplifier. This cutoff turns out to be at about 10 Hz, or $|s| = 20\pi$, so that the dc gain is about 10^5 as we suggested earlier.

8.9. Frequency filtering with op-amps

In many signal-processing systems, some examples of which we will explore later, it is necessary to attenuate those Fourier components of a signal outside of a certain band of frequencies while passing with uniform gain those Fourier components within the band. It is impossible to do this precisely. Yet, approximations to an ideal filter can be achieved that satisfy the requirements closely enough to be useful.

Figure 8.22 illustrates two examples of first-order, low-pass filters. Except for sign (the circuit of Fig. 8.22b inverts), the system functions of both the "passive" filter in Fig. 8.22a and the "active" filter in Fig. 8.22b

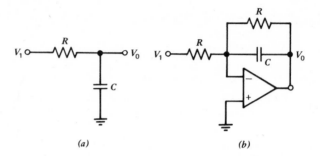

Fig. 8.   22. Low-pass filters: (a) passive and (b) active.

are given by

$$H(s) = \frac{V_0(s)}{V_1(s)} = \pm \frac{1}{1 + sCR} \qquad (8.28)$$

Why then would we want to use an active filter with an op-amp when we could equally well use a simple RC network? The answer to this question lies in the ability of each circuit to drive a succeeding circuit. Any current drawn from the passive circuit's output changes its frequency response characteristic. This is not true for the active filter; any current drawn from its output comes from the output stage within the op-amp and not from the RC part of the circuit. Thus, whenever we want to cascade a first-order, low-pass filter with another filter, or with any circuit with a relatively low input impedance, the active filter of Fig. 8.22b is the circuit of choice.

Next suppose that we want a second-order, low-pass filter; that is, one whose system function is proportional to $1/s^2$ at high frequencies. One way to achieve this is to use two active first-order, low-pass filters in series or in cascade, that is, one after the other. However, it turns out that a second-order frequency characteristic can be obtained with a single op-amp. Such a second-order, low-pass filter is shown in Fig. 8.23. In order to calculate the output voltage when the input voltage is unity, in a very uncluttered manner, we choose the unit of s to be $1/RC$. This is equivalent to setting $R = 1$, $C = 1$ or to defining a new variable s as RC times the old s. Now we use the Kirchhoff's current law at the node V to write:

$$1 - V = \frac{3Vs}{b} + V + (V - V_0) \qquad (8.29)$$

Application of the KCL to the node at the inverting input of the op-amp itself leads to

$$V = \frac{-V_0 sb}{3} \qquad (8.30)$$

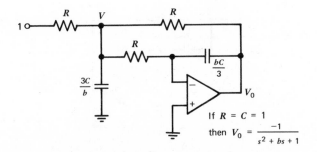

Fig. 8.   23. Second-order, low-pass active filter.

Now substitute from Eq. 8.30 into Eq. 8.29 and solve for V_0:

$$V_0 = \frac{-1}{s^2 + bs + 1} \tag{8.31}$$

Example.   Design, that is, choose resistor and capacitor values to make a second-order, low-pass filter with system function having equal real roots and a break frequency of 1000 Hz.

The condition of equal roots forces us to choose $b = 2$ and, for this choice of b, Eq. 8.31 becomes

$$V_0 = \frac{-1}{(s + 1)(s + 1)} \tag{8.32}$$

The break frequency occurs at $s = j$ so that $\omega = 1$ in units of $1/RC$ or at $\omega = 2\pi \cdot 1000 = 6280$ rad/sec in ordinary units. Thus $1/RC = 6.28 \times 10^3$. One possible choice is $R = 1.6$ kΩ and $C = 0.1$ μF. This value of C, however, is not the actual value of either capacitor in the filter circuit; these are given by $3C/b$ and $bC/3$, which, with $b = 2$, work out to 0.15 and 0.067 μF.

A second-order, high-pass filter can be designed by using the previously discussed low-pass filter as a basis. One simply replaces all resistors by capacitors and all capacitors by resistors. The resistors are chosen to have the same ratio as the ratio of impedances of the capacitors in Fig. 8.23. Thus, in Fig. 8.24, the three capacitors all have the same value C while the resistor from V to ground has the value $bR/3$ and the feedback resistor has the value $3R/b$. The equations analogous to (8.29) and (8.30) (with $R = 1$, $C = 1$) are

$$(1 - V) \cdot s = 3V/b + sV + (V - V_0) \cdot s$$

and

$$V = -V_0 b/3s \tag{8.33}$$

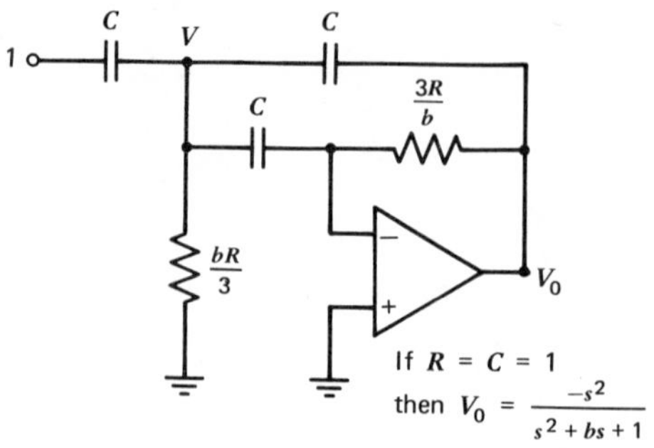

If $R = C = 1$

then $V_0 = \dfrac{-s^2}{s^2 + bs + 1}$

Fig. 8.   24. Second-order, high-pass active filter.

Upon dividing through the first of these by s, we see that these equations are identical to Eqs. 8.29 and 8.30 except that wherever s appears in the former equations, $1/s$ appears in Eqs. 8.33. The solution for V_0 must be

$$V_0 = \frac{-1}{(1/s)^2 + b/s + 1} = \frac{-s^2}{s^2 + bs + 1} \tag{8.34}$$

The high-pass filter has a gain of -1 at high frequencies while at low frequencies the gain falls off quadratically with s; on a (logarithmic) Bode plot, the slope of the low-frequency falloff would be 2, or a factor of 100 decrease in gain for each factor of 10 decrease in frequency.

Notice that, in Eq. 8.31 for the second-order, low-pass filter, it is possible to choose b so that the poles of the system function are complex. If we choose b in this manner, the low-pass circuit simulates a tuned circuit. Suppose, for example, that we make the left-hand capacitor in Fig. 8.23 equal to $1\ \mu F$, the right-hand capacitor equal to $0.0001\ \mu F$, and choose $R = 1\ k\Omega$. Then $3C/b = 10^{-6}$ and $bC/3 = 10^{-10}$, so that $C = 0.01\ \mu F$ and $b = 0.03$. Now the poles of the system function of Eq. 8.31 are found by setting the denominator equal to zero. Thus the poles in this example are at:

$$s = -\frac{b}{2} \pm \frac{\sqrt{b^2 - 4}}{2} \approx -0.015 \pm j \tag{8.35}$$

These poles are plotted roughly to scale in Fig. 8.25. Also shown in the figure are the vectors $s - s_1$ and $s - s_2$ where s_1 and s_2 are the locations of the poles given in Eq. 8.35. If Eq. 8.31 is written in factored form as:

$$V_0 = \frac{-1}{(s - s_1)(s - s_2)} \tag{8.36}$$

we can visualize the magnitude of the response of the filter circuit to an input sinusoid (pure imaginary value of s) by referring to Fig. 8.25. The magnitude of V_0 is given by the reciprocal of the product of the lengths

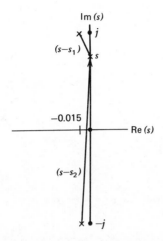

Fig. 8.   25. Pole-zero plot for LPF with complex poles: a tuned filter.

of the vectors drawn from the poles to s. As s approaches j, the length of the vector $s - s_1$ reaches a minimum; the peak value of V_0 therefore occurs very near a frequency $\omega = 1$ or $f = 1/2\pi$ (in units of $1/RC$). The actual peak frequency in hertz in the present example is about

$$f_{\text{peak}} = \frac{1}{2\pi RC} = \frac{1}{2\pi} \cdot 10^{-3} \cdot 10^8 = 1.6 \times 10^4 \text{ Hz} \qquad (8.37)$$

We can summarize the frequency response of the second-order, low-pass filter with complex poles with a Bode plot. Fig. 8.26 shows such a Bode plot for the example above. The independent variable is chosen to be ω,

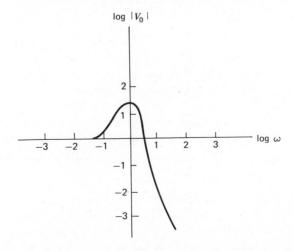

Fig. 8.   26. Bode plot for tuned circuit.

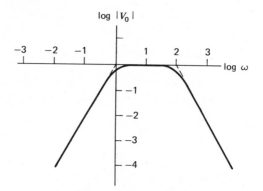

Fig. 8.   27. Bode plot for cascaded high- and low-pass filters.

measured in units of $1/RC$, so that the peak of the Bode plot occurs at approximately $\omega = 1$, or $\log \omega = 0$. At low frequencies, the output for unity input approaches 1, while for high frequencies the output falls off with an eventual slope of -2 on the log-log Bode plot. The maximum value of V_0 in this example is approximately 30.

A filter in which the response is attenuated both at high frequencies and at low frequencies is called a band-pass filter. These can be achieved by cascading high- and low-pass filters; for example, if second-order, high- and low-pass filters are cascaded, the Bode plot of the resultant system might look as in Fig. 8.27. It should be clear that the width and location

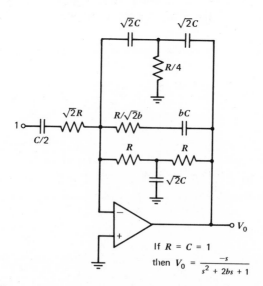

Fig. 8.   28. Single op-amp version of band-pass filter.

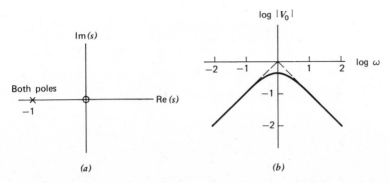

Fig. 8.   29. Pole-zero and Bode plots for circuit of Fig. 8.28 with $b = 1$.

of the passband are controlled by the location of the break frequencies of the high-pass and low-pass filters. If the break frequencies of the high- and low-pass sections of the filter are close to one another, we have what is called a narrow-band filter.

Another approach to a band-pass filter is to make both a high-pass and low-pass filter with a single op-amp. Figure 8.28 shows one method of achieving a first-order, low- and high-pass characteristic. The analysis, though not difficult, is unwieldy and is left for the reader. As is so often the case in an op-amp circuit, the —input is at virtual ground, and the input current flows entirely through the three feedback paths. The output that results from a unit input voltage is given by

$$V_0 = \frac{-s}{1 + 2bs + s^2} \tag{8.38}$$

in which the unit of s is $1/RC$. If b is equal to unity, the roots of the denominator are real and equal. For this special case the pole-zero plot of the system function and the Bode plot appear in Fig. 8.29; the filter would be considered to be narrow-band.

8.10. Op-amp sinusoidal oscillators
Sinusoidal oscillators, in addition to their obvious use as a source of test signals for other circuits, have a number of other important applications in electronic instruments. These applications include frequency (and therefore time) references, and local oscillators in frequency-shifting or heterodyning systems (see Chapter 12). Here we present two examples of sinusoidal oscillators built around op-amps.

The basic idea in any sinusoidal oscillator is to create a circuit in which the only signal of the form e^{st} that can exist is one for which s is purely imaginary. As we learned earlier (in our discussions of the stability of a feedback system in Chapter 5) if the s value for a nonzero signal has a negative real part, any such signal will die away exponentially; if the s

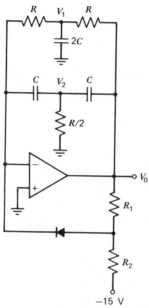

Fig. 8.   30. Twin-tee oscillator.

value has a positive real part, the signal will grow exponentially (and, in any real system, will result in an oscillation that is limited in amplitude by the nonlinearities that are eventually reached); and if the s value has a real part that is exactly zero, a sinusoidal oscillation will exist.

As an example of the application of these ideas about stability and instability, look at the twin-tee oscillator of Fig. 8.30 and attempt to find the conditions under which sinusoidal oscillations will occur. This oscillator is so named because of the two R–C tee networks in the feedback loop. The diode and associated resistors R_1 and R_2 are used to limit the amplitude of the oscillations and to insure that they are nearly sinusoidal; we neglect this part of the circuit in our initial analysis. The usual virtual ground approximation holds and, as long as the output voltage V_0 is not too positive, the diode does not conduct. Then the sum of the currents flowing from the nodes V_1 and V_2 toward the −input terminal of the op-amp must be zero. We choose to measure s in units of $1/RC$ so that, in effect, $R = 1$ and $C = 1$. The condition on the currents is then

$$V_1 + sV_2 = 0 \qquad (8.39)$$

We also apply Kirchhoff's current law to the nodes V_1 and V_2 and obtain the following two equations:

$$V_1(2 + 2s) = V_0$$

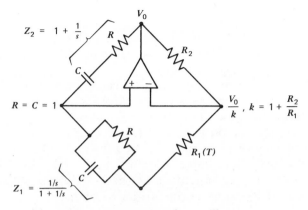

Fig. 8.   31. Wien-bridge oscillator.

and

$$V_2(2 + 2s) = V_0 s \qquad (8.40)$$

We now substitute for V_1 and V_2 in Eq. 8.39 to arrive at the condition on the output voltage:

$$V_0 \cdot \frac{(s^2 + 1)}{2(s + 1)} = 0 \qquad (8.41)$$

For this equation to be satisfied, for a signal $V_0 e^{st}$, *either* V_0 must be precisely zero *or* $s = \pm j$. Thus only a sinusoidal signal of angular frequency $\omega = 1$ can be present at the output. Notice that Eq. 8.41 depends on the fact that the resistor in the lower-tee network has exactly half the resistance of the resistors in the upper-tee network. We leave for the reader to show that if the ratio is either larger or smaller than one-half, the critical value of s that satisfies Eq. 8.41 will lie either in the right-half or the left-half s plane. In the latter case no signal will appear at the output but in the former case an exponentially growing oscillation will occur. Since the possibility of choosing the resistor values to have *exactly* the right ratio is clearly impossible, we are, in practice, always faced with the choice of no oscillations or growing oscillations.

The diode and the associated resistor chain connected from output to the negative supply voltage are in the circuit to limit the buildup of the oscillations. This network comes into play whenever the actual instantaneous value of output voltage is positive enough to raise the midpoint of R_1 and R_2 above zero (actually the critical voltage is a few tenths of a volt because the diode is not an ideal diode). When the diode conducts, it supplies any current demanded by the tee networks, and the output voltage is momentarily clamped at a fixed value. After a small fraction of a cycle the output voltage falls below the clamping value, and the oscillation proceeds as if the diode were not there. If the distortion due to the clamping network is excessive, the output of this oscillator can be put through a low-pass filter, which passes only the fundamental sinusoidal frequency generated by the oscillator and blocks the higher harmonics that are responsible for the distortion.

Figure 8.31 shows another oscillator and another method that can sometimes be used to prevent distortion. This oscillator is called a Wien bridge oscillator because the network is in the form of a bridge circuit. The op-amp actually amplifies any difference voltage across the bridge and drives the bridge with its output. To find the condition for oscillation we employ the virtual short approximation; V_+ and V_- are very closely at the same potential. The potentials at the two inputs of the op-amp are found in terms of V_0 by using the divider relation

$$V_- = [R_1/(R_1 + R_2)] \cdot V_0 = V_0/k$$

$$V_+ = V_0 \cdot \frac{Z_1}{Z_1 + Z_2} = \frac{V_0 \dfrac{1/s}{1 + 1/s}}{1 + 1/s + \dfrac{1/s}{1 + 1/s}} = \frac{V_0 s}{s^2 + 3s + 1} \qquad (8.42)$$

In order for these two potentials to be equal, either $V_0 = 0$ and there are no oscillations, or:

$$s^2 + (3 - k)s + 1 = 0 \qquad (8.43)$$

The roots of this equation depend on the value of k and are found with the aid of the quadratic formula:

$$s = \frac{-(3 - k) \pm \sqrt{(3 - k)^2 - 4}}{2} \qquad (8.44)$$

For $k \geqslant 5$, the roots are real and are both positive. Thus, for such a choice of k, a growing exponential wave (actually a pair of them) will appear at the output, and in practice the output of the op-amp will switch rapidly between + and − saturation levels. For $1 < k < 5$, the roots are complex, with the real part being positive (leading to growing oscillations) if $k > 3$, and with the real part being negative (leading to decaying oscillations) if $k < 3$. When k is exactly equal to 3 the roots lie on the imaginary axis at $s = \pm j$, and sinusoidal oscillations will occur. The problem, of course, in avoiding distortion is to make k equal to 3, exactly. This problem is solved in the Wien bridge oscillator circuit by using a temperature-dependent resistance for R_1 such as a small incandescent lamp, or a fuse (e.g., a 1/100 A instrument fuse for which $R_1 \approx 150\,\Omega$). The idea is to choose the value of the other resistance R_2 so that when R_1 has no current through it, k is slightly larger than 3. As oscillations build up, the temperature of R_1 rises, its resistance increases, and k decreases. A steady state is reached when $k = 3$, because if the amplitude of the oscillations were to increase any further, k would fall below 3 and the oscillations would then die away.

Another solution to the distortion problem in a Wien bridge oscillator is to use a resistor with a negative temperature coefficient for R_2 (i.e., a resistor whose resistance falls when the temperature rises). Any semiconductor in which more electrons and holes exist at higher temperatures will do; these are commercially available under the name of "thermistors."

Varying the frequency of oscillation of either of the two oscillators presented above requires changing all three resistors (or all three capacitors) in the twin-tee oscillator and both resistors (or both capacitors) in the Wien bridge oscillator. "Ganged" potentiometers are commercially available, in which a pair of pots is adjusted simultaneously with one shaft. These make the Wien bridge oscillator a particularly simple one to build. Generally, the pots are used to vary the frequency over a little more than a decade; to change the "range" a multiposition switch is used to vary the capacitors by factors of ten.

In concluding this section we should note that we have derived the conditions for oscillations in the twin-tee and Wien bridge oscillators by making use of the virtual ground and virtual short approximations, which depend for their validity on the gain of the op-amp being large. Obviously the gain will not be large if we attempt to make oscillations occur at a frequency near the unity gain frequency of the op-amp, around a megahertz for a type-741 op-amp. Another limitation is the maximum slew rate of the op-amp; as long as the amplitude of the oscillations remains small this will be no problem, but if the frequency and amplitude are high enough for the slew rate limitation to come into play, the oscillations will tend to become triangular, rather than sinusoidal.

Problems for Chapter 8

8.1. In the op-amp circuit shown in Fig. P8.1 calculate the gain V_2/V_1 under the usual assumptions of very high open-loop gain and input

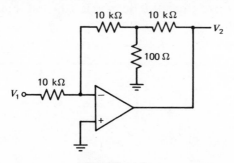

Fig. P8.   1

impedance for the op-amp itself. What single feedback resistor could replace the tee network in the feedback path and give the same voltage gain?

8.2. Find an expression for the output voltage V_0 in terms of V_1 and A in the op-amp circuit of Fig. P8.2 if the op-amp is an ideal one with $V_0 = A(V_1 - V_2)$. Your answer must not contain V_2.

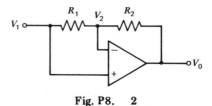

Fig. P8.   2

8.3. Calculate the closed-loop gain expression V_2/V_1 for the integrator circuit (Fig. P8.3) if the open-loop gain of the op-amp is $A(s) = 10^6/(s/10 + 1)$. What is the smallest value of the feedback capacitor for which the overall circuit is stable?

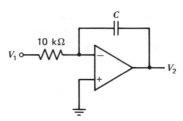

Fig. P8.   3

8.4. Stray capacitance in a circuit is unavoidable. In an op-amp feedback circuit, stray capacitance from the inverting input to ground has the effect of changing the location in the s plane of the solutions of the equation $F(s)G(s) = -1$.

(a) For an op-amp whose gain is $F(s) = [A/(s/a + 1)]$, find A and a if it is known that the dc gain is 10^5 and that the magnitude of $F(s)$ is unity at $\omega = 10^6$ rad/sec.

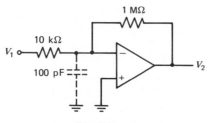

Fig. P8.   4

(b) In this feedback circuit (Fig. P8.4), which includes a stray capacitance of $100\,\mathrm{pF}$ ($1\,\mathrm{pF} = 10^{-12}\,\mathrm{F}$), find the feedback factor $G(s)$.

(c) Calculate the s-plane solutions of $F(s)G(s) = -1$.

8.5. Compute the system function $H(s) = V_2(s)/V_1(s)$ for the operational amplifier circuit in Fig. P8.5. The gain of the op-amp itself is given by $A(s) = [10^5/(s/10 + 1)]$.

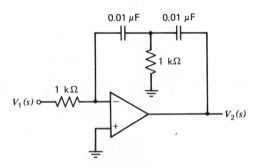

Fig. P8.   5

8.6. Prove that in a voltage-follower circuit (output directly connected to inverting input) with an op-amp with a common-mode gain of A_C and differential gain A_D, the voltage gain is $(A_D + A_C)/(1 + A_D)$.

8.7. In a noninverting amplifier with closed-loop gain of 100, the op-amp has a gain characteristic of $A(s) = [10^5/(s/10 + 1)]$. Find the time constant of the step response, assuming the amplifier remains linear.

8.8. What factors limit the use of a very small fraction of the output voltage being fed back to the inverting input in order to "amplify" the input-offset voltage measurement?

8.9. For an op-amp with gain $A(s) = Aa/(s + a)$ show that the input admittance I_{in}/V_{in} is given by $1/R[1 + Aa/(s + a)]$ when connected in the circuit in Fig. P8.9. Thus if $Aa/(s + a) \gg 1$ we have $Y_{in} \doteq (s + a)/RAa$.

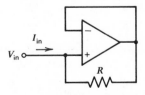

Fig. P8.   9

8.10. Show that, if $RC = 1$, the circuit (Fig. P8.10) has gain $V_o/V_i = (s/2)/\{1 + [1/A(s)](2 + s)\}$ and that if $A(s) = A_0 a/(s + a)$, the circuit

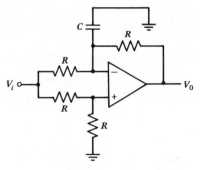

Fig. P8.   10

represents a band-pass filter with system function $(A_0 a \cdot s/2)/[s^2 + s(2 + a) + A_0 a + 2a]$. Also, make a pole-zero plot for the case $A_0 = 10^5$, $a = 10$.

8.11. A "gyrator" is a two-port network in which $I_1 = gV_2$ and $I_2 = -gV_1$. Show that the following network[†] (Fig. P8.11) performs as a gyrator, with $g = 1$.

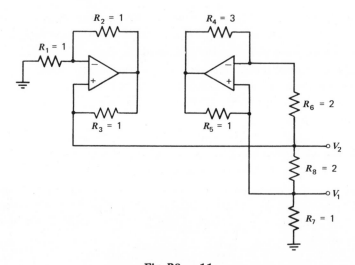

Fig. P8.   11

8.12. In the circuit in Fig. P8.12, which includes an offset voltage V_{os} and a bias current I_-, express the output voltage in terms of R_1, R_2, V_{os}, and I_-. Show for $R_1 = 100 \ \Omega$, $R_2 = 100 \ k\Omega$, $V_{os} \approx 1 \ mV$, and $I_- \approx 100 \ nA$ that the contribution to V_0 from I_- is negligible compared with that from V_{os}.

[†] R. M. Inigo, *IEEE J. Sol. State Circuits*, SC6, 88–89 (1971).

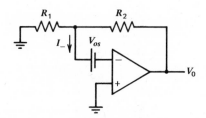

Fig. P8.   12

8.13. In the circuit used to measure the differential gain of an op-amp:
   (a) What would happen if R_2 were a short circuit ($R_2 = 0$)?
   (b) If, with $R_1 = R_2 = R_3 = 10\ k\Omega$ and $R_4 = 100\ \Omega$, one measured V_0 to be 1 V and V_1 to be $-10\ mV$, what would the input voltage, V_{in}, be?

8.14. In the section describing the type-741 op-amp, the effect of the output amplifier was neglected in obtaining the transfer characteristic of the frequency-compensation stage. Suppose that the output amplifier can be modeled by a load resistance R_L, so that the compensation circuit looks like (Fig. P8.14). Assume that $K \gg 1$ and that point A is at ground potential, and find an expression for v_B in terms of C, R_L, s, and K. If $K = 5000$, what must R_L be in order to have a break frequency of 10 Hz?

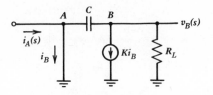

Fig. P8.   14

8.15. The op-amp circuit (Fig. P8.15) is called a negative-impedance converter.
   (a) Show that the input impedance $Z = v/i$ is $-R_1 R_L / R_2$.
   (b) If $R_1 = R_2$ and a series R, L, C circuit is connected between input and ground, find R_L in terms of R in order for sinusoidal oscillations to occur.
   (c) To limit the oscillations, a thermistor may be used in place of R_1 or R_2 so that if the amplitude of the oscillations becomes too large the thermistor heats up, its resistance gets smaller, and the conditions for oscillation are not met. Which of the two resistors, R_1 or R_2, should be a thermistor?

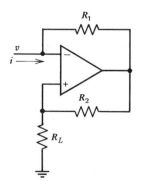

Fig. P8.   15

8.16. A filter for which $| V_{out}/V_{in} | = (1 + \omega^6)^{-1/2}$ is called a third-order Butterworth filter. P. R. Geffe has shown (see references at the end of the chapter) that such a filter can be built from a single op-amp. The circuit (see Fig. P8.16) has $R = 1$, $C_3 = 0.20245$, $C_2 = 3.5468$, and $C_1 = 1.3926$.

(a) Show that, with input voltage e^{st}, $V = 1/(1 + 2s + 2s^2 + s^3)$.

(b) Show that for $s = j\omega$ the magnitude of the voltage transfer function is indeed given by $(1 + \omega^6)^{-1/2}$.

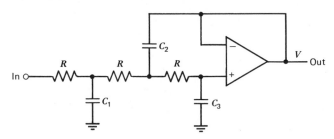

Fig. P8.   16

REFERENCES

Bleuler, E., R. O. Haxby, Eds., *Methods of Experimental Physics, Vol. 2, Electronic Methods*, Chapter 6, Academic Press, New York, 1975.

Geffe, P. R., "How to Build High-Quality Filters out of Low-Quality Parts," *Electronics*, 49, 111–113 (Nov. 11, 1976).

Graeme, J. G., *Applications of Operational Amplifiers: Third Generation Techniques*, McGraw-Hill, New York, 1973.

Graeme, J. G., G. E. Tobey and L. P. Huelsman, *Operational Amplifiers: Design and Applications*, McGraw-Hill, New York, 1971.

Hnatek, E. R., *Applications of Linear Integrated Circuits*, Wiley–Interscience, New York, 1975.

Philbrick/Nexus Research, *Applications Manual for Operational Amplifiers*, Philbrick/Nexus Research, Dedham, Mass., 1968.

Solomon, J. E., "Monolithic Op-Amp: A Tutorial Study." *IEEE J. Solid State Circuits*, SC9, 314–332 (December 1974).

CHAPTER 9

Logic Devices and
Combinational Circuits

9.1. Introduction We turn now to a discussion of digital systems, which are characterized by inputs and outputs that can take on one of only two values at any time, rather than a continuous range of values. Since digital systems are *not* LTI, the analytical methods of convolution, Laplace transforms, and Fourier transforms that we developed earlier are of no help in the analysis and synthesis of these systems. Instead we usually take advantage of the limited number of input configurations and enumerate all of the resulting outputs.

Digital systems may be divided into two classes — combinational, in which the outputs at a given instant are functions of the inputs at that instant, and sequential, in which the outputs depend on both the present inputs and earlier inputs. In sequential systems the earlier inputs must be remembered in some fashion; it is the introduction of memory that makes possible instruments such as counters, calculators, and digital computers.

In this chapter we develop the mathematical notation and elementary tools used in the analysis and synthesis of digital systems. We also describe two of the several commercial families of integrated-circuit logic gates. These integrated circuits, which themselves may perform quite complex digital functions, can serve as building blocks for even larger systems.

9.2. Logical functions, truth tables, and Boolean algebra Of primary importance in both combinational and sequential systems is the notion of a logical function, in which an output variable is a function of one or more input variables. As in ordinary algebra, we denote logical variables by letters, with or without subscripts, except that the values of the variables are confined to the two values 1 and 0. In electronic representations of logical functions, the 1 and 0 are represented by definite values or a definite range of values of current or voltage; for example, any voltage between 0 and 1.5 V might represent a zero, while any voltage between 2 and 3.5 V might represent a one.

The simplest logical function is the NOT function, represented by a bar over a variable. Thus, $y = \bar{x}$ (read y equals NOT x) is a shorthand notation that says $y = 1$ when $x = 0$ and $y = 0$ when $x = 1$. The function is summarized by a truth table, as in Fig. 9.1, which is merely a list of all

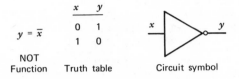

$$y = \bar{x}$$

NOT
Function   Truth table   Circuit symbol

Fig. 9.   1. The NOT function.

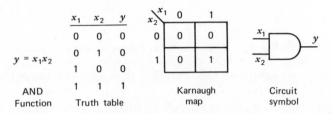

$$y = x_1 x_2$$

AND
Function   Truth table   Karnaugh   Circuit
                                        map   symbol

Fig. 9.   2. The AND function.

possible inputs and the corresponding outputs. The circuit symbol is
useful when various simple logical functions are interconnected to form a
complex system.

Functions of two input variables require a truth table with four rows,
since each of the two variables, for example, x_1 and x_2, can independently be 1 or 0. Two of the two-variable functions are so important that
they have special names and symbols. These functions are the AND
and the OR. The AND function, $y = x_1 x_2$ or $y = x_1 \cdot x_2$ (read y
equals x_1 AND x_2), is illustrated in four equivalent ways in Fig. 9.2. The
Karnaugh map is merely an alternate representation of the truth table;
each square corresponds to a unique set of input variables. For example,
the upper right-hand square in the Karnaugh map conveys the same
information as the third row of the truth table. The name of the AND
function comes from the fact that the output is 1 only if x_1 AND x_2
are 1.

Figure 9.3 shows the OR function, in which the output is 1 if either x_1
OR x_2 is 1, OR both. Sometimes this is called the inclusive OR function.

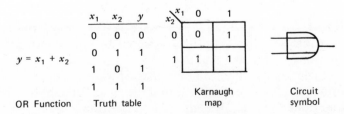

$$y = x_1 + x_2$$

OR Function   Truth table   Karnaugh   Circuit
                                          map   symbol

Fig. 9.   3. The OR function.

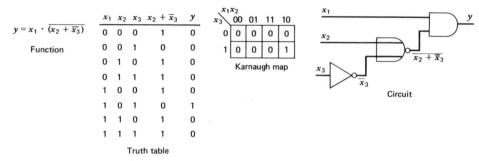

$$y = x_1 \cdot \overline{(x_2 + \overline{x}_3)}$$

Function

x_1	x_2	x_3	$x_2 + \overline{x}_3$	y
0	0	0	1	0
0	0	1	0	0
0	1	0	1	0
0	1	1	1	0
1	0	0	1	0
1	0	1	0	1
1	1	0	1	0
1	1	1	1	0

Truth table

$x_3 \backslash \begin{array}{c} x_1 x_2 \\ \end{array}$	00	01	11	10
0	0	0	0	0
1	0	0	0	1

Karnaugh map

Circuit

Fig. 9.   4. A function of three logical variables.

(There is also an exclusive OR function in which the output is 1 if either input is 1 and the other is 0.)

The ideas above can be extended to include combinations of the basic NOT, AND, and OR functions. For example, the function $y = x_1 \cdot \overline{(x_2 + \overline{x}_3)}$ is illustrated in Fig. 9.4. Several aspects of its various representations deserve comment. In the Karnaugh map the various squares are labeled in a particular way, in anticipation of an important analytical technique to be described later. The idea is that, in going from one square to a horizontally or vertically adjacent square, only one of the input variables changes. It would be incorrect to have the 01 and 10 columns next to each other.

In the circuit diagram of Fig. 9.4, we have employed a useful notation that combines the NOT function with the OR function to represent the function $x_2 + \overline{x}_3$. The circuit ⟶ performs the NOR function.

A similar notation can be used for the NAND function. Sometimes it is useful to place the small circle that indicates inversion at the input of a circuit symbol. Thus ⟶ represents the function $y = x_1 \overline{x}_2$.

However, by far the most important thing we can learn from Fig. 9.4 is that function $y = x_1 \cdot \overline{(x_2 + \overline{x}_3)}$ can be represented by more than one combination of the elementary functions. For example,

$$y = x_1 \overline{x}_2 x_3 \tag{9.1.}$$

and

$$y = \overline{\overline{x}_1 + x_2 + x_3} \tag{9.2.}$$

are equivalent to $y = x_1 \cdot \overline{(x_2 + \overline{x}_3)}$ because they have the same truth table. The reader should verify this statement by constructing truth tables for the functions of Eqs. 9.1 and 9.2.

Since a single logical function may take on various forms, it is useful to have a formalism for deciding, without using truth tables, whether two logical functions are equivalent. An important application of this formalism is in simplifying a logical function so that its circuit representation contains fewer elements. Such a formalism, known as Boolean algebra, was developed originally for describing the "Laws of Thought" and published by George Boole in the mid-nineteenth century. Boolean algebra is a set of axioms, definitions, and theorems based on variables that can take on only discrete values such as 1 and 0, and the AND, OR, and NOT operations. Although we will not introduce Boolean algebra in a formal way, we will, perforce, arrive at some of its conclusions by a graphical analysis of Karnaugh maps. Also there is an important theorem from Boolean algebra that we will find useful, known as DeMorgan's theorem. DeMorgan's theorem states that

$$\overline{x_1 x_2 \ldots} = \bar{x}_1 + \bar{x}_2 + \ldots \tag{9.3}$$

or equivalently

$$\overline{x_1 + x_2 + \ldots} = \bar{x}_1 \bar{x}_2 \ldots$$

These relations are easily proved by constructing truth tables for each side of the equation. In words, DeMorgan's theorem states that the NOT of an AND or OR function can be constructed by changing the function from AND to OR or vice versa and inverting its arguments.

Example. Prove that $\overline{x_1 x_2 + x_3} = (\bar{x}_1 + \bar{x}_2)\bar{x}_3$ by successive applications of DeMorgan's theorem.

First we find $\overline{x_1 x_2 + x_3} = \overline{(x_1 x_2)} \cdot \bar{x}_3$ by using the second version of DeMorgan's theorem. Then we apply the first version to the expression $\overline{x_1 x_2}$ to arrive at $(\bar{x}_1 + \bar{x}_2) \cdot \bar{x}_3$. Note that proof could have been done with a truth table, but the original equality could be "invented" easily with DeMorgan's theorem, while considerable insight might have been necessary to invent if from a truth table.

Example. Construct a circuit that performs the exclusive OR function from elementary AND, OR, and NOT elements.

The required Karnaugh map is shown in Fig. 9.5a and one form of the circuit in Fig. 9.5c. We see that the output is 1 if either $x_1 \cdot \bar{x}_2$ OR

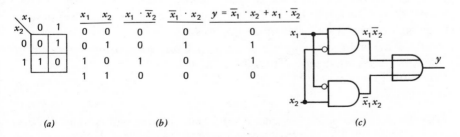

(a)   (b)   (c)

Fig. 9.   5. The exclusive-OR function.

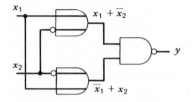

Fig. 9.   6. Another version of the exclusive-OR circuit.

$\bar{x}_1 \cdot x_2$. An alternate version of the circuit can be constructed with the help of DeMorgan's theorem. Thus $\bar{y} = (\bar{x}_1 \cdot x_2) \cdot (x_1 \cdot \bar{x}_2) = (x_1 + \bar{x}_2) \cdot (\bar{x}_1 + x_2)$. The circuit based on this form appears in Fig. 9.6.

9.3. The TTL logic family While AND and OR functions can be generated with simple diode-resistor circuits, the NOT function requires, basically, an inverting amplifier. Several families of logic circuitry that differ in the configuration of the basic "gates" have been produced commercially as integrated circuits. Perhaps the most popular of the several families that use the silicon bipolar transistor as the amplifying device is known as the TTL (for transistor-transistor logic) family. The basic gate in the TTL family is the NAND gate, one form of which is illustrated in Fig. 9.7. A characteristic feature of a TTL gate (except for a simple inverter) is a multiple-emitter input transistor.† Also the supply voltage for this family is standardized at a nominal 5.0 V (usually ±0.5 V).

The operation of the TTL NAND gate is as follows. If the voltage at both inputs is above about 2.0 V (logic 1), the emitter-base junctions of Q_1 are reverse biased. However, current flows from the 5-V supply down through the 4-kΩ resistor through the forward-biased, base-collector diode of Q_1 and into the base of Q_2, turning Q_2 on. The resultant emitter current in Q_2 flows partly through the base of Q_4, which turns Q_4 on and clamps the output at essentially ground potential. Since Q_2 is on, its collector voltage is approximately equal to its emitter voltage, that is, about 0.6 V as a result of the base-emitter diode drop in Q_4. The base of Q_3 is not at a high enough voltage to turn Q_3 on, and therefore the only current that flows through Q_4 is that flowing from the external circuit into the output terminal of the NAND gate. This output terminal may be connected to the input terminals of several other NAND gates; the number of other inputs that can be driven from a single output is called the fan-out of the logic gate. The fan-out is at least 10 for a TTL gate.

Now consider the operation of the TTL NAND gate when either (or

---

†The diodes in the emitter circuits of the input transistor prevent any negative transient voltage < -0.6 V from reaching the emitter-base junction and burning out the input transistor.

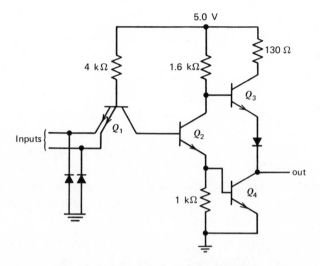

Fig. 9.   7. The basic TTL NAND gate.

both) of its inputs are below about 0.8 V (logic 0). Now the current through the 4-kΩ resistor (approximately 1 mA) flows out through one or both emitters of Q_1. There is now no base current in Q_2 so that it is off and provides no base current to Q_4. Base current does tend to flow in Q_3, and a sizeable collector current can flow in Q_3. Now the output will be at about 3.8 V.

In most applications the output of a NAND gate is connected to the inputs of other gates; thus, when the output of a gate is high, it need not provide a steady current to succeeding gates. However, during the switching process, the output of a NAND gate must supply current to charge the stray capacitance that inevitably exists in the output circuit. Although the speed of a system based on TTL gates is often limited by this stray capacitance, when this capacitance is negligible the switching times of the transistors within the TTL gate still result in a finite "propagation delay" on the order of 20 nsec.

There are several different forms of the basic TTL gate that differ principally in the circuitry used to drive the output transistors Q_3 and Q_4. In some versions, Q_3 is omitted entirely so that the collector of Q_4 is connected only to the output terminal. The resulting gate is called an open-collector NAND gate. Although the different versions of TTL gate may have somewhat different input-output characteristics, Fig. 9.8a shows a typical input-output transfer characteristic that applies for slowly varying input voltages and Fig. 9.8b shows a typical output response (noncapacitive load) to rapid changes in input voltage. Later we will see that in some applications, particularly those involving sequential systems,

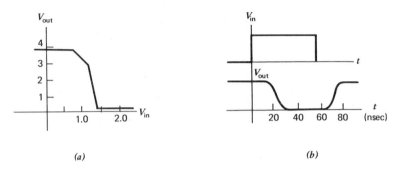

Fig. 9.   8. TTL gate properties: (*a*) voltage transfer characteristic and (*b*) square pulse response.

proper account must be taken of the finite slope of the transfer function or the nonzero propagation delay.

9.4. The CMOS logic family   A logic family that is finding increasing applications is the Complementary Metal-Oxide-Semiconductor family.† The basic gates in this family are made from enhancement-mode, *p*-type and *n*-type MOSFETs, with substrates shorted to sources. Figure 9.9*a* shows the basic CMOS inverter, whose operation can be explained in terms of the characteristic curves of Fig. 9.9*b* and 9.9*c*. The supply voltage is shown as +10 V but CMOS logic gates can operate with a wide range of supply voltages, from as low as 3 V up to about 15 V.

Suppose, first, that the input voltage is zero. In the *n*-type MOSFET (with *n*-type source and drain regions as in Fig. 6.26), there is no induced *n*-type channel because the gate-source voltage is less than the three volts or so (see the characteristic curve of Fig. 9.9*b*) required to induce electrons into the *p*-type substrate between source and drain. On the other hand, the gate-source voltage in the *p*-type MOSFET is −10 V, which is more than negative enough to form a *p*-type channel by inducing holes into the *n*-type substrate between the *p*-type source and drain regions. Thus the *p*-type MOSFET conducts, while the *n*-type MOSFET does not. If the CMOS inverter has a capacitive load, stray wiring capacitance plus the input capacitances of other CMOS gates, then this capacitance is quickly charged to approximately +10 V.

If the input voltage is +10 V the situation is the reverse of that described above, with the *p*-type MOSFET nonconducting ($V_{GS} = 0$) and the *n*-type MOSFET conducting ($V_{GS} = 10$), and the load capacitor discharged to 0 V.

NOR and NAND gates can be constructed by using series and parallel

† See J. R. Burns, *RCA Review*, 25, 627−661 (1964).

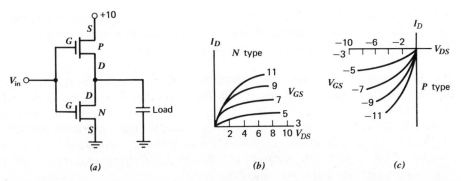

Fig. 9.   9. CMOS inverter: (*a*) circuit configuration, (*b*) characteristics of *n*-channel device, and (*c*) characteristics of *p*-channel device.

combinations of MOSFETs. Figure 9.10 shows a NOR gate; the output is low when either input is high (or both) so that at least one of the parallel *n*-type MOSFETs is conducting and at least one of the series *p*-type MOSFETs is nonconducting. Only if both inputs are near zero will both *p*-type MOSFETs be conducting and both *n*-type MOSFETs be noncon-ducting so that the output is high.

An important advantage of CMOS logic over other types of logic, including TTL, is its very low power consumption. Virtually no power is required when the CMOS output is in its high or low state because no current is necessary to maintain the charge on the load capacitance. Only during switching, and the associated charging and discharging of the load capacitance, does appreciable current flow. Actually there is some current flow due to reverse-bias leakage in the nonconducting MOSFET in the steady state, but this current is measured in fractions of a nanoampere. Because of the charging times of its output voltages, CMOS logic is

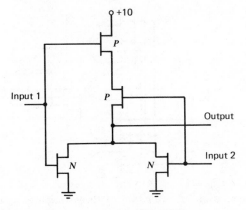

Fig. 9.   10. CMOS NOR gate.

generally slower than other forms of logic, with the maximum switching rate depending on the number of inputs driven by a particular output. The lowest attainable propagation delay is about 100 nsec.

9.5. Synthesis of combinational circuits   One is often faced with the problem of designing a combinational logic circuit to implement a desired truth table. It is always possible to produce the desired output with a "sum of products" circuit. For example, consider the 3-input truth table of Fig. 9.11. Each row that contains a 1-output can be represented by a 3-input AND circuit. The outputs of the five AND circuits are then ORed together to give the required logic function.

The circuit of Fig. 9.11, while correct, is not the simplest circuit that can implement the truth table. It is possible to achieve the same function by simplifying the original expression that represents the truth table. This simplification is most easily done with the aid of a Karnaugh map. Figure 9.12 shows the three-variable Karnaugh map for the truth table of Fig. 9.11. We can represent all four 1's in the top row by the single logical function $\bar{x}_3$, and we can represent both 1's in the second column by the function $\bar{x}_1 x_2$. Thus the output is 1 when either of these functions is 1, or both, as occurs when $x_1 = 0$, $x_2 = 1$, and $x_3 = 0$. The resulting simplified function is given in Fig. 9.12. You can verify that this function has the same truth table as the more complex function of Fig. 9.11. The resulting simplification in the circuit is dramatic!

The key idea in simplifying a logical expression by means of the Karnaugh map method is to group together 2, 4, 8, etc. 1's in a square or

x_1	x_2	x_3	y
0	0	0	1
0	0	1	0
0	1	0	1
0	1	1	1
1	0	0	1
1	0	1	0
1	1	0	1
1	1	1	0

$$y = \bar{x}_1\,\bar{x}_2\,\bar{x}_3 + \bar{x}_1 x_2 \bar{x}_3 + \bar{x}_1 x_2 x_3$$
$$+\; x_1 \bar{x}_2 \bar{x}_3 + x_1 x_2 \bar{x}_3$$

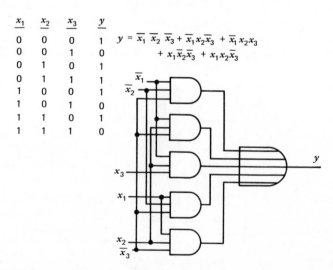

Fig. 9.   11. Implementation of a function of three variables with a sum-of-products circuit.

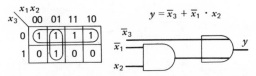

Fig. 9.   12. Karnaugh map and simplified circuit for the function of Fig. 9.11.

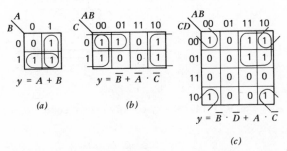

Fig. 9.   13. Examples of two-, three-, and four-variable Karnaugh maps.

rectangular array. In doing the grouping one mentally "bends" each edge of the map to join the opposite edge. Figure 9.13 gives examples of two-, three-, and four-variable maps with the 1's grouped together. As in the example of Fig. 9.12, a given 1 may be in more than one group. Thus, in Fig. 9.13a, the 1 that represents $A \cdot B$ is in the group that represents A and in the group that represents B. Similarly, the upper left-hand 1 in Fig. 9.13b and the upper right-hand 1 in Fig. 9.13c are in two groups. Notice that, in the three-variable map, the first and last column are grouped together because the squares are "adjacent" and contain 1's. (Recall that "adjacent" squares means that only one variable changes in going from one to the other.) Similarly, the four corners in the four-variable map are all adjacent and, since they all contain 1's, they are represented by $\bar{B} \cdot \bar{D}$.

As an example of the application of Karnaugh maps to the simplification of logic functions, let us design a binary-coded-decimal (BCD) to seven-segment-display decoder. In particular, our job is to design a combinational circuit that takes as inputs the 4-bit binary representations of any of the ten decimal digits and produces seven outputs. Each output drives one segment of the display, for example, a set of light-emitting diodes. (We represent a lighted segment by a logic 1 output and a dark segment by a logic 0.) Figure 9.14 shows the BCD code and the standard labeling scheme for a seven-segment display. Figure 9.15 shows, for each of the seven segments, which of the ten decimal digits must cause that segment to light.

A useful first step in the design is to make a Karnaugh map with the decimal digits entered in the squares (Fig. 9.16a). Notice that there are six squares that do not have a decimal representation; since the four binary

Decimal	Binary X_8 X_4 X_2 X_1
0	0 0 0 0
1	0 0 0 1
2	0 0 1 0
3	0 0 1 1
4	0 1 0 0
5	0 1 0 1
6	0 1 1 0
7	0 1 1 1
8	1 0 0 0
9	1 0 0 1

Seven—segment display

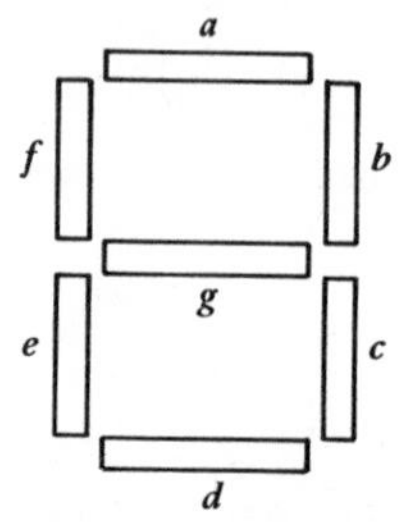

Fig. 9.   14. Binary-coded-decimal code.

Segment	Decimal digits for which lighted
a	0, 2, 3, 5, 7, 8, 9
b	0, 1, 2, 3, 4, 7, 8, 9
c	0, 1, 3, 4, 5, 6, 7, 8, 9
d	0, 2, 3, 5, 6, 8
e	0, 2, 6, 8
f	0, 4, 5, 6, 8, 9
g	2, 3, 4, 5, 6, 8, 9

Fig. 9.   15. The seven-segment display for numerals.

bits that represent a BCD number will never exist simultaneously in any of these six "states," they are called "don't care" states and are represented by an x in the Karnaugh map. When we simplify the map for each of the seven output functions we can make each x a 1 or a 0, whichever makes the function simpler.

Karnaugh maps for each of the seven segments are illustrated in Fig. 9.16b to 9.16h. In each case we have circled 1's and x's to produce the smallest number of OR and AND operations. Study one or two of the maps and corresponding output functions, and then, on some of the others, practice by drawing the map, circling 1's and x's, and writing down the simplest logic expression.

9.6. Multiplexors and demultiplexors

A multiplexor is a device that has several inputs and one output. Any one of the inputs can be connected to the output, depending on the state of a set of "address" inputs. A demultiplexor is similar except that there is only one input and several outputs. The separate "address" inputs determine which of the outputs is

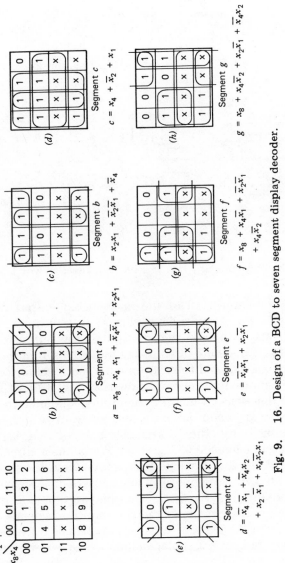

Fig. 9. 16. Design of a BCD to seven segment display decoder.

215

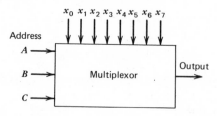

Fig. 9.   17. An 8-input logic multiplexor.

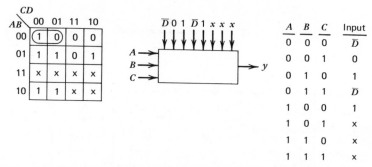

Fig. 9.   18. Programming a four-variable logic function with a multiplexor.

connected to the input. All of the nonselected outputs remain at, for example, logic 1. Some multiplexors have an additional "enable" input that, in one state, connects the input(s) to the output(s) as described above, and in the other state disconnects the inputs from the outputs.

Multiplexors can be used to perform combinational logic. For example, consider the 8-input multiplexor shown in Fig. 9.17. There are three address inputs that select which of the inputs is connected to the output according to a binary code. Suppose that we want to program the logical function described by the Karnaugh map of Fig. 9.16g, which is reproduced for convenience in Fig. 9.18. We connect three of the inputs, say A, B, and C, to the three address inputs. Now for each of the eight combinations of ABC, we choose an appropriate logic signal to be applied to the corresponding input of the multiplexor. For $ABC = 000$, we see from the circled entries in the Karnaugh map that when $D = 0$, $y = 1$ and when $D = 1$, $y = 0$. To achieve this output behavior we set $x_0 = \bar{D}$. For $ABC = 001$, we have $y = 0$, independent of D, so we set $x_1 = 0$. For $ABC = 010$ we set $x_2 = 1$ and so on. For the three don't care configurations of ABC, the corresponding inputs x_5, x_6, and x_7 can be anything; in practice one would connect them all to logic 1 or logic 0, rather than to D or $\bar{D}$ in order to reduce the load seen by the gates that produce D and $\bar{D}$.

The 8-input multiplexor can be used to "program" any 4-input logic function. By an extension of the ideas above it can even help program functions of five or more variables. For example, consider the function:

$$y = ABCDE + ABC\bar{D}\bar{E} + A\bar{B}CDE + A\bar{B}C\bar{D}E \qquad (9.4)$$

This can be "factored" as in ordinary algebra (to see this construct a truth table) to give

$$y = ABC \cdot (DE + \bar{D}\bar{E}) + A\bar{B}C(DE + \bar{D}E) \qquad (9.5)$$

In order to implement this function with the 8-input multiplexor we connect ABC to the select inputs, the logic function $DE + \bar{D}\bar{E}$ to input 7, the logic function $DE + \bar{D}E = E$ to input 5, and logic 0 to all the other inputs. Thus, the output will be 1 if and only if one of the required combinations of the five input variables occurs.

9.7. The read-only memory: the ultimate in combinational logic   A read-only memory (ROM) is a combination of multiplexors, demultiplexors, and an interconnection "matrix" all formed on a single semiconductor chip. As an example, consider the 64-bit ROM illustrated schematically in Fig. 9.19. The memory occurs at the intersections of horizontal and vertical wires. If a connection (indicated by a dot) is made, the bit is a 1; if no connection is made, the bit is a 0. The address inputs x_1, x_2, and x_3 select which one of the 8 demultiplexor outputs is 1; all others are 0.

One way to look at the ROM of Fig. 9.19a is that it stores eight 8-bit words. Each stored word can be made to appear on the eight output lines,

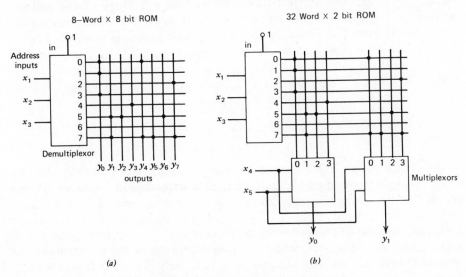

(a)   (b)

Fig. 9.   19. Examples of read-only memories.

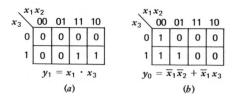

Fig. 9.   20. Karnaugh maps for the y_0 and y_1 outputs of the ROM of Fig. 9.19a.

$y_0, \ldots, y_7$ by choosing the proper address word; for example, the address word $x_1 x_2 x_3 = 101$ evokes the 8-bit output word 0110 0010. This view of a ROM might be termed a computer-oriented view, and indeed there are many applications in computers in which a memory containing an unalterable set of words is useful. One example of a ROM application is in subroutines for computing transcendental functions, such as e^x, $\sin x$, etc. in a pocket calculator.

There is another productive way of looking at a ROM, which might be called the logic-oriented view. Thus, the ROM of Fig. 9.19a may be viewed as storing eight different functions of the three variables $x_1 x_2 x_3$. For example, consider output line y_1. Since $y_1 = 1$ if either demultiplexor line 5 or 7 is 1, which requires $x_1 x_2 x_3 = 101$ or 111, y_1 can be represented by the Karnaugh map of Fig. 9.20a. We see from the map that y_1 represents the logical function $x_1 \cdot x_3$; the function y_0 is illustrated by the Karnaugh map of Fig. 9.20b.

Other organizations of a 64-bit ROM are possible. For example, Fig. 9.19b shows a 32-word, 2-bit organization. Here 5 address bits are required to select two output functions y_0 and y_1. Three of the address inputs are decoded to select one of the eight rows; the other two inputs select which one of the four inputs to each multiplexor is connected to its output. For example, suppose the multiplexor inputs 0,1,2,3 are selected by $x_4 x_5 = 00, 01, 10, 11$, respectively. Then the output function y_1 is 1 if $x_4 x_5 = 00$ and $x_1 x_2 x_3 = 000$ or 111, or if $x_4 x_5 = 01$ and $x_1 x_2 x_3 = 111$, etc. The complete output equation for y_1 is

$$y_1 = \bar{x}_1 \bar{x}_2 \bar{x}_3 \bar{x}_4 \bar{x}_5 + x_1 x_2 x_3 \bar{x}_4 \bar{x}_5 + x_1 x_2 x_3 \bar{x}_4 x_5$$
$$+ x_1 \bar{x}_2 x_3 x_4 \bar{x}_5 + \bar{x}_1 x_2 \bar{x}_3 x_4 x_5 + x_1 x_2 x_3 x_4 x_5 \quad (9.6)$$

Unfortunately, a ROM cannot be made simply by making connections at the intersections of an array of wires as is suggested in Fig. 9.19. To see the difficulty, imagine that demultiplexor output line 5 is a 1. As a consequence, output lines y_1, y_2, and y_6 are 1, as desired, but, in addition, because of the connection at the intersection of horizontal line 7 and y_1, we would have y_4 and $y_5 = 1$, and because of the interconnection at line 0 and y_4, we would have $y_0 = 1$. Clearly we must find a way to isolate the various output lines if we are to fabricate a ROM.

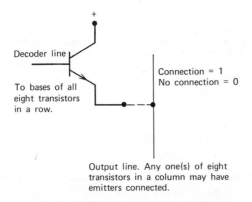

Fig. 9.   21.   A single cell of a ROM.

One method of isolation is to use a bipolar transistor at each intersection with all 64 collectors tied to the same supply voltage. Each demultiplexor output is connected, in parallel, to the bases of the eight transistors in a row. On the other hand, only the emitters at which a 1 is to be stored are connected to a vertical output line. This scheme is illustrated in Fig. 9.21 for an n-p-n bipolar ROM. The unselected state of a decoder output is low, so that none of the transistors in the row is turned ON. To select a row, the decoder output line goes high, and each of the eight transistors in the row will conduct — provided that its emitter is connected to an output line — and a current will flow in the output line. None of the current in an output line can flow back through to the base of an unselected row transistor because that would constitute current through a reverse-biased, emitter-base diode.

The connections between the emitters and output lines of an integrated-circuit, bipolar ROM are made at the time of fabrication of the integrated circuit and thus are unchangeable. Because the initial setup expense for the manufacture of an integrated circuit is high, a ROM that is "programmed" at the factory can be justified only if a large number of identical units is required, such as in pocket calculators, digital watches, or automotive applications. There exists, however, another class of read-only memory called programmable read-only memories (PROM) that not only can be programmed in the field but also can be erased. Thus, a PROM can be manufactured as a general purpose logic device, and each user can load the memory, or program it, with his or her own desired pattern of 1's and 0's.

The technology of a PROM is based on the fact that charge can be introduced into an insulated region consisting of silicon dioxide (SiO_2) and will not decay away for a long time. A typical half-life for storage of charge at room temperature in SiO_2 might be hundreds of years, a time longer than the usefulness of most man-made information.

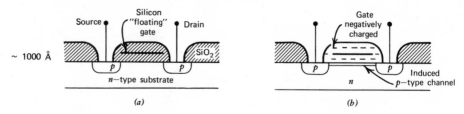

Fig. 9. 22. Basic "memory element" of an electrically programmable read-only memory (PROM): (a) nonconducting state and (b) conducting state.

Figure 9.22 shows one method of constructing a PROM bit. As in an ordinary p-channel, enhancement-mode MOSFET, heavily doped p-type source and drain regions are formed by diffusion of impurities into an n-type silicon substrate. An insulating SiO_2 coating containing a patch of pure silicon called a floating gate is then grown over the area between source and drain. After manufacture, the MOSFET is nonconducting because there is no p-type channel. However, such a channel can be induced if sufficient negative charge is placed on the floating gate. This is accomplished by applying a brief (~1 msec) pulse of about 50 V across the source-drain leads; strong electric fields are induced in the 1000 Å thick insulator between the floating gate and the negative contact, and electrons flow through the SiO_2 to charge the gate.

In a PROM one of these floating gate MOSFETs would be formed at each intersection point between an emitter and an output line in Fig. 9.21, and, in addition, all transistors in the demultiplexor and multiplexor circuits would be replaced by MOSFETs.

One way of "erasing" a PROM is to expose it to X rays or ultraviolet light, which excite electrons to conduction states in the SiO_2 and render its conductivity high enough so that the charge on the floating gate can leak off. Other PROM devices can be made electrically erasable by the inclusion of additional electrodes within the SiO_2, which are brought out to external wires. Then a positive voltage pulse applied to these electrodes can create breakdown paths in the SiO_2 through which the floating gates can discharge.

Problems for Chapter 9

9.1   A circuit employed as part of a certain TTL integrated logic circuit is shown in Fig. P9.1
    (a) Determine the truth table for outputs Q_1 and Q_2 in terms of the eight possible states of the three inputs A, B, and C.
    (b) Represent the truth table by a Boolean function, using ORs, ANDs and NOTs, and by a logic circuit diagram. *Hint.* It is

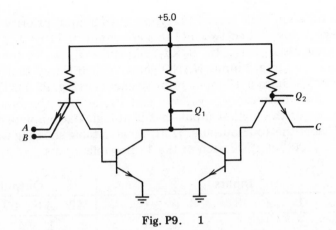

+5.0

Q_1

Q_2

A
B

C

Fig. P9.   1

possible to do this for Q_1 with one OR, one AND, and one NOT operation.

9.2   Devise a circuit, using only 2-input NANDS and inverters, and with inputs A, B, C, and D that produces $Y = A + B + C + D$.

9.3.   Find the simplest logical circuitry for a full adder. Both "Sum" and "Carryout" outputs are required. These are functions of three inputs, "A," "B," and "Carryin." (*Hint.* Use three-variable Karnough maps.)

9.4.   Write truth tables and devise logic circuits using only NAND gates for the following functions.
(a) $A \cdot (B + C)$
(b) $(A \cdot B) + C$
(c) $A \cdot (B + \overline{C})$

9.5.   Devise a circuit that uses only 2-input NAND gates to yield $Y = (\overline{A + B}) + (\overline{C + D})$. Assume that only A, B, C, and D are available as inputs.

9.6.   For each of the Karnaugh maps in Fig. P9.6, find the simplest logic expression, being careful to choose the "don't cares" properly in (*b*).

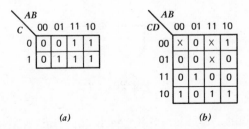

(a)   (b)

Fig. P9.   6

9.7. The function table for a TTL-type 74148 8-input priority encoder is shown below. When two 74148's are connected together as shown, the outputs A_2, A_1, A_0, B_2, B_1, B_0, and EO_B can be connected by means of three 2-input NAND gates in such a way that the entire circuit becomes a 16-input priority encoder (Fig. P9.7). That is, the four final outputs, y_4, y_2, y_1, and y_0, assume the binary coded number of the highest input that is at logic 0. Complete the circuit and label the four outputs. *Hint.* Make a table showing the state of A_2, A_1, A_0, etc., for each of the 17 possible inputs.

Inputs									Outputs			
EI	0	1	2	3	4	5	6	7	$A2$	$A1$	AO	EO
1	x	x	x	x	x	x	x	x	1	1	1	1
0	1	1	1	1	1	1	1	1	1	1	1	0
0	x	x	x	x	x	x	x	0	0	0	0	1
0	x	x	x	x	x	x	0	1	0	0	1	1
0	x	x	x	x	x	0	1	1	0	1	0	1
0	x	x	x	x	0	1	1	1	0	1	1	1
0	x	x	x	0	1	1	1	1	1	0	0	1
0	x	x	0	1	1	1	1	1	1	0	1	1
0	x	0	1	1	1	1	1	1	1	1	0	1
0	0	1	1	1	1	1	1	1	1	1	1	1

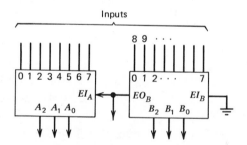

Fig. P9. 7

9.8. In the circuit in Fig. P9.8, each of the NAND gates has a propagation delay of 20 nsec. Find the quiescent levels of outputs

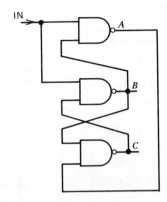

Fig. P9.   8

A, B, and C when the input is logic 0. Then sketch the time course of the three outputs after the input goes to logic 1 at time $t = 0$.

9.9. Program an 8-input multiplexor to produce the logical functions $y = ABC\bar{D} + AB\cdot(C + D) + \bar{A}\bar{D}$.

9.10. Program an 8-input multiplexor to produce the logical function $y = AB + A\cdot(\bar{B}C + \bar{C}D) + A\bar{B}\bar{C}D$.

9.11. The spot on an oscilloscope is to be moved in a block figure-eight pattern (Fig. P9.11a), in response to the contents of a 3-bit binary counter that is driven by an oscillator. The idea is simply to drive an integrator with either a positive, negative, or zero signal when the spot is to be driven rightward, leftward, or to remain stationary.

Fig. P9.   11a

The output of the integrator drives the x-deflection plates of a scope. Similarly, a y integrator is driven by a positive, negative, or zero signal. For example, in Fig. P9.11b, voltages A and B have the form shown during the eight time segments. Resistor R is chosen so that when A and B are zero, I is negative and a positive-going ramp occurs. When A is positive and B is 0 or vice versa, $I = 0$ and V_x remains constant, and when both A and B are positive, I is positive and a negative-going ramp occurs.

Devise decoding circuits for obtaining the logic voltages A and B

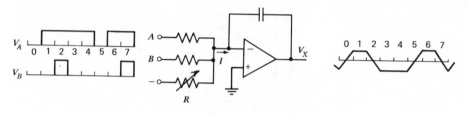

Fig. P9.   11*b*

from the Q and $\bar{Q}$ outputs of a 3-bit binary counter. For example, $B = 1$ when the counter reads 010 or 111. *Hint*. Use three variable Karnaugh maps.

9.12. Repeat problem 11, but find decoding circuits for a pair of voltages that can be used to drive a y integrator.

9.13. When entering data into a certain programmable ROM, the "spec sheet" requires that the *complement* of each address bit A_0 through A_7 be presented to the address inputs $A_0{}'$ through $A_7{}'$ and then, in response to a logic signal E, the actual address bits are presented to the address inputs. In other words the signal at $A_n{}'$ is a logical function of A_n and E. What is this logical function? Implement it with NAND's and inverters.

REFERENCES

Burns, J. R., "Switching Response of Complementary-Symmetry MOS Transistor Logic Circuits," *RCA Review*, 25, 627—661 (1964).

Eaton, S. S., "Complementary MOS Logic Design and Applications," *The Electronic Engineer*, 29, 52—57 (May 1970).

Garrett, L. S., "Integrated Circuit Digital Logic Families," *IEEE Spectrum* (Oct., Nov., Dec. 1970).

Karstad, K., "CMOS for General-Purpose Logic Design," *Computer Design*, 12, Number 5, 99—106 (May 1973).

Lenk, J. D., *Handbook of Logic Circuits*, Reston Publishing Co., Reston, Va., 1972.

Texas Instruments Inc. Engineering Staff, *T.T.L. Data Book for Design Engineers*, Texas Instruments, 1973.

CHAPTER 10

Sequential Logic

10.1. Introduction   As important and powerful as combinational logic is, the real potentials of digital electronics are realized only by sequential logic circuits. Here we will show how sequential logic may be built from simple combinational circuits by adding feedback and trace the development of sequential logic from the basic memory element, through various flip-flops and counters, to the ultimate in sequential devices — the microprocessor and its handmaiden — the random access memory.

10.2. The latch: a basic memory element   A latch is a device that uses feedback to drive its output to one of two stable states, called 1 and 0, in response to an input signal and to keep it there until a different input signal drives it to the other stable state. Several forms of latches are shown in Fig. 10.1 along with their "quiescent" input states. The latch of Fig. 10.1a is based on a trigger circuit. In the quiescent state the input is at 0 and the output is at either its positive or its negative limit. If the output is positive, then the feedback to the positive input assures that the differential input is positive so that the positive output is reinforced. Similarly if the output is negative the feedback assures that the differential input is negative. A brief positive input pulse is all that is necessary to switch the output from positive to negative and keep it there, while a brief negative input pulse switches the output from negative to positive.

Figure 10.1b shows a latch built from a pair of logical NOR circuits. The quiescent state is with both inputs at logic 0. Then one of the outputs must be 0 and the other must be 1. (To see this try both outputs at 0 or both at 1; a contradiction will occur.) If the upper output is in its 1 state it can be made 0 by applying a 1 at A. The 0 at C forces $D = 1$, which sets the latch so that C remains at 0 even after A returns to 0. Similarly, a brief 1 at B sets the latch to the $D = 0$, $C = 1$ state. Notice that either C or D tells us the state of the latch; if one is considered to be the output, then the other is the complement of the output.

The double-NAND latch of Fig. 10.1c works like the double-NOR latch, except that the quiescent state of the inputs is $A = B = 1$. Taking either input to 0 sets the corresponding output to 1. An unsymmetrical latch built from a NAND and a NOR is shown in Fig. 10.1d. The quiescent

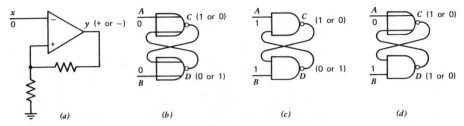

Fig. 10. 1. Different forms of latches.

input state is $A = 0$, $B = 1$; both outputs are then either 0 or 1. The outputs are both set to 0 by a 0 at B and are both set to 1 by a 1 at A.

A latch exhibits memory in the sense that the output state depends on which of two events occurred most recently at the inputs. If the input events occur simultaneously, for example, $A = B = 0$ in Fig. 10.1c, both outputs will respond, in this case by going to 1; but if the inputs return to 0 simultaneously, we cannot predict which of the two stable output states will occur. This situation is called a "race" between the two inputs, the "winner" being the one that, because of a slightly different trigger level or random fluctuations of charge carriers in the transistors, is the *last* to become 0.

10.3. The *D* flip-flop: a clocked latch   Sequential logic systems may be classified as synchronous or asynchronous. In a synchronous system, the states of latches and other sequential devices derived from them, such as flip-flops and counters, change only when a particular logic signal called CLOCK makes a transition, say from 0 to 1. Thus, all of the memory elements change together, and certain timing problems that can occur in asynchronous, or unclocked, circuits are avoided. Another advantage of clocked systems is ease of analysis; we usually do not have to worry about the state of the inputs between the clock transitions, since the system cares about these inputs only at the sharply defined transition times.

The *D* flip-flop is an elementary example of a clocked system and a building block for more complex sequential circuits. It has a *D* (for "data") input, a *C* (for "clock") input, and a latch. The state of the latch can change only at the clock transitions; at these times the latch is set to 1 if $D = 1$ and to 0 if $D = 0$.

To construct a *D* flip-flop from a basic latch, we must find a way to apply the appropriate input signals at the clock times. For example, consider the double-NAND latch of Fig. 10.1c, the natural latch of the TTL logic family. We must devise a control circuit that makes one input 1 and the other 0, in response to the *D* signal, at the clock time. At other times the latch inputs can remain constant or both can be 1, but they may not both be 0. One approach is to develop a control circuit that itself is a

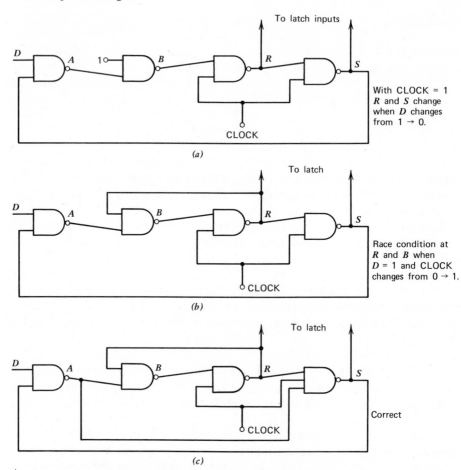

Fig. 10. 2. Evolution of a D flip-flop.

sequential circuit. We show a sequence of development steps for such a control circuit in Fig. 10.2.

Figure 10.2*a* pictures a feedback system in which the feedback loop is broken by a CLOCK = 0 signal. With CLOCK = 0 the circuit obeys the equations $D = \bar{A} = B$ and $R = S = 1$ and, since both latch inputs are 1, the latch itself (not shown) cannot change state. When CLOCK = 1 there are two possibilities. If $D = 0$ at the instant CLOCK becomes 1, then $A = 1$, $B = 0$, $R = 1$, and $S = 0$ and, even if D later changes to 1, the states of A, B, R, and S will not change because the 0 at S effectively blocks out the D input. Since R and S are inputs to the NAND latch of Fig. 10.1*c*, that latch is set to one of its stable states by the $S = 0$ input.

Now look at the other possibility, $D = 1$, when CLOCK becomes 1. We must have $A = 0$, $B = 1$, $R = 0$, and $S = 1$. As a result, the latch is set to its other stable state by the $R = 0$ input. However, the control circuit of Fig. 10.2a is not quite correct because D may go to 0 after CLOCK has become 1, which sets $A = 1, B = 0, R = 1$, and $S = 0$ and changes the latch back to its original state. This behavior does not conform to our definition of a D flip-flop. To avoid this difficulty, we must prevent the postclock $1 \rightarrow 0$ transition of D from affecting R and S. Fortunately this is easy to do; we simply add another feedback loop from the output of R to the unused input of B(Fig. 10.2b). Since R is 0 just after CLOCK goes to 1, this 0 forces $B = 1, R = 0$, and $S = 1$ and effectively blocks out any later change in D. This new feedback loop does not affect the operation (described earlier) in which D was 0 at the clock transition because, at that time, $R = 1$, and the inputs of B are the same as in the circuit of Fig. 10.2a.

However, a new and subtle difficulty has arisen in the $D = 1$ case. Recall that just before the $0 \rightarrow 1$ clock transition we have, with $D = 1$:

$$A = 0, B = 1, R = 1, S = 1 \qquad (10.1)$$

Now CLOCK goes to 1. It is possible that, because of a very slow clock transition, S could become 0 and force $A = 1, B = 0$ before R has a chance to become 0. Then R would remain 1 and S would remain 0, which is *not* correct. In other words, there is a race between R and B; if $R \rightarrow 0$ before $B \rightarrow 0$, the circuit works as desired, but if $B \rightarrow 0$ before $R \rightarrow 0$ there is an error. This race can be eliminated by the addition of a third input to NAND gate S, which is driven by A. The final configuration of the latch controller is shown in Fig. 10.2c. This new connection has no effect on the operation if $D = 0$ when CLOCK becomes 1 because A is 1 at that time. But when $D = 1$ and $A = 0$, the new connection prevents S from going to 0 as CLOCK changes from 0 to 1.

Let us summarize the operation of the latch driver of Fig. 10.2c. When CLOCK $= 0$, $R = S = 1$, and the latch remains in whatever state it was in. If $D = 1$, then $A = 0$, $B = 1$ while if $D = 0$ then $A = 1$, $B = 0$. In other words, during the time CLOCK $= 0$, A and B are free to respond to changes in D. Now when CLOCK makes a 0 to 1 transition, S assumes the state of D *at the time of the transition*, and R assumes the opposite state. Thereafter B, R, and S do not respond to changes in D; A can change with D in the case $S = 1$, but then $R = 0$ and the change in A does not affect B. As a result of these operations, we may say that the D flip-flop "remembers" the state of the D input at the time of a 0 to 1 clock transition.

The symbol for a D flip-flop is shown in Fig. 10.3a. Of course a reference terminal (ground) and a supply voltage terminal must be included in a real device, but these are omitted for clarity in Fig. 10.3. The little vertex at the C input suggests that CLOCK is effective only when it makes a transition from 0 to 1. Often a D flip-flop contains an

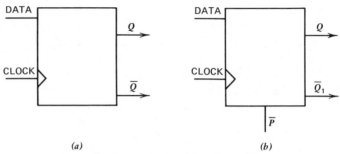

(a)   (b)

Fig. 10.   3. Symbols for D flip-flops.

additional unclocked $\overline{\text{PRECLEAR}}$ input that sets $Q = 0$, $\overline{Q} = 1$ regardless of the states of the DATA and CLOCK inputs. This feature, shown in Fig. 10.3b, can be added to the latch driver circuit of Fig. 10.2c by including additional inputs to NAND gates A and R and to the latch NAND gate driven by S. Nothing happens when these inputs are all 1, but when they are all forced to 0 by $\overline{\text{PRECLEAR}}$, A and R are forced to 1, B is forced to 0, and the latch is set to the same state that would be set by $S = 0$. Then, when the $\overline{\text{PRECLEAR}}$ returns to 1, the feedback loops retain the state $A = 1$, $B = 0$, $R = 1$, and $S = 0$ if CLOCK = 1, while if CLOCK = 0, then $R = S = 1$, and the latch is ready to be set by the next $0 \rightarrow 1$ clock transition.

10.4. Shift registers and binary counters.   The D flip-flop may be made the basis of a number of other useful sequential devices. For example, Fig. 10.4 shows a shift register. A shift register contains a sequence of D flip-flops connected in such a way that, at clock time, each output assumes the state of the previous output. Thus each 1 or 0 logic signal that is present at the D input of the first D flip-flop is shifted through the shift register on successive clock signals.

    One aspect of this shift register that deserves discussion is the fact that the input to each flip-flop may change as that flip-flop is responding to

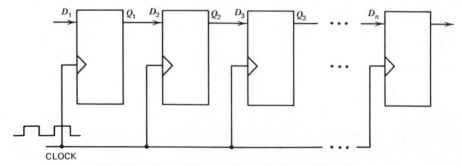

CLOCK

Fig. 10.   4. A shift register made from D flip-flops.

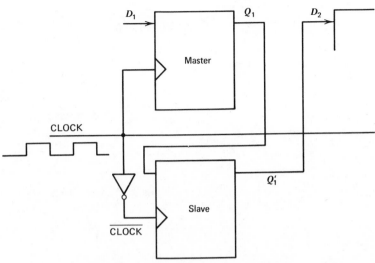

Fig. 10.   5. The master-slave principle.

the clock transition. For example, if the D input to the third flip-flop changes from 1 to 0 at the clock transition, then that flip-flop may be set, ambiguously, to either 1 or 0. This problem is exacerbated by slight differences in the trigger levels of the several flip-flops. Fortunately, there is a solution. We simply expand each D flip-flop into two flip-flops and invert the clock pulse to the second flip-flop (Fig. 10.5). The result is sometimes called a master-slave flip-flop. At the upward transition of CLOCK, the "master" flip-flop output, Q_1, assumes the state of its input, but nothing changes in the "slave" flip-flop because its clock input is falling. Then when CLOCK goes to 0, $\overline{\text{CLOCK}}$ goes to 1 and the "slave" output, $Q_1{}'$, follows the "master" by assuming the state of Q_1. In the master-slave arrangement, the input to each D flip-flop is constant during the time its output is changing, and there is no ambiguity. The user must remember, however, that when the slave output changes, it assumes the state of the D input at an earlier time.

The master-slave flip-flop can be made the basis of a simple binary counter. All that is necessary is to feed back an inverted version of Q' to input D (see Fig. 10.6a). The operation of the system is shown in the timing diagram of Fig. 10.6b. Each time CLOCK makes an upward transition Q changes state. The complete system of Fig. 10.6a is called a T (for toggle) flip-flop; its circuit symbol, Fig. 10.6c, is like that of the D flip-flop but without the D input.

Counters that cycle through 2^N states (N an integer) may be built by cascading T flip-flops. A 16-counter is illustrated in Fig. 10.7, along with its sequence of states. The state table is derived from the fact that each output changes only when the preceding output makes a 0 to 1 transition.

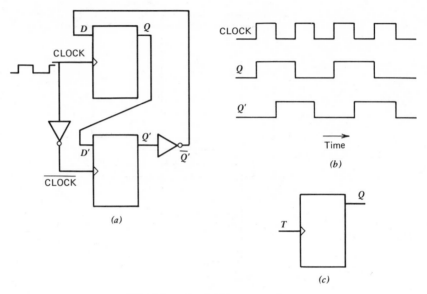

Fig. 10. 6. A binary counter.

Although each of the four T flip-flops in Fig. 10.7 is a clocked, or synchronous circuit, the overall system is asynchronous because each flip-flop must wait for the preceding one to go to 0. For example, when the counter goes from its 1111 state, back to its 0000 state, the operation is like a row of dominoes; Q_A goes to 0, then Q_B, then Q_C, and finally Q_D, all in rapid succession. Even though the "dominoes" fall very rapidly (~50-nsec intervals for TTL circuitry), the overall delay time in a long 2^N counter of this type can limit its usefulness, since the true state of the counter does not occur until after N-1 delays. Synchronous counters can be constructed from D flip-flops, but we will wait to make them from the type of flip-flop discussed next.

10.5. The J-K flip-flop   The J-K flip-flop is a generalized flip-flop that can mimic the behavior of either the D or the T flip-flops discussed in section 10.4. The J-K flip-flop is a clocked flip-flop with two control inputs, labeled J and K. What the flip-flop does at the clock transition time is determined by the state of its J and K inputs at that time. Figure 10.8 shows a J-K flip-flop and its table of operation. When $J = K = 0$, the output is "frozen"; it does not change at the clock transition time. When $J = K = 1$, the flip-flop behaves like a T flip-flop, with Q changing state on each upward clock transition. When $J = 1$ and $K = 0$ the output becomes 1 at the next upward clock transition, and when $J = 0$ and $K = 1$ the output becomes 0 at the clock transition. All that is necessary to

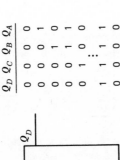

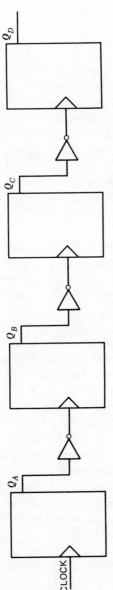

Fig. 10. 7. A 4-bit binary counter.

233

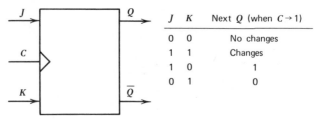

Fig. 10.   8.   A J-K flip-flop.

J	K	Next Q (when $C \rightarrow 1$)
0	0	No changes
1	1	Changes
1	0	1
0	1	0

mimic a D flip-flop is to connect D to J and $\bar{D}$ to K; the output then assumes the state of D at the transition time.

A J-K flip-flop always contains a built-in delay feature so that the flip-flop works properly even when the J and K inputs change at the same time as the clock signal. This delay feature is sometimes achieved by the master-slave technique. For very rapid operation, it is necessary to know the "setup" time of the J-K inputs. This is the minimum time interval prior to a clock transition during which the J and K inputs must remain constant in order that the flip-flop can decide unambiguously how it is to behave at clock time.

10.6 Synchronous counters J-K flip-flops can be used as building blocks in the design of synchronous counters that sequence through any desired number of states. Such counters are achieved by driving the inputs of J-K flip-flops with combinational logic that is a function of the present state of the counter.

Example.   Use J-K flip-flops to design a 5-counter, whose three outputs, Q_C, Q_B, and Q_A, cycle through the five states: 000,001,010,011,100.

We begin by noting that if a flip-flop is to be changed from 0 to 1, its inputs at clock time can be *either $J = 1$, $K = 1$ or $J = 1$, $K = 0$.* If it is to be changed from 1 to 0, its inputs can be either $J = 1$, $K = 1$ or $J = 0$, $K = 1$. If it is to remain at 1 its inputs can be either $J = 0$ $K = 0$ or $J = 1$, $K = 0$, and if it is to remain at 0 its inputs can be either $J = 0$, $K = 0$ or $J = 0$, $K = 1$. These conditions are noted at the right of Table 10.1, in which are listed the states of each of the three pairs of J-K inputs that are required for the 5-counter to progress to its next output state.

For example, look at the behavior of Q_A. When $Q_C Q_B Q_A = 000$, we want Q_A to change to 1 on the next clock transition. Thus, when $Q_C Q_B Q_A = 000$ we must have $J_A = 1$ and K_A either 1 or 0 in order to effect the required change in Q_A. When $Q_C Q_B Q_A = 001$ we must have $J_A = 0$ or 1 and $K_A = 1$ in order to prepare for the next transition of Q_A, and so on.

Table 10.1 shows that each of the six J-K inputs is a logical function of the three variables Q_C, Q_B, Q_A. A three-variable Karnaugh map analysis

TABLE 10.1. Design of Synchronous 5-Counter

Present Outputs			Inputs Required to Prepare for Next State								Effect			
Q_C	Q_B	Q_A	J_A	K_A	J_B	K_B	J_C	K_C			Effect	J	K	(x = don't care)
0	0	0	1	x	0	x	0	x			$0 \rightarrow 1$	1	x	
0	0	1	x	1	1	x	0	x			$1 \rightarrow 0$	x	1	
0	1	0	1	x	x	0	0	x			Remain 1	x	0	
0	1	1	x	1	x	1	1	x			Remain 0	0	x	
1	0	0	0	x	0	x	x	1						

235

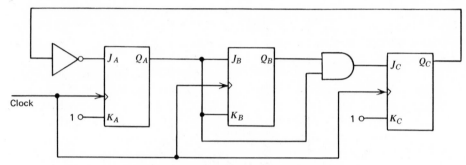

Fig. 10.   9. A synchronous 5-counter.

can be performed in order to express the six functions as simply as possible. (There are don't cares in states $Q_C Q_B Q_A$ = 101,110,111 in all of the maps as well as the don't cares in Table 10.1.) The result of this analysis is

$$J_A = \overline{Q_C} \qquad K_A = 1$$
$$J_B = Q_A \qquad K_B = Q_A$$
$$J_C = Q_A \cdot Q_B \qquad K_C = 1$$

and the circuit for the 5-counter is shown in Fig. 10.9.

Several types of counters, which include the J-K flip-flops and all of the necessary combinational feedback circuitry, are available in integrated-circuit form. Examples of these are a 16-counter, a combination 2-counter and 5-counter that can be connected as a binary-coded-decimal (BCD) counter (see Table 10.2), and a BCD counter that can count up or down.

TABLE 10.2.   Binary Coded Decimal Code

Decimal Digit	Q_8	Q_4	Q_2	Q_1
0	0	0	0	0
1	0	0	0	1
2	0	0	1	0
3	0	0	1	1
4	0	1	0	0
5	0	1	0	1
6	0	1	1	0
7	0	1	1	1
8	1	0	0	0
9	1	0	0	1

10.7 The microprocessor — a generalized sequential device   In section 10.5 we saw that the output of a J-K flip-flop can assume different states at the time of a clock transition depending on the state of the J and K inputs at that time and on the state of the output at that time. The J and K inputs can be thought of as a 2-bit input data "word," an extension of the 1-bit data input of a D flip-flop. The other new feature of the J-K flip-flop was that the output of the J-K flip-flop just prior to when CLOCK goes to 1 can affect the new output after CLOCK goes to 1, perhaps because the previous output was stored in an internal "slave" flip-flop when CLOCK went from $1 \to 0$.

We can imagine a more general device in which there are more than two inputs, more than one output, and more than one slave flip-flop for remembering previous results. (A set of these slave flip-flops is called a register.) At a $0 \to 1$ CLOCK transition the outputs would assume values (0 or 1) depending on the inputs and the contents of the internal registers.

What we have described is called a microprocessor. These devices, usually designed to mimic the central processing unit (CPU) of a digital computer, were first introduced in the early 1970s by Intel Corporation in the form of a single integrated circuit. Since then, prices have fallen dramatically, from several hundred dollars to about $20, because of the mass-production concept. Microprocessors are truly general purpose logic devices, therefore, they are widely applied, and the large production volume that results allows the cost of development and other fixed costs to be distributed over many units.

Let us look at the kinds of operations that can be performed by a microprocessor that is designed to be a computer CPU. The great idea of a computer, in addition to its speed in performing logic operations, is that the CPU can select its own inputs. This input almost always comes from some sort of memory (not included on the microprocessor chip) where data is organized into "*words*" with a fixed number of bits. Word lengths in microprocessor systems are usually 8 bits, although there are commercial microprocessors based on 4-bit words and 16-bit words. (Large digital computers typically are based on 32-bit words.) An 8-bit word is called a *byte*; we assume that the memory is organized into bytes in the following discussion.

Each byte in memory is labeled by an address, which itself is a set of binary digits. For example, if the memory contains 2^{16} (or 65,536) bytes, then 16 binary digits (2 bytes) are required to address any byte in memory. Our microprocessor will then contain a 16-bit register, called an address register (AR). All 16 outputs of the AR must be available at the pins of our integrated-circuit microprocessor. In addition, the microprocessor must provide a READ signal and a WRITE signal at two more pins. When READ goes to 1, circuitry external to the microprocessor must make available to eight "input" pins of the microprocessor the contents of the location in memory specified by the AR. Similarly, when WRITE goes

TABLE 10.3.   Pin Assignments in Typical
Microprocessors

Number of Pins	Use
16	AR (output)
8	DATA (input/output)
1	READ (output)
1	WRITE (output)
1 or 2	CLOCK (inputs)
1, 2, or 3	Supply voltages
1	Ground
29 to 32	

to 1, the information at eight output pins on the microprocessor is stored in the memory location specified by the AR. It is possible to use the same eight pins for output as are used for input. The complement of pins through which the microprocessor communicates with the external world, is shown in Table 10.3. A typical microprocessor is packaged in a standard 40-pin, dual-in-line integrated circuit.

Next we describe how a microporcessor selects the memory addresses that provide its input and how it can process that data before sending information back out. At some point in its cycle of operations, a CLOCK input causes an address to be loaded into AR. The next CLOCK input causes a READ output that brings a byte from memory and stores it in an 8-bit internal register called an instruction register (I). On succeeding clock pulses the microprocessor takes action based on the contents of I. A typical set of actions might be to: (1) MOVE a pair of bytes stored in two internal registers to the AR, and (2) READ the contents of the addressed memory location into the microprocessor, at the same time adding this byte to the contents of an internal register, distinct from the instruction register, called the accumulator (AC). Notice that the entire sequence required 4 operations, two to fetch an instruction from memory and two to execute it.† An example of a 3 operation sequence would be one that did not involve the contents of memory in the execution phase: (1) LOAD AR, (2) READ Instruction from memory into I, and (3) add 1 to the contents of AC. Microprocessors generally have several internal registers, and typical instructions can move the contents of one register to another, clear, complement or increment the contents of registers, and add, AND, or OR the contents of one internal register with another.

Thus far we have avoided the question of where the microprocessor obtains the address that is loaded into the AR prior to the fetching of an

---

† In a microprocessor each of the operations usually requires several sub-steps; thus, several CLOCK pulses may be required for each operation.

instruction from memory. Usually this is obtained from an internal 16-bit register called a program counter (PC). Each time an instruction is read into I from memory, PC is incremented by 1, so that PC continually contains the address of the next instruction to be fetched from memory.

To achieve its true potential, the microprocessor must be able to decide where its next instruction is coming from on the basis of its own calculations. Thus some of the instructions will allow modification of the contents of PC. Not only could PC be changed by adding to it, subtracting from it, or replacing it by a pair of bytes from internal registers but also those two bytes could come from the next two memory locations following the one that contained the actual instruction. Even more flexibility is provided by requiring certain conditions to be met in order for PC to be changed. Typical conditions would be that an internal register such as AC was 0, or that two internal registers were equal. The effect of these conditional changes of the PC is to create what are called conditional jump instructions. When the conditions are met, the normal sequential order of instructions is broken and the next instruction is obtained from an entirely different memory location.

The foregoing very brief introduction to microprocessors is intended merely to advertise their existence and to give some indication of their power and flexibility as sequential logic devices. Many applications in the consumer area have been forecast for microprocessors, including monitoring conditions in a automobile engine, determining cycles in home appliances such as ovens and washing machines, and in automatic vending machines to choose products, make correct change, and so on. There will undoubtedly be numerous applications in electronic instrumentation. Oscilloscopes are a prime target for microprocessor applications because of the huge market. One can imagine, for example, setting a knob to represent a multiple of the sweep time/cm and obtaining an instantaneous digital display of the vertical deflection voltage at that time. Another knob might even be set to tell the microprocessor to average the readings at a particular time during each of several traces before updating the display.

To obtain a working knowledge of microprocessors, the interested reader should consult specialized books on the subject† as well as the device application notes published by the various manufacturers of microprocessors.

10.8. Semiconductor memories   Although the set of instructions to be obeyed by a microprocessor could be and often is stored in a ROM, the most flexible microprocessor systems demand that the program be

---

† For example, Adam Osborne, *An Introduction to Microcomputers*: Vol. I, *Basic Concepts* and Vol. II, *Some Real Products*, Adam Osborne and Associates, Inc., Berkeley, CA, 1976.

changeable. Also if the system requires any storage of data, one must be able to store information in memory as well as to read it out. A memory in which any word can be changed at any time is called a read-write random-access memory, or RAM for short.

For many years the most economical memory device for systems larger than a few hundred bits was the magnetic core, in which data is stored as the direction of residual magnetization in a small (~1 mm diameter) ferrite toroid. While magnetic cores are still used in very large computers for their fast-access memory, the economic advantage of cores for memories of a few thousand bits or less has been surpassed by memories based on bipolar or field-effect transistors. The time required to get information in and out of a semiconductor RAM can be as low as about 100 nsec, which is about five times faster than the fastest core memories. One reason for this speed advantage is that a core is a destructive read-out device while semiconductor memory devices are not. In order to tell if a core is magnetized in one direction, one must try to change its direction of magnetization with a brief pulse of current in a wire that passes through the toroid; if a current is induced in another "sense" wire, then the core had a 1 and if no current was induced the core had a 0. If the contents of the memory are to remain the same after readout, then the stored information must be immediately read back in. This process requires time and adds to the complexity of the memory control circuits.

Although semiconductor RAMs have nondestructive readout, they do have the disadvantage of requiring power, even when the contents are not being changed. Thus it is necessary either to supply continuous power to a system, or to have a portion of the memory stored in a ROM, or to reenter information, for example, from punched paper tape each time the power is turned on.

Semiconductor memories are classified as dynamic or static, depending on whether information "leaks off" after a certain time. Dynamic memories must be "refreshed," usually every few milliseconds. While this requires additional control logic and adds a little to the average time for reading and writing, a dynamic memory is more compact, that is, has more bits per integrated circuit package and is cheaper. As of the middle 1970s the typical number of bits per package in a dynamic RAM is 2048 as against 1024 in a static RAM.

The fastest semiconductor RAMs are based on bipolar transistors, but these use considerably more power than FET memories. Since FET memories are fast enough for microprocessor applications, we will concentrate on them.

The heart of any RAM is the basic storage cell. Figure 10.10 shows three such cells. The 6-transistor cell of Fig. 10.10a is a typical static cell. This particular one, made of n-channel, enhancement-mode MOSFETs, requires a positive supply voltage V_{DD} and a supply V_{GG} that for the

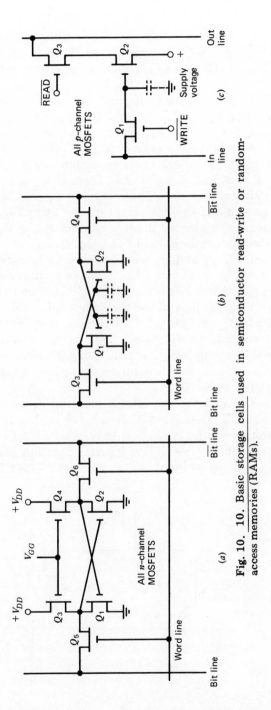

Fig. 10. 10. Basic storage cells used in semiconductor read-write or random-access memories (RAMs).

241

moment we will take to be equal to V_{DD}. Recall that the n channel is induced only if the gate voltage is sufficiently positive; thus, the MOSFETs in Fig. 10.10a are ON (conducting) if their gates are Hi and are OFF (not conducting) if their gates are Lo.

To understand the operation of this static cell, suppose first that the *word* line is Lo and V_{GG} is Hi, so that Q_5, Q_6, are OFF and Q_3, Q_4 are ON. (The reason for the names "word line" and "bit line" will become apparent when we discuss the organization of cells into a larger memory.) With Q_5 and Q_6 OFF, there are two stable states of the other four MOSFETs. A 1 is stored if Q_2 is ON. Then, if Q_2 is designed to have a greater ON conductance than Q_4, the gate of Q_1 will be Lo, and Q_1 will be OFF thus confirming the Hi voltage on the gate of Q_1 necessary for it to be ON. Since by symmetry there is another stable state with Q_1 ON, we can recognize the combination of Q_1, Q_2, Q_3, Q_4 as a basic latch. This circuit would work if Q_3 and Q_4 were replaced by resistors, but it actually requires less space on a silicon chip to make a transistor. Also we have another advantage if the loads are MOSFETs rather than resistors. It is feasible to take V_{GG} Lo part of the time while information is being stored so that Q_3 and Q_4 are turned OFF. While they are OFF, the stray capacitance, between the Hi gate of Q_1 or Q_2 and other structures, remains charged, but the only currents that flow are very tiny leakage currents. In this way the stand-by power drain due to current supplied by V_{DD} can be greatly reduced.

In order to READ from or WRITE into the cell of Fig. 10.10a, the word line is taken Hi. Then, to WRITE a 1, the bit line is made Hi and the $\overline{\text{bit}}$ line Lo, which sets the latch. Similarly, a zero is written if the bit line is taken Lo and the $\overline{\text{bit}}$ line Hi. A READ operation is accomplished by connecting the bit lines as inputs to the gates of MOSFETs. The MOSFET

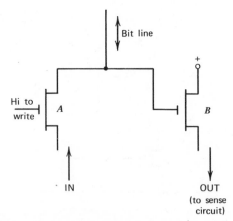

Fig. 10.   11. Read-write circuitry for each bit line of memories with cells shown in Figs. 10.10*A* and 10.10*B*.

connected to the Hi bit line will conduct and the other will not. How the same bit line can be used for writing and reading is illustrated in Fig. 10.11. To WRITE, the gate of MOSFET A is taken Hi so that A conducts. The Hi or Lo voltage at the input then appears on the bit line. To READ, A is turned OFF, and the Hi or Lo voltage on the bit line turns B ON or OFF, thus providing current or no current to a sensing circuit (not shown).

Next, look at the cell of Fig. 10.10b, which is identical to the cell of Fig. 10.10a except that the load transistors have been omitted. Now information is stored *only* on the stray gate capacitances. Since this charge can leak off through the reverse-biased junctions of MOSFETs, this cell is a dynamic one and needs to be refreshed every few milliseconds. Refreshing is accomplished by taking *both* bit lines Hi and then raising the word line so that Q_3 and Q_4 conduct. Now Q_3 and Q_4 act as load transistors, and the regenerative action of the latch restores the charge on the gate that was Hi to its maximum. To complete the restore cycle, the word line is taken Lo before the bit lines are returned Lo. Although reading and writing are accomplished in exactly the same way as in the static cell of Fig. 10.10a, it should be clear that additional logic is required for the refresh cycle, and a few microseconds must be reserved for this purpose every few milliseconds.

Additional saving of chip space is achieved with the 3-transistor, dynamic RAM cell of Fig. 10.10c. This cell, implemented with p-channel, enhancement-mode transistors, is used in the type-1103 1024-bit dynamic RAM, popular in the early 1970s. Here, information is stored as charge or as no charge on the gate capacitance of Q_2. Both Q_1 and Q_3 are OFF during the storage mode; this requires Hi voltages on the gates of Q_1 and Q_3 because they are p-channel devices. To WRITE into this cell, the WRITE line is taken low and a Hi or Lo voltage on the IN line charges or discharges the gate capacitance of Q_2 through Q_1. To READ, the READ line is taken Lo so that Q_3 conducts, and the supply voltage is or is not connected to the OUT line depending on whether a 0 or 1 is stored in the cell. Since storage is dynamic, a refresh cycle must occur every couple of milliseconds. Refresh is accomplished by gating on an inverting amplifier whose input is the OUT line and whose output is the IN line. Although these inverting amplifiers add complexity to the memory chip, one amplifier can service several cells; in the type-1103 memory there is one refresh amplifier for every 16 cells — thus 64 amplifiers on the chip.

Next we focus on the way in which memories are organized. One of the most important properties of a multibit memory is the number of connecting wires required to get information in and out. For example, consider a 1024-bit static memory constructed of the cells in Fig. 10.10a. It would be possible to have a bit and a bit line to every cell (2048 lines in all) and one word line to all cells simultaneously. Such an arrangement would completely defeat any size advantage of a RAM by requiring an enormous number (2049 plus power supply and ground) of inter-connections to the memory. An improvement would be to send one word line to each cell (1024 in all) and a common bit and bit line to all cells, but still 1028 connections would be required. A significant improvement occurs if we think in terms of a square array (indeed, this is the way cells are arranged on a chip) with 32 rows and 32 columns. Now we can use 32 word lines, one to each row, and 64 bit lines, two to each column. When we take a word line Hi, we can READ or WRITE all 32 cells in a row at the same time, each with its own 1 or 0.

With the row-column organization, we have reduced the number of lines that are required to connect to our 1024-bit memory to about 100. But we can do better! Instead of sending in 32 word lines, we can send in 5 (because $2^5 = 32$) if we include a "5 in to 1 of 32 out" demultiplexor right on the memory chip. And we can use a multiplexor on the bit lines too. We send in 5 lines with binary signals that tell the multiplexor which bit line to connect to a *single* DATA IN/OUT line. (Only a single IN-OUT line is necessary if we use some combinational logic on the chip to produce the proper input signals to the bit and $\overline{\text{bit}}$ line and if we use a separate input line to specify READ and WRITE, for example, READ when the signal on this line is 1 and WRITE when it is zero.

The final version of our 1024-bit memory is shown in Fig. 10.12. In addition to 10 address lines, a DATA line, a READ/WRITE line, power, and ground, we have added a CHIP SELECT line. When CHIP SELECT is 0, all word lines are disabled and the DATA IN/OUT line is disabled within the memory so that no reading or writing can occur. When CHIP SELECT = 1 all lines are enabled, and DATA is READ (if READ/WRITE = 1) or WRITTEN (if READ/WRITE = 0). Notice that this memory "fits" in a standard 16-pin IC package.

The CHIP SELECT line is important for two reasons. First, it allows the

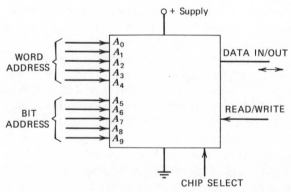

Fig. 10.   12. External connections necessary for a 1024-bit memory.

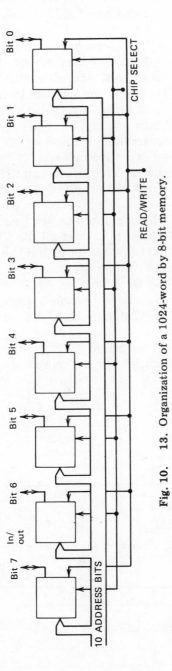

Fig. 10.   13. Organization of a 1024-word by 8-bit memory.

memory to be clocked, so that we are not forced always either to READ or WRITE, and second, it permits many chips to be connected in a large memory.

Figure 10.13 illustrates a very simple multichip memory. This organization would be called a 1024-word by 8-bit memory. All 10 address signals are sent in parallel to all eight chips. Also all eight CHIP SELECT inputs and all eight READ/WRITE inputs are connected together. The DATA IN-OUT lines are brought out separately. Thus, when CS = 1, one bit is READ from or WRITTEN into one cell on each chip. The address of the cell on each chip is the same, and it is specified by the 10 address bits.

The next, and final, order of complexity is to use a decoder on additional address bits to choose which of several CHIP SELECT lines are made 1. A 4096-word by 8-bit memory requiring 32 chips is shown in Fig. 10.14. Ten of the twelve address lines go, in parallel, to all 32 chips. Also the R/W line goes to all 32 chips. The last two address lines go to a "2 in, 1 of 4 out" demultiplexor, so that all the CHIP SELECT inputs in a row, but *only* those in that row are set to 1 at the same time. The four DATA IN/OUT pins in each column are connected in parallel; only the chip in the selected row can be READ or WRITTEN at any one time. Finally, if the demultiplexor has an "enable" input, so that all 4 CHIP SELECT lines are 0 unless MEMORY ENABLE = 1, we can retain our clocked memory feature.

Additional memory could be added to Fig. 10.14 in blocks of 1024 words by using a larger demultiplexor and more address lines. However, it

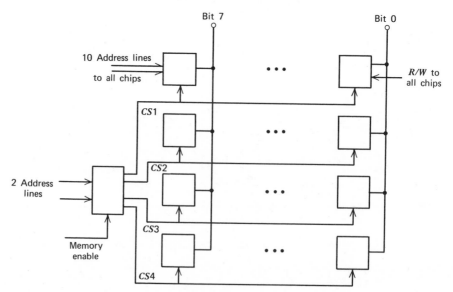

Fig. 10.   14. Organization of a 4096-word by 8-bit memory.

should be apparent that we have already "built" a memory system of considerable flexibility that could be used with an 8-bit microprocessor and a relatively small amount of additional interface logic to produce a true "microcomputer."

Problems for Chapter 10

10.1. What is wrong with each of the following circuits (Fig. P10.1) as potential latches? In each case describe the operation for all four combinations of the two inputs.

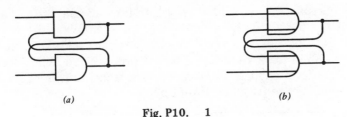

(a)   (b)

Fig. P10.   1

10.2. In the circuit in Fig. P10.2, when $x = 0$, output y is also observed to be 0.
(a) Find the logic level at the output of each gate.
(b) Now x becomes 1. Find the new state of all seven gate outputs. If there is a critical race (one NAND input changing from $0 \rightarrow 1$ while the other changes from $1 \rightarrow 0$) assume that the output of that NAND remains at 1.

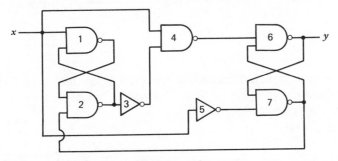

Fig. P10.   2

10.3. One version of a TTL flip-flop is shown in Fig. P10.3. Begin with the state CLOCK = 0, $Q = 0$, and $\bar{Q} = 1$ and make a table that shows the output state of each gate as the clock continually switches between 0 and 1. Verify the flip-flop operation.

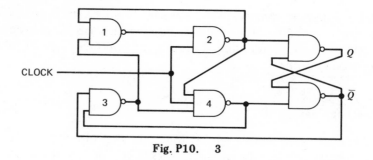

Fig. P10.   3

10.4. The circuit in Fig. P10.4 is a flip-flop. Begin with the state: CLOCK = 0, $Q = 0$, $\bar{Q} = 1$. Make a table showing the changes of each gate output as the clock switches back and forth between 0 and 1. Keep track of the number of propagation delays that occur before a level changes after each clock transition. Is this a positive- or negative-edge triggered flip-flop?

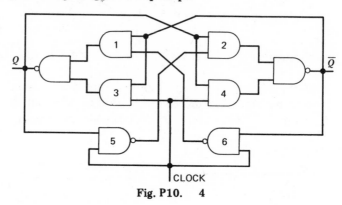

Fig. P10.   4

10.5. Draw the Karnaugh maps for the six J-K inputs of the 5-counter in the example of section 10.6, and verify that the circuit of Fig. 10.9 gives the desired sequence of states.

10.6. A 6-counter is to be built from type-D flip-flops, the desired order of states being shown in the table. *Part* of the circuit is drawn (see Fig. P10.6). On the downstroke of T, outputs Q_4, Q_2, and Q_1 are shifted into Q_4' Q_2', and Q_1' and then on the upstroke of T, the proper functions of Q_4', Q_2', and Q_1' are shifted back into Q_4, Q_2, and Q_1. Find the functions of Q_4', Q_2', and Q_1' that must be connected to D_4, D_2, and D_1.

Q_4	Q_2	Q_1
0	0	0
0	0	1
0	1	0
0	1	1
1	0	0
1	0	1
0	0	0
	etc.	

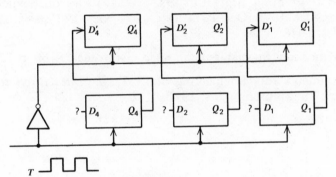

Fig. P10.   6

10.7. Design a binary-coded 12-counter using *J-K* edge-triggered flip-flops and NAND gates.

REFERENCES

Electronics, Special issue on microprocessors, April 15, 1976.

Hodges, D. A., "Review and Projection for Semiconductor Components for Digital Storage," *Proc. IEEE*, 63, 1136—1147 (1975).

Hodges, D. A., Ed., *Semi-Conductor Memories*, IEEE Press, 1972.

IEEE Journal of Solid State Circuits, Special issues on Semiconductor Memories, Oct. 1970, Oct. 1971, Oct. 1972, Oct. 1973, Oct. 1974, Oct. 1975, Oct. 1976.

Scientific American, Special issue on microelectronics, Sept. 1977.

Texas Instruments Inc. Engineering Staff, *T.T.L. Data Book for Design Engineers*, Texas Instruments, 1973.

CHAPTER 11

Digital-to-Analog and
Analog-to-Digital Conversion

11.1. Introduction   Because of the flexibility in data analysis afforded by digital systems such as special-purpose computers, there is often a need in modern instrumentation systems to convert an analog signal to a form appropriate to a digital device. For example, if the analog signal is a voltage that varies between 0 and +10 V, we might convert it once every 20 μsec to a succession of binary numbers, each with 12 binary digits. Since a change in the least significant digit of a 12-bit binary number represents a fractional change of 1 part in $2^{12} = 4096$ of the largest possible number that can be represented by the 12 bits, our digital number may not be exactly proportional to the input analog signal. In this example, if the binary number 0000 0000 0000 represents 0.0 V and if 1111 1111 1111 represents +10.0 V, the digitizing error† or discrepancy between the actual analog voltage at a particular time and the voltage represented by the digital "equivalent" may be as large as half the voltage corresponding to the least significant bit, or $\pm$ ½ $\cdot$ (10/4096) = $\pm$ 1.2 mV in this example.

The opposite process, that of converting a digital number to its analog equivalent, is equally important not only because output display devices such as x-y plotters and oscilloscopes respond to analog signals but also because the fastest methods of converting a signal from analog to digital form use a digital-to-analog converter. Because of the latter fact, and the fact that D-to-A converters are inherently simpler to understand than A-to-D converters, we describe D-to-A conversion methods first.

11.2. Digital-to-analog conversion   We learned earlier that there are different ways, for example, straight binary and binary-coded-decimal, of representing a number in binary form. As long as the relative weight of each binary digit is independent of the other digits, the D-to-A conversion process is quite straightforward. Thus, in the 8421 BCD representation, a 1 in the 8-bit must represent eight times the analog signal that is represented by a 1 in the 1-bit, a 1 in the 4-bit must represent four times the analog signal of the 1-bit, and so on.

† Effects introduced by the ability to sample a continuous analog signal only at discrete times are discussed in Chapter 12.

251

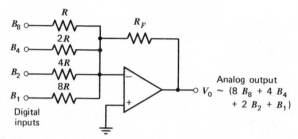

Fig. 11.   1. A 4-bit D-to-A converter made from an op-amp.

One way of achieving the required proportionality between the digital and analog representations is with an op-amp circuit in the form of an adder. In Fig. 11.1 there are four input resistors, in the ratio 1:2:4:8. A 1 in the 8-bit of the digital number results in eight times the current to the inverting input as a 1 in the 1-bit. The output voltage is therefore proportional to the binary-coded-decimal input number. The constant of proportionality depends on the resistor ratio, R_F/R. Care must be taken that the maximum binary input, in this case 1001 or decimal 9, gives an output voltage within the range of the op-amp's output voltage swing.

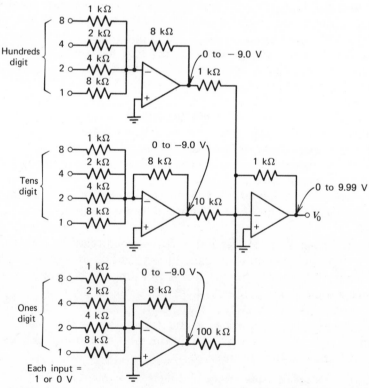

Fig. 11.   2. D-to-A converter for three decimal BCD digits.

Binary-coded-decimal signals with several digits can be converted to analog form by using a D-to-A converter of the type shown in Fig. 11.1 for each BCD digit and then adding the outputs in another op-amp. Figure 11.2 shows a three-decimal-digit, binary-coded-decimal, D-to-A converter, designed to give an output between 0 and 9.99 V. In practice, several of the resistors would have to be made variable to allow for accurate calibration. The need for variable resistors and for several op-amps makes this circuit somewhat expensive. Another disadvantage of this circuit is that the binary input voltages, in this example a nominal 0 to 1 V, may vary among themselves or may vary with the load current being drawn from the supply voltage for the digital registers whose outputs drive the D-to-A inputs. Thus if there is a 1-mV change in supply voltage to the 8-bit of the hundreds digit, an 8-mV change in final output voltage results. This is practically equal to the output (10 mV) caused by a 1 in the 1-bit of the ones digit. For high accuracy it is best to use some form of transistor switch at the input such as that shown in Fig. 11.3. The dc collector reference voltage should be regulated separately from the supply voltage to other digital circuitry. When the transistor input voltage is low, the transistor is cut off, and current from the supply flows through 1 kΩ to the op-amp input; when the transistor input voltage is high, the current is shunted to ground through the saturated transistor. The fraction of a volt that exists across even a well-saturated transistor may cause a slight current to the op-amp and produce an output offset voltage. We can compensate for this with a small, negative bias current to the op-amp.

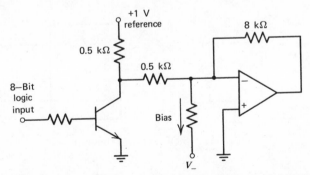

Fig. 11. 3. D-to-A input circuitry featuring a separate reference voltage.

The need for several values of precision resistor, (1, 2, 4, and 8 kΩ) in the op-amp adder circuit can be eliminated by using a precision ladder network of resistors. The ladder network of Fig. 11.4 is appropriate for 4 digits of binary input, but it can be extended merely by adding more "rungs" to the ladder. Note that only two values of precision resistor are needed, and that these values differ by a factor of two.

We will analyze the ladder network of Fig. 11.4 with the aid of an

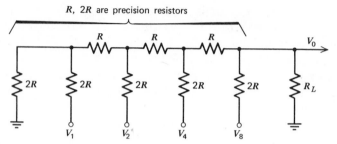

Fig. 11.   4. A 4-bit ladder-type D-to-A converter.

important theorem from network theory known as Thévenin's theorem. (For a proof of this theorem see problem 11.1.) Thévenin's theorem is illustrated in Fig. 11.5. It states that any linear network, however complex, that contains any number of voltage sources, any number of current sources, and any number of resistors can be represented by a circuit with just two elements — a series combination of an "equivalent" voltage source V_T and an "equivalent" resistance R_T. That is to say that the effect on any external circuit, here represented by the load resistance, R_L, is simply that of a voltage source in series with a resistance. [This is the dc form of Thévenin's theorem; the theorem also applies when the network contains voltage and current generators of the form e^{st}, and resistors, capacitors, and inductors. In this case the equivalent circuit contains a voltage source $V_T(s)e^{st}$ and an impedance $Z_T(s)$.]

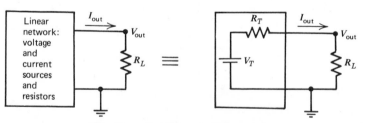

Fig. 11.   5. Thévenin's theorem.

In practice, or in theory, if one wishes to determine the magnitude of the voltage generator in the Thévenin equivalent circuit, one can simply replace R_L by an open circuit and measure or calculate the output voltage of the network. The Thévenin equivalent voltage, V_T, is clearly equal to this *open-circuit output voltage*. Similarly, one can determine the value of the resistor, R_T, in the Thévenin equivalent circuit by replacing R_L with a short circuit and measuring or calculating the output current. Clearly R_T equals the Thévenin equivalent voltage divided by the *short-circuit output current*. These relations are summarized in Fig. 11.6 and Eqs. 11.1.

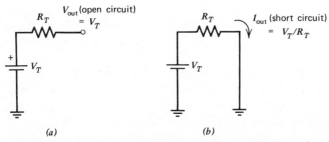

Fig. 11.   6. Calculations of the Thévenin equivalent voltage and resistance.

$$V_T = V_{out} \text{ (open-circuit)}$$

$$R_T = \frac{V_{out} \text{ (open circuit)}}{I_{out} \text{ (short circuit)}} \quad\quad (11.1)$$

Next we use Thévenin's theorem to find the output voltage in Fig. 11.4 when any one of the input voltages, V_1, V_2, V_4, and V_8, is 1 and the rest are 0. According to the *superposition theorem* we can then find the output when more than one of the input voltages is 1 by adding the individual responses. In our analysis we assume that $R = 1$. This does not mean that we are necessarily choosing R to be 1 Ω; we are simply choosing units so that, whatever R is, that is the fundamental unit of resistance for this problem.

First, let $V_8 = 1$ and $V_1 = V_2 = V_4 = 0$. By successive use of the formulas for series and parallel resistors, the circuit of Fig. 11.7a reduces to that of Fig. 11.7b. (The parallel combination of resistors a and b is equivalent to a single resistor of value 1, which is in series with resistor c. This series combination of two unit resistors is replaced by a single resistor of value 2, and so on.) We can apply Thévenin's theorem to the reduced circuit of Fig. 11.7b by considering that everything to the left of the dotted line is "inside" the circuit and that R_L is "outside." According to the theorem, the inside circuit can be replaced by a series combination of voltage source and resistor, whose values can be found from Eqs. 11.1. Thus, to find the value of the equivalent voltage source, we replace R_L in

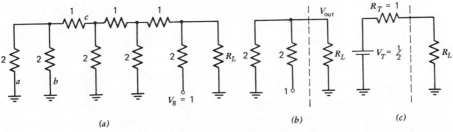

Fig. 11.   7. Reduction of the ladder network.

Fig. 11.7b by an open circuit. We see immediately that the voltage divider action of the two equal resistors results in V_T = ½. To find the Thévenin resistance of the circuit, we replace R_L by a short circuit and determine the short-circuit output current. Since the unit voltage source is now shorted to ground through a resistor of value 2, the current is ½. The second of Eqs. 11.1 shows that R_T = ½/½ = 1. The final Thévenin equivalent circuit appears in Fig. 11.7c.

Now to find the response to a unit value of V_4, let V_4 = 1 and $V_1 = V_2 = V_8 = 0$, (Fig. 11.8a). The circuit is reduced to that of Fig. 11.8b by the series-parallel resistance formulas. The circuit is now reduced by *two* applications of Thévenin's theorem; first, the "inside" is considered to be to the left of dotted line a (we have already worked out the Thévenin equivalent voltage source and resistance for this little circuit) and reduces to that of Fig. 11.8c. Then the circuit is broken at dotted line b and the Thévenin equivalent voltage source and resistance calculated as before. The result is shown in Fig. 11.8d. Note that the Thévenin resistance of the circuit is unity, whether V_8 = 1 or V_4 = 1, but that the equivalent voltage source is half as large when V_4 = 1 as it was when V_8 = 1.

The reader is invited to carry out the analysis for the cases of V_2 = 1 and V_1 = 1. In each case the circuit can be reduced to the form of Fig. 11.7c or 11.8d, with a Thévenin equivalent resistance of 1. When V_2 = 1 the Thévenin equivalent voltage source is 1/8, and when V_1 = 1 it is 1/16. The overall effect of the ladder circuit is to weight the effects on the output of V_8, V_4, V_2, and V_1 respectively, by the factors 1/2, 1/4, 1/8, and 1/16. Additional rungs placed on the left end of the ladder would

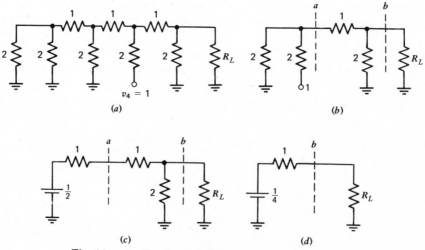

Fig. 11.   8. Further reduction of the ladder network.

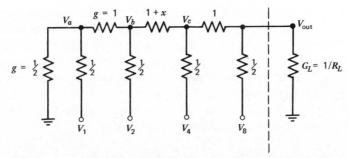

Fig. 11.   9. Ladder network with an erroneous resistor value.

result in the additional inputs being weighted by the factors 1/32, 1/64, etc. The circuit is clearly a D-to-A converter for a binary-coded signal.

No discussion of a ladder-type D-to-A converter is complete without considering the effect of slight errors in one or more resistance values. For example, consider the circuit of Fig. 11.9, in which we have shown the conductances instead of the resistances (conductance = 1/resistance). One of the unit conductances in the top row has the erroneous value $1 + x$. The problem is to find how the output depends on all the input voltages and x.

Now, with less symmetry than before, a straightforward nodal analysis is less cumbersome than Thévenin's theorem analysis. We write down four nodal equations in terms of the input voltages, V_1, V_2, V_4, and V_8, and the four nodal voltages, V_a, V_b, V_c, and V_{out}. These are:

$$
\begin{aligned}
V_1/2 &= V_a \cdot (\tfrac{1}{2} + \tfrac{1}{2} + 1) - V_b \\
V_2/2 &= -V_a + V_b \cdot (1 + \tfrac{1}{2} + 1 + x) - V_c \cdot (1 + x) \\
V_4/2 &= -V_b \cdot (1 + x) + V_c \cdot (1 + x + \tfrac{1}{2} + 1) - V_{out} \\
V_8/2 &= -V_c + V_{out} \cdot (1 + \tfrac{1}{2} + G_L)
\end{aligned}
\tag{11.2}
$$

Equations 11.2 are to be considered as four linear equations in four unknowns: V_a, V_b, V_c, and V_{out}. We are interested only in the value of V_{out}; it can be obtained by a determinant solution:

$$
V_{out} = \frac{
\begin{vmatrix}
2 & -1 & 0 & V_1/2 \\
-1 & (\tfrac{5}{2} + x) & -(1 + x) & V_2/2 \\
0 & -(1 + x) & (\tfrac{5}{2} + x) & V_4/2 \\
0 & 0 & -1 & V_8/2
\end{vmatrix}
}{
\begin{vmatrix}
2 & -1 & 0 & 0 \\
-1 & (\tfrac{5}{2} + x) & -(1 + x) & 0 \\
0 & -(1 + x) & (\tfrac{5}{2} + x) & -1 \\
0 & 0 & -1 & \tfrac{3}{2} + G_L
\end{vmatrix}
}
\tag{11.3}
$$

Two aspects of this solution are noteworthy; the input voltages appear only in the numerator, and G_L appears only in the denominator. The latter fact means that G_L only affects the magnitude of the output but *not* the relative effects of the various inputs on the output. In fact, we need only compute the numerator determinant in order to find the relative weights of the input voltages. Accordingly, we expand the numerator determinant in terms of the elements of the last column and denote the denominator determinant simply by D.

$$V_{out} = -\frac{V_1}{2D} \begin{vmatrix} -1 & (5/2 + x) & -(1+x) \\ 0 & -(1+x) & (5/2 + x) \\ 0 & 0 & -1 \end{vmatrix} + \frac{V_2}{2D} \begin{vmatrix} 2 & -1 & 0 \\ 0 & -(1+x) & (5/2+x) \\ 0 & 0 & -1 \end{vmatrix}$$

$$-\frac{V_4}{2D} \begin{vmatrix} 2 & -1 & 0 \\ -1 & (5/2 + x) & -(1+x) \\ 0 & 0 & -1 \end{vmatrix} + \frac{V_8}{2D} \begin{vmatrix} 2 & -1 & 0 \\ -1 & (5/2 + x) & -(1+x) \\ 0 & -(1+x) & (5/2+x) \end{vmatrix}$$

$$(11.4)$$

Expansion of the third-order determinants leads, after some manipulation, to the final expression:

$$V_{out} = \frac{V_1}{2D} \cdot (1+x) + \frac{V_2}{2D} \cdot 2 \cdot (1+x) + \frac{V_4}{2D} \cdot 4 \cdot \left(1 + \frac{x}{2}\right) + \frac{V_8}{2D} \cdot 8 \cdot \left(1 + \frac{5}{8}x\right)$$

$$(11.5)$$

Notice, in Eq. 11.5, that if $x = 0$, the relative effects of V_1, V_2, V_4, and V_8 on the output are in the ratio 1:2:4:8 as was found in the Thévenin's theorem analysis. However, when $x \neq 0$, the ratios are slightly different. For example, when $x = 0.01$, a 1% error in the conductance value, the ratios are 1.01:2.02:4.02:8.05. The effect of V_8 is seen to be 3 parts in 800 lower, and the effect of V_4 is 4 parts in 800 lower than they should be if the output voltage is to be an accurate analog version of the digital inputs. We may conclude that the values of resistance in the ladder network, at least in the higher-order "rungs" of the ladder, must be held to extremely close tolerances in order to obtain an accurate conversion.

11.3. Analog-to-digital conversion

There are numerous ways of obtaining a digital number whose value is closely proportional to an analog signal. In one method, a linear ramp voltage is generated, starting from 0 V. At the same time a counter is incremented at equal intervals by a periodic signal. The linear ramp rises until its voltage becomes equal to the analog input voltage, at which time the input to the counter is removed. Assuming the counter started at zero, the digital number now in the

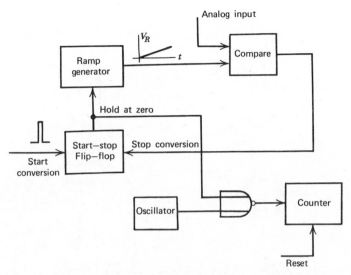

Fig. 11.   10. Ramp-type A-to-D converter.

counter is proportional to the analog signal. A block diagram of the instrumentation needed to accomplish this ramp-type A-to-D conversion is shown in Fig. 11.10. The ramp generator is held off until a flip-flop is set by a "start conversion" pulse. The flip-flop output also controls a gate that can prevent the oscillator output, which runs continually, from reaching the counter. As the ramp voltage crosses the analog input voltage level the comparator output resets the flip-flop. After the number in the counter has been stored in some register (not shown), or otherwise served its useful life, a reset pulse clears the counter and the system is ready for another conversion.

There are two disadvantages of the simple ramp converter we have described. The first is that the time required for a conversion may be longer than in other methods. For example, if the counter contains three decimal digits and can count at a maximum rate of 20 MHz, then a time of approximately 1000×50 nsec = 50 μsec is required to convert the largest signal. A second disadvantage is that the long-term accuracy of the converter depends on both the oscillator frequency and the slope of the ramp; since these quantities can vary with temperature, humidity, and component aging, high accuracy is difficult to maintain. Despite the need for frequent calibration, this type of A-to-D converter has been widely used in pulse-height analyzers in nuclear physics applications, where the peak size of a voltage pulse is converted to a digital number and one count is stored in a memory location to denote that one pulse of a particular amplitude has occurred.

Many modern digital voltmeters use a variation of the simple ramp-type conversion scheme, called the dual-ramp method. In this method,

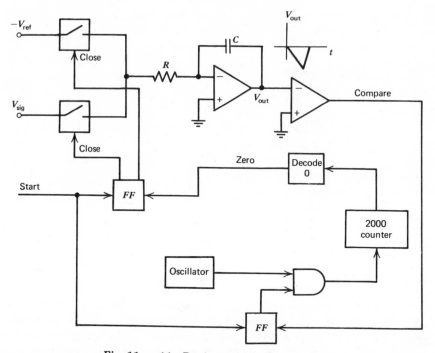

Fig. 11.   11. Dual-ramp A-to-D converter.

illustrated in Fig. 11.11, a measurement cycle begins when the analog input is applied to an integrator. In a typical "3½ digit" or 0 to 1.999-V meter, the same signal that closes the integrator switch also enables a 2000 counter to count pulses from an oscillator. When the counter has cycled through 2000 counts and reads 0 again, the analog input to the integrator is replaced by a stable reference voltage of −2.000 V. The integrator returns toward 0 while the counter continues to count. When the integrator output reaches 0, as determined by a comparator, the counter is stopped, with an accumulated count proportional to the signal voltage.

The great advantage of the dual-ramp, A-to-D conversion method is

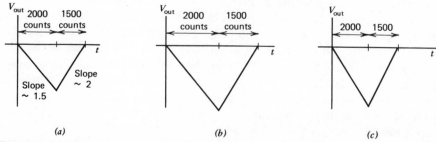

(a)   (b)   (c)

Fig. 11.   12. Ramp output in a dual-ramp converter: (a) normal situation, (b) oscillator running slow, and (c) reduction in RC time constant.

that the conversion accuracy is quite independent of the oscillator frequency and the proportionality factor between the input voltage to the integrator and the slope of its output voltage (see Fig. 11.12). Fig. 11.12a shows the integrator signal when a 1.500-V input signal is converted. The final output of the counter is 1500. Figure 11.12b shows the situation when the oscillator frequency is smaller than in Fig. 11.12a. The counter takes longer to reach 0 but V_{out} also takes longer to return to 0 by the same factor, and the final count is still 1500. Finally, Fig. 11.12c shows the situation when the slope factor is increased, say because a temperature change reduced R or C in Fig. 11.11. Now the voltage reached by the integrator is more negative than before, but it still takes 1500 counts for the integrator to return to 0 at its increased rate.

The dual-ramp converter is seen to be a method of comparing an unknown voltage with a reference voltage. Only the reference voltage needs to have long-term stability. The availability of many excellent integrated circuit voltage references makes the dual-ramp, A-to-D converter practical.

Another A-to-D conversion method similar to the simple ramp method utilizes a D-to-A converter to provide an analog voltage that is compared with the analog signal. Figure 11.13 illustrates this technique in block diagram form. The only real difference between this counter-type, D-to-A converter and the ramp-type described above is that the ramp generator is replaced by the D-to-A converter, whose output is a psuedoramp — a staircase waveform. As in the ramp-type converter, the conversion time is limited by the oscillator frequency and the speed of the counter. A slight advantage of the staircase method is that it avoids the problem of generating a very accurately linear ramp voltage. On the other hand, the

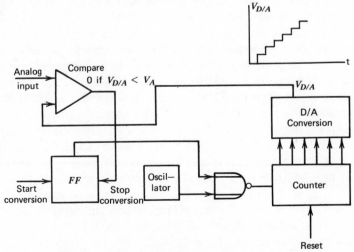

Fig. 11.   13. Counter-type A-to-D converter.

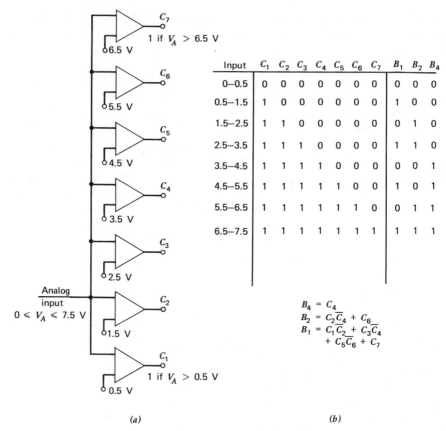

Input	C_1	C_2	C_3	C_4	C_5	C_6	C_7	B_1	B_2	B_4
0–0.5	0	0	0	0	0	0	0	0	0	0
0.5–1.5	1	0	0	0	0	0	0	1	0	0
1.5–2.5	1	1	0	0	0	0	0	0	1	0
2.5–3.5	1	1	1	0	0	0	0	1	1	0
3.5–4.5	1	1	1	1	0	0	0	0	0	1
4.5–5.5	1	1	1	1	1	0	0	1	0	1
5.5–6.5	1	1	1	1	1	1	0	0	1	1
6.5–7.5	1	1	1	1	1	1	1	1	1	1

$$B_4 = C_4$$
$$B_2 = C_2\overline{C_4} + C_6$$
$$B_1 = C_1\overline{C_2} + C_3\overline{C_4} + C_5\overline{C_6} + C_7$$

(a)   (b)

Fig. 11.   14. Multiple comparator A-to-D converter.

D-to-A converter must have a "settling time" that is shorter than the period of the oscillator. Also the dependence of conversion accuracy on oscillator stability is no better than in the simple ramp converter.

A very fast way of performing an A-to-D conversion, but one that utilizes considerably more circuitry than either of the two previous methods, is exhibited in Fig. 11.14a. In the particular example shown, an analog signal in the range from 0 to 7.5 V is to be converted into a 3-bit binary number. Seven comparators are used, with their trigger levels set at 0.5, 1.5, 2.5, 3.5, 4.5, 5.5, and 6.5 V. The table in Fig. 11.14b shows the state of the comparator outputs for different ranges of analog input voltage. If one comparator's output is 1, then so are all those with lower reference voltages. Also shown in the table are the binary bits that are to correspond to each set of comparator states. Inspection of the table reveals that each binary bit can be generated by a logical function of the comparator outputs. These logical functions are listed in the table and can be produced by conventional logic circuitry.

Although this multiple-comparator-type, A-to-D converter can be made very fast because its conversion time is essentially the time required for the comparators to switch plus a few nanoseconds propagation delay in the logic circuits, it suffers from the fact that a small error in one of the trigger levels can make a larger relative error in the width of not one, but two, analog voltage ranges. For example, suppose the reference voltage to comparator 3 is 2.55 V instead of 2.50 V. The voltage range over which $C_3 = 1$ and $C_4 = 0$ is now reduced from 1.00 to 0.95 V while the voltage range over which $C_2 = 1$ and $C_3 = 0$ is increased from 1.00 to 1.05 V. For this reason, the comparator-type converter is seldom used in devices such as pulse-height analyzers in which there is need for strict equality of the width of each input range.

The last type of A-to-D converter we describe is the successive approximation type. The basic idea is to try a "1" in the most signficant bit of a digital register. Then this trial digital number is converted to an analog signal y and compared with the input signal x that is to be converted to digital form. If $x > y$ then the trial "1" in the most significant bit is retained, but if $x < y$, then the trial "1" is erased. The same algorithm is then performed on the next most significant bit, and so on until a "1" has been tried in the least significant bit and either retained or erased. Since the erasure of a bit can coincide with the setting of the next, less significant bit, the operation may be summarized as in Table 11.1 for an 8-bit, A-to-D converter.

TABLE 11.1   Time-Table for Successive-Approximation-Type A-to-D Converter

Time	Operation
t_0	Set bit 7=1 (most significant bit)
t_1	Erase bit 7 if $x < y$; set bit 6=1
t_2	Erase bit 6 if $x < y$; set bit 5=1
t_3	Erase bit 5 if $x < y$; set bit 4=1
t_4	Erase bit 4 if $x < y$; set bit 3=1
t_5	Erase bit 3 if $x < y$; set bit 2=1
t_6	Erase bit 2 if $x < y$; set bit 1=1
t_7	Erase bit 1 if $x < y$; set bit 0=1 (least significant bit)
t_8	Erase bit 0 if $x < y$

In this method the number in the 8-bit digital register becomes a successively better approximation to x as the successive bits are tried. An example of the operation, shown in Table 11.2, may help in understanding this technique. Suppose that the analog voltage, x, is just barely greater than $139/256$ of the full-scale output voltage of the D-to-A converter. The

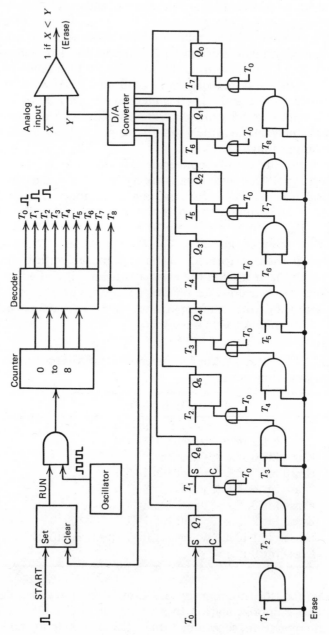

Fig. 11.   15. Successive approximation A-to-D converter.

TABLE 11.2   Development of the Digital Output in a Successive-Approximation Converter

Time	Digital Register	y	x	Comment
t_0	1000 0000	128/256	139/256	
t_1	1100 0000	192/256	139/256	$192 = 128 + 64$
t_2	1010 0000	160/256	139/256	$160 = 128 + 32$
t_3	1001 0000	144/256	139/256	$144 = 128 + 16$
t_4	1000 1000	136/256	139/256	$136 = 128 + 8$
t_5	1000 1100	140/256	139/256	$140 = 128 + 8 + 4$
t_6	1000 1010	138/256	139/256	$138 = 128 + 8 + 2$
t_7	1000 1011	139/256	139/256	$139 = 128 + 8 + 2 + 1$
t_8	1000 1011	139/256	139/256	

contents of the digital register and the D-to-A output, y, are shown at each time in the sequence. In the comment column we show how the numerator of the y column is composed.

Figure 11.15 is a block diagram of the complete A-to-D converter. The squares represent latches, which can be set to 1 by a brief positive pulse at S and cleared to 0 by a brief positive pulse at C. The conversion sequence begins when a START pulse sets the RUN latch to 1, thereby allowing the oscillator to begin incrementing the counter. The decoder produces short pulses at each of its nine outputs in the order $T_0, T_1, \ldots T_8$. The pulse at T_0 sets $Q_7 = 1$ and all the other Q's to 0. The pulse at T_1 will clear Q_7 only if the comparator output is 1, indicating that $x < y$ and, at the same time, will set $Q_6 = 1$. Pulses at T_2 through T_8 act in a similar way, except that T_8, in addition to clearing Q_0 if the comparator is 1, resets the RUN latch.

11.4. Multiplying D-to-A converters

If the reference voltage in any of the D-to-A converters discussed earlier is made to vary in time, the result is an output that "multiplies" the reference voltage by a fraction that depends on the digital input. In its simplest form, Fig. 11.16, a set of digitally controlled switches, made, for example, with MOSFETs, allows different currents to flow to the inverting input of an op-amp. When only the switch corresponding to the most significant bit (MSB) of the digital number is closed, $V_{out} = -\frac{1}{2} V_{in}$; when only the next switch is closed, $V_{out} = -(1/4) V_{in}$; etc. The summing property of the op-amp permits the value of K to be chosen in 2^N ways where N is the number of bits in the digital number.

The multiplying D-to-A converter in Fig. 11.16 is known as a "two-quadrant" converter because the constant K cannot be negative. It is possible to achieve four-quadrant conversion in which K can be

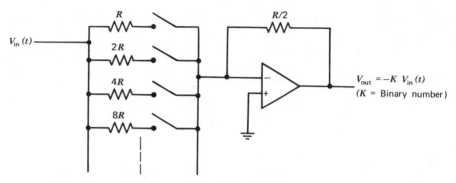

Fig. 11.   16. Two-quadrant multiplying D-to-A converter.

positive or negative by using another op-amp and single-pole double throw switches (each achieved, for example, with a pair of MOSFETs) as in Fig. 11.17.

If a given switch is in the "1" position, it supplies current to the upper op-amp, tending to produce a negative voltage at V' and an equal positive voltage at V_{out}. If a switch is in the "0" position it supplies current to the lower op-amp, tending to produce a negative voltage at V_{out}. The summing property gives a range of voltages at V_{out} from nearly $-V_{in}$ (when all switches are set at 0) to nearly $+V_{in}$ (when all switches are set at 1). Actually the lower bound for K is $-[1-(1/2^N)]$, where N is the number of bits in the binary number.

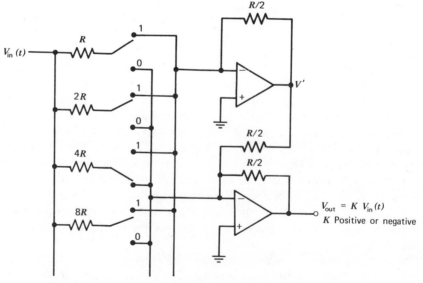

Fig. 11.   17. Four-quadrant multiplying D-to-A converter.

Applications of multiplying D-to-A converters include computer-controlled amplification and attenuation (as in psychoacoustic experiments where sounds of different intensities are presented to a subject), amplitude modulation (as in communication systems or phase-sensitive detectors), and the multiplication of one analog signal by another (in this case, an A-to-D converter is used to derive the digital input to the multiplying D-to-A converter).

Problems for chapter 11

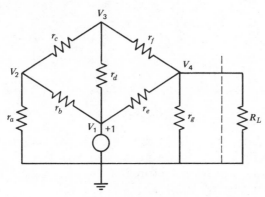

Fig. P11.   1*a*

11.1. Write the four nodal equations for nodes 1, 2, 3, and 4 in Fig. P11.1*a*. Rearrange these so that the three unknown voltages, V_2, V_3, and V_4 appear on the left sides of the equation, and the known voltage, $V_1 = 1$, appears on the right side. Solve for V_4 by determinants, and carry the evaluation of the determinants far enough to show that

$$V_4 = \frac{k \cdot V_1 - R_L}{R + R_L}$$

where k and R depend on the values of the internal resistance only. Note that V_4 is the same as in the Thévenin equivalent circuit (Fig. P11.1*b*):

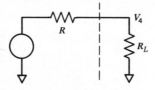

Fig. P11.   1*b*

11.2. Analyze the ladder network of Fig. 11.4 for the two cases: (a) $V_2 = 1$, $V_1 = V_4 = V_8 = 0$, and (b) $V_1 = 1$, $V_2 = V_4 = V_8 = 0$. Show that in case (a) the Thévenin equivalent voltage is 1/8 and the Thévenin equivalent resistance is 1, while in case (b) the corresponding values are 1/16 and 1.

11.3. Analyze the ladder network of Fig. 11.4 when $V_1 = V_2 = 0$, $V_4 = V_8 = 1$ and show that the Thévenin equivalent voltage source is $1/2 + 1/4 = 3/4$ and the Thévenin equivalent resistance is 1.

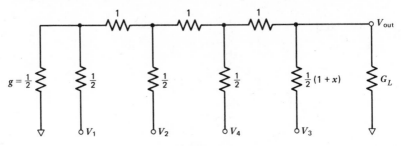

Fig. P11.   4

11.4. Make a nodal analysis of the "erroneous" ladder network in Fig. P11.4 to find V_{out} in terms of V_1, V_2, V_4, V_8 and x. By how much do the weighting factors for V_1, V_2, V_4, and V_8 deviate from the desired ratio of 1:2:4:8?

11.5. Design a ladder-type, D-to-A network (as in Fig. P11.5) whose inputs are unit *current* sources $a_0, a_1, a_2 \ldots$, and whose output voltage v_0 is given by the following expression: $v_0 = \alpha(a_0 + ka_1 + k^2 a_2 + \ldots k^N a_N)$; $k < 1$, $a_i = 0$ or 1. Use the fact that the equivalent resistance looking to the left from each node must be the same, to show that R_N and R must obey the relation $R = R_N^2/(1 + R_N)$. Then find the values for the ladder resistance, R, and the termination resistor, R_N for (a) $k = 1/2$, (b) $k = 1/10$, and (c) $k = 1/16$.

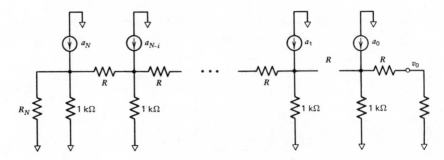

Fig. P11.   5

11.6. One of the ways negative numbers can be represented in binary notation is with a "two's complement" code. For positive numbers the most significant bit (MSB) is 0, for negative numbers it is 1. To change the sign of a number one simply complements it and adds 1. Thus, in an 8-bit number, decimal 3 is represented by 0000 0011 and −3 by 1111 1101. (Note that their sum is 0.)

(a) What are the most positive and most negative decimal numbers that can be represented by 8 bits in the two's complement code?

(b) A successive approximation A-to-D converter must be designed somewhat differently if it is to properly digitize both positive and negative analog signals. Find the algorithm that is comparable with that in Table 11.1. What changes must be made in the circuit of Fig. 11.15?

(c) If the analog signal is *negative* and 47/128 of the full scale negative value, construct a table, analogous to that in Table 11.2, showing how the A/D's digital output changes during the conversion process.

11.7. Analog-to-digital conversion can be accomplished by measuring the frequency of the output of a voltage to frequency (V-F) converter, such as is shown in Fig. P11.7. Beginning with the integrator capacitor discharged, the current generator, I_0, is connected to the − input of the op-amp for a fixed time τ, during which the output of the integrator rises linearly. After time τ, the current generator is disconnected and the op-amp output falls linearly. When it reaches 0 the comparator is triggered, the current generator is reconnected, and the cycle begins again.

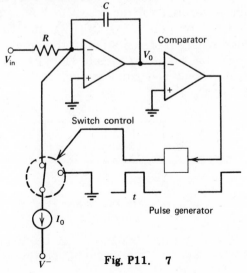

Fig. P11. 7

(a) Sketch and dimension a graph of V_0 versus t.
(b) Show that for input voltages that are not too large, the frequency of the sawtooth voltage is $V_{in}/\tau R I_0$.
(c) Show that the maximum allowed input voltage is $I_0 R$.
(d) Why is the scheme not a good one for very small input voltages?

11.8. Charge redistribution A-to-D conversion.[†] In this technique the switch S_1 is first connected to ground and all capacitors' lower terminals are connected to V_{in}. Then S_1 is opened and all lower terminals are connected to ground (see Fig. P11.8a).

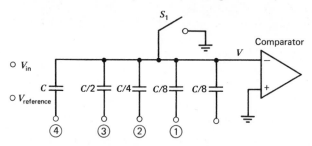

Fig. P11.   8a

(a) Show that $V = -V_{in}$.
(b) Now point ④ is connected to $V_{reference}$. Show that V_{in} rises by an amount $V_{ref}/2$. (See hint below.)
(c) If $V > 0$, point ④ is reconnected to ground; otherwise it remains connected to V_{ref}. Next point ③ is connected to $V_{reference}$. Show that V_{in} rises by an amount $V_{ref}/4$. Again, if $V > 0$ point ③ is returned to ground, and so on.
(d) Sketch V versus t for the case $V = .7 \ V_{ref}$ and show the final connections of points ④, ③, ②, and ①. Does the circuit exhibit a binary representation of the input signal?
    Hint. In the following circuit (Fig. P11.8b) show that when point A is switched from ground to V_{ref}, V rises by $V_{ref} \times [C_1/(C_1 + C_2 + C_3)]$.

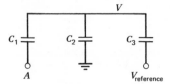

Fig. P11.   8b

[†]See J. L. McCreary, P. R. Gray, "All Mos Charge Redistribution A/D Conversion Techniques — Part I," *IEEE J. Sol. State Circuits* 371 (1975).

REFERENCES

Bleuler, E., R. O. Haxby Eds, *Methods of Experimental Physics, Vol. 2, Electronic Methods*, Part A, Chapter 8, Academic Press, New York, 1975.

Hoeschele, D. F., *Analog-to-Digital/Digital-to-Analog Conversion Techniques*, Wiley, New York, 1968.

Schmid, H., *Electronic Analog/Digital Conversions*, Van Nostrand Reinhold, New York, 1970.

Sheingold, D. H., Ed., *Analog-Digital Conversion Handbook*, Analog Devices, Norwood, Mass., 1972.

Hnatek, E. R., *User's Handbook of D/A and A/D Converters*, Wiley-Interscience, New York, 1976.

CHAPTER 12

Multiplication, Modulation, and Sampling

12.1. Introduction Until now the analog systems we have discussed have been for the most part linear, time-invariant systems. The systems to be discussed in this chapter are nonlinear by their very nature because they involve the multiplication of one time signal by another; nevertheless, the frequency domain techniques developed for LTI systems are still found to be useful. Multiplication, in the form of modulation of a signal, is an important function in nearly all communications systems, and sampling is becoming increasingly important as digital techniques become more widely available for instrumentation systems. In addition, multiplication is of direct importance in analog computation where it is not uncommon to require the instantaneous product of two time-varying signals.

12.2. Frequency-domain description of the multiplication process We are concerned here primarily with multiplication of a signal that may be nonperiodic by another signal that is periodic, such as a sinusoid or a pulse train. The first signal, $x(t)$, is assumed to have a frequency spectrum $X(f)$, as is illustrated in the general multiplier system diagram in Fig. 12.1. Being periodic, the multiplying signal, $s(t)$, has a frequency spectrum that is a set of impulses; for example, a pair of impulses at frequencies $\pm f_0$ for a sinusoid of frequency f_0. Earlier, we learned that multiplication in the time domain is represented by convolution in the frequency domain. Thus, if $s(t)$ is a sinusoid, the spectrum of the product wave, $x(t) \cdot s(t)$, consists of a set of replicas, or images, of the spectrum $X(f)$ shifted in frequency both up and down by an amount equal to the frequency of the multiplying sinusoid. Figure 12.1 refers specifically to the situation of a cosine-multiplying signal whose frequency components are real. Multiplication by a sine wave would produce rotations by $\pm 90°$ in the real-imaginary plane of the Fourier transform of $x(t)$ in addition to a frequency shift. The basic function of shifting a signal to other frequencies has several extremely important applications in electronics, some of which will be described in the following sections.

12.3. Frequency-domain multiplexing Multiplexing is the process that enables more than one signal at a time to be sent over a channel. The channel

272

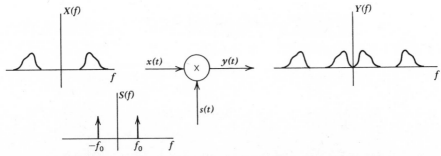

Fig. 12.   1. Frequency-domain description of multiplication.

may be a transmission line, a microwave link, a telephone circuit, a laser beam, or an electromagnetic pathway in empty space. In frequency-domain multiplexing, several signals, all occupying approximately the same frequency band, such as the audio band (20—20,000 Hz) or the video band (0—6 MHz), are each multiplied by a sinusoid of a different frequency, and the products added so that the resulting spectrum, $Z(f)$, consists of images of the original Fourier spectra shifted to different frequencies. Figure 12.2 illustrates the frequency-domain multiplexing of two input signals, $x_1(t)$ and $x_2(t)$. The images can be "stacked" quite close together in the frequency domain; clearly the wider the bandwidth of the communication channel (which includes the multiplying devices, the medium through which the information is sent, and the receiving devices that must separate out the various signals), the more different signals can be sent. This observation explains the continual research and development in high-frequency circuits and devices. For example, the frequency range from 1 to 10 GHz (1 GHz = 10^9 Hz) can contain 10 times as many television channels as the frequency range from 100 MHz to 1 GHz.

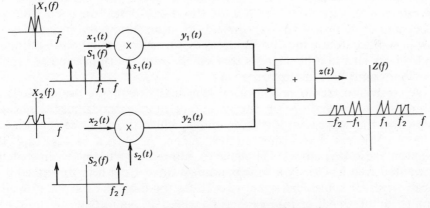

Fig. 12.   2. Frequency-domain multiplexing.

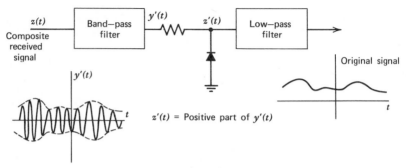

Fig. 12.   3. Diode-detector receiver.

Once several signals have been multiplexed and sent over a single communications channel, the job of a receiver is to separate them again. One way to do this, though not the most common way in public communications systems such as radio and television broadcasting, is simply to multiply the signal by a sinusoid that is in phase with the sinusoid (properly delayed to account for transmission time) originally used to shift the signal to higher frequency. The result of this multiplication is to shift the images of the original signal both up and down in frequency. The upshifted images of the desired signal and the upshifted and downshifted images of the other signals are blocked by a low-pass filter, while the downshifted image of the desired signal, which appears near zero frequency, is passed.

Another type of receiver is illustrated in Fig. 12.3. Here the composite received signal is put through a band-pass filter that blocks all the undesired images. The desired image is then shifted down to its original location in the frequency domain by a simple diode detector. One way to look at the operation of the diode detector is to note that it multiplies its input signal by a square wave that is 1 for, say, positive half cycles of the input and 0 for negative half cycles. Since a square wave is a superposition of sine waves, one of the results of the diode detection is to shift the spectrum $Y(f)$ down to its original location near zero frequency. A low-pass filter following the diode detector selects this "correct" image and rejects all the other images that result from multiplication of $y(t)$ by the "harmonics" of the square wave.

A receiver of radio or television broadcasts for home use employs an interesting combination of the two receiving methods just described, known as the superheterodyne technique, which proves to be cheaper and more troublefree than either method alone. As shown in Fig. 12.4, the received radio-frequency (r-f) signal is filtered to eliminate some of the unwanted frequencies by a wide band-pass filter. Next it is multiplied by a local oscillator sinusoidal signal that shifts the desired r-f signal down (or up), to an intermediate frequency (i.f.) where it is amplified and filtered by

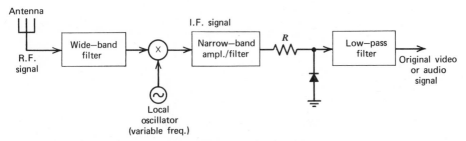

Fig. 12.   4. Superheterodyne radio receiver.

two or three stages of narrow-band i.f. amplifiers. At this point only the
desired information remains, but it is still at a high frequency. The i.f. output
is next shifted down to the original signal frequency, for example, audio or
video, by a diode detector and low-pass filter. A different signal in the
received r-f composite signal can be selected merely by changing the local
oscillator frequency, the i.f. frequency response remaining constant.

12.4. Chopper-type dc amplifier   For several reasons it has always
proved difficult to design a good amplifier that has a frequency response
extending to zero frequency. One problem is with the very slow drifts in
characteristics of the amplifying devices such as vacuum tubes or
transistors. Another problem is with drift caused by changes in temperature of certain critical components. These drifts are a form of low-
frequency noise, which is separate and distinct from thermal noise and
shot noise (see Chapter 13). In addition, most solid-state devices exhibit a
form of noise known as $1/f$ noise, in which the spectral density of the
noise produced by the amplifying element itself increases at low
frequencies according to a $1/f$ law.

An often used technique for amplifying very low-frequency or slowly
varying signals is to chop them at a frequency f_c; that is, multiply them by
a square wave, *before* amplification. Two of the images so produced lie
around $\pm f_c$; these images are selected by a band-pass filter and tuned
amplifier, and later shifted back down to zero frequency by another
square-wave multiplication followed by low-pass filtering. This chopper-
type amplifier avoids the problems associated with drift and $1/f$ noise by
performing the amplification at a frequency at which these noise sources
are negligible. A block diagram of a chopper-type dc amplifier is
illustrated in Fig. 12.5.

12.5. Frequency synthesis and calibration   The shifting property of a
multiplier can be used to produce sinusoids of many frequencies from just
a few oscillators. For example, if we already have oscillators at frequencies
f_1 and f_2, we can, by multiplication and appropriate band-pass filtering,

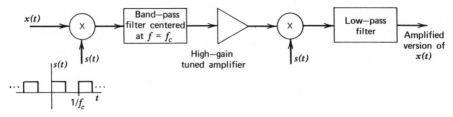

Fig. 12.   5. Chopper-type dc amplifier.

produce sinusoids at frequencies $2f_1$, $2f_2$, $f_1 + f_2$, and $f_1 - f_2$. In this way one avoids having to build an oscillator that can be adjusted in frequency over a wide range and, more importantly, it allows one to adjust accurately a small number of oscillators and be confident of the accuracy of all of the derived frequencies. Nowadays, with the advent of inexpensive digital gates and counters, it is easy to synthesize frequencies that are lower than that of a master oscillator. Together, these two techniques for generating new frequencies can yield oscillators of excellent precision.

As an example, suppose we have an oscillator that has been adjusted to run at 1.00 MHz, and we wish to obtain frequencies of 100 KHz to 1 MHz in 100-KHz steps. One way to do this is to use a decade counter and some decoding circuitry to produce square waves at frequencies of 100, 200, and 500 KHz. Band-pass filtering is then used to eliminate the third, fifth, etc. harmonics of each wave, and then a single multiplier can be used to generate the remaining 6 of the 10 desired frequencies. To get 300 KHz one multiplies the 200-KHz wave by the 100-KHz wave and filters out the difference frequency of 100 KHz. To get 400 KHz, one multiplies the 200-KHz signal by itself and filters out the dc difference frequency. Signals at 600, 700, 800, and 900 KHz are obtained by similar multiplications followed by band-pass filtering. In every case the accuracy of the frequency that is synthesized is determined by that of the master oscillator. For example, suppose the master oscillator is really running at 1.001 MHz instead of 1.000 MHz. Then the 100- and 200-KHz signals are really at 100.1 and 200.2 KHz, and the derived 300-KHz signal is at 300.3 KHz. Note that the fractional error in the frequency of all of the derived frequencies is the same (0.1%) as it is for the master oscillator.

Calibration of frequencies against a standard is accomplished by similar methods. For example, suppose one has available a 1.000-MHz frequency standard and wishes to adjust a local oscillator to exactly 700 KHz. One approach is to use the system of Fig. 12.6. The frequency standard is first divided down to 200 KHz and 500 KHz. Next the 500 KHz and the local oscillator signal are multiplied, and the difference frequency, which should be close to 200 KHz, is passed by a band-pass filter. Finally, this wave and the accurate 200-KHz signal are multiplied to give a difference

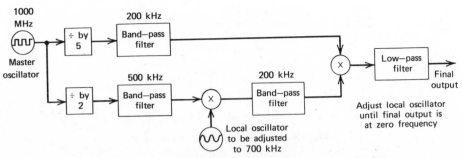

Fig. 12.   6. Adjustment of oscillator against a standard.

frequency near zero. A low-pass filter blocks the 400-KHz sum frequency, and the local-oscillator frequency is adjusted until the output of the low-pass filter is at zero frequency.

12.6. Multiplier circuits   Over the years many ingenious circuits have been designed to accomplish the multiplication of two time-varying signals. Some of these take advantage of a physical process that depends on the product of two quantities. One such multiplier utilizes the Hall effect, in which a current in a thin piece of semiconductor flows at right angles to a magnetic field. The Lorentz $\vec{v} \times \vec{B}$ force on the current carriers (holes or electrons) forces the carriers to one side of the sample and results in a small, transverse Hall-effect voltage across the sample (see Fig. 12.7). The magnitude of the Hall-effect voltage is exactly proportional to the product of the longitudinal current and the current that produces the magnetic field.

Another multiplier system produces a train of equally spaced rectangular pulses, at a repetition rate of, say, 10 KHz. One of the signals to be multiplied controls the amplitude of the pulses, the other controls their duration. The pulse train is sent through a smoothing filter whose output is proportional to the area of a pulse and therefore to the product of the signals that were to be multiplied. This type of multiplier has the

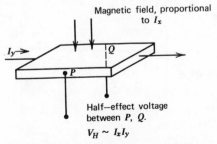

Fig. 12.   7. Hall-effect multiplier.

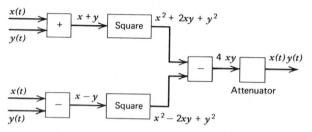

Fig. 12.   8. Quarter-square multiplier.

disadvantage that it produces a series of samples of the true product wave rather than a continuous and instantaneous product wave; however, as we will learn in the next section, the set of samples may be entirely adequate in certain situations.

In recent years three types of multipliers have emerged as the simplest and most reliable. Some of these modern multipliers can accurately multiply signals from zero frequency to many megahertz. One type, known as a quarter-square multiplier, operates by analogy with the algebraic identity:

$$xy = \tfrac{1}{4} \, | \, (x + y)^2 - (x - y)^2 | \qquad (12.1)$$

A quarter-square multiplier system is shown in Fig. 12.8. In this system, the adder and the subtractors can be made from operational amplifiers. The squaring devices are usually networks of biased diodes, connected to give a piecewise-linear approximation to the square of an input voltage. A diode squaring circuit for positive input voltages is shown in Fig. 12.9. If

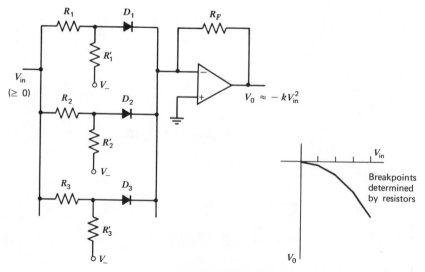

Fig. 12.   9. Diode squaring circuit.

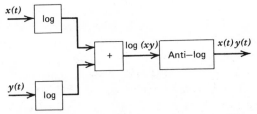

Fig. 12.   10. Logarithmic-amplifier multiplier.

$V_{in} = 0$ the left sides of all diodes are at a negative voltage, and no current flows to the inverting input of the op-amp. As V_{in} increases, a point is reached at which each diode in turn begins to conduct. For example, diode D_1 begins to conduct when the input voltage exceeds $R_1 V_- /R_1'$ if the diode is ideal. For a nonideal diode the input voltage has to be somewhat higher to initiate conduction because of the diode voltage drop, and also the turn-on is somewhat gradual, which happily rounds off the otherwise sharp corners in the curve of V_0 versus V_{in}.

A quarter-square multiplier is a fairly complex instrument, especially when one considers that a diode squaring circuit responds only to positive or negative input voltages but not both. Thus each squaring circuit in Fig. 12.8 requires two diode networks, one arranged to square positive input voltages, the other arranged to square negative input voltages.

Another type of multiplier utilizes logarithmic amplifiers. The fundamental idea here is to form log x and log y, add these signals in an op-amp, and take the antilog to arrive at xy (see Fig. 12.10). Figure 12.11 shows how the nonlinear characteristic of a transistor can be used as the feedback element in an op-amp circuit to provide a logarithmic amplifier characteristic. As is typical of op-amp circuits, the inverting input and therefore the collector voltage is at virtual ground. The minority carrier concentrations in the collector, base, and emitter regions of the transistor are shown in the Figure. Because the collector-base voltage is zero, the electron concentration at the collector edge of the base remains at the equilibrium concentration, n_0, regardless of the emitter-base voltage. The

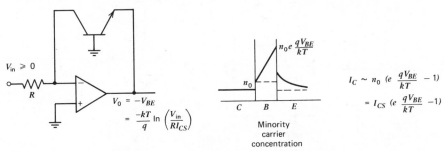

Fig. 12.   11. Op-amp logarithmic amplifier.

current of electrons injected from the emitter into the base is proportional to the slope of the electron concentration curve in the base. If the base current is negligible in comparison with the emitter current, the collector current is nearly equal to the emitter current and is proportional to $e^{q V_{BE}/kT} - 1$. For $V_{BE} > 100\,\text{mV}$, the 1 can be neglected, since $kT/q = 26\,\text{mV}$ at room temperature, so that I_C reduces to the form $I_{CS} e^{q V_{BE}/kT}$. We note that $I_C = V_{\text{in}}/R$ in Fig. 12.11 and $V_{BE} = -V_{\text{out}}$, which enables us to solve for V_{out}. The result is

$$V_{\text{out}} = -\frac{kT}{q} \ln\left(\frac{V_{\text{in}}}{R I_{CS}}\right) \tag{12.2}$$

The antilog device utilizes a similar circuit, except that the transistor is in the input circuit of the op-amp with its collector connected to the inverting input, its base grounded, and with the input voltage applied to its emitter. Note that the logarithmic circuit of Fig. 12.11 operates correctly only for positive input voltages. If multiplication of voltages that can be either positive or negative is desired then one must use a separate channel for negative inputs.

Except for the Hall-effect multiplier, none of the schemes we have discussed so far provides four-quadrant multiplication, that is, multiplication in which both quantities to be multiplied can be either positive or negative. The translinear multiplier (TLM) we are about to describe not

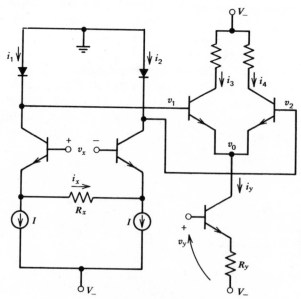

Fig. 12.   12. Simplified translinear multiplier.

only provides four-quadrant capability but it is easily frabricated as an integrated circuit. The operating principle of the translinear multiplier is best understood by considering the simplified version sketched in Fig. 12.12. This circuit consists of two somewhat different differential amplifiers. The differential amplifier on the left has diodes in its collector circuits and constant-current generators in its emitter circuits. The differential amplifier on the right has a variable-current generator, i_y, in its common emitter circuit. A differential input voltage, v_x, produces a proportional difference current: $i_1 - i_2 = i_x$. We next show that the function of this circuit is to create a difference current $i_3 - i_4$ in the right-hand differential amplifier that is proportional to the product of i_x and i_y.

In order to simplify the analysis we make three assumptions:

1. All transistors and diodes are sufficiently forward biased that their currents are given by $i = Ae^{qv/kT}$, where, in the case of the transistors, v is the base-emitter voltage. This expression is accurately true for transistors as long as the collector junction is reverse biased by more than 0.1 V (so that the minority-carrier concentration at the collector edge of the base is zero).

2. All transistors have sufficiently high current gain that we can neglect their base currents. This means that $i_E = i_C$.

3. Devices that are symmetrically located in the circuit are identical, and operate at the same temperature. This temperature condition can be quite accurately met when the devices are formed on a single integrated-circuit chip.

First, we prove that the circuit requires

$$\frac{i_1}{i_2} = \frac{i_4}{i_3} \tag{12.3}$$

This relation comes from the following voltage-current relations:

$$
\begin{aligned}
i_1 &= Ae^{-qv_1/kT} \\
i_2 &= Ae^{-qv_2/kT} \\
i_3 &= A'e^{q(v_1 - v_0)/kT} \\
i_4 &= A'e^{q(v_2 - v_0)/kT}
\end{aligned}
\tag{12.4}
$$

Direct substitution of Eqs. 12.4 into Eq. 12.3 leads to an identity.

Next, we write the KCL equations:

$$
\begin{aligned}
i_1 &= I + i_x \\
i_2 &= I - i_x \\
i_y &= i_3 + i_4
\end{aligned}
\tag{12.5}
$$

We substitute the first two of these into Eq. 12.3 and rearrange to obtain

$$(i_3 - i_4) \cdot I = -(i_3 + i_4) \cdot i_x \qquad (12.6)$$

Substitution from the third of Eqs. 12.5 leads immediately to the desired result:

$$i_3 - i_4 = -\frac{i_x i_y}{I} \qquad (12.7)$$

In order for this result to be useful, we must show that i_x and i_y are each proportional to an applied voltage. This will be the case if the resistors R_x and R_y are large enough. To get an idea of how large these resistors must be, we recall that a change in the base-emitter voltage v of a transitor results in a change of emitter current as follows:

$$i_E = A e^{qv/kT}$$

$$di_E = \frac{q}{kT} \cdot A e^{qv/kT} dv = \frac{q i_E dv}{kT} \qquad (12.8)$$

Thus, for *changes* in voltage and current, the transistor acts like a resistor of value $dv/di_E = kT/q i_E$. For emitter currents of 1 mA and more, at room temperature this resistance is 26 Ω or less (because kT/q = 26 mV at $T = 300°$K). Therefore, as long as R_x and R_y are much larger than 26 Ω, these incremental resistances can be neglected and we have the simple relations:

$$v_x = i_x R_x$$

$$v_y = i_y R_y \qquad (12.9)$$

In view of Eqs. 12.7 and 12.9, and the fact that the differential current $i_3 - i_4$ produces a differential voltage between the collectors of the right-hand pair of transistors in the circuit of Fig. 12.12, we have a circuit that, within certain limitations, develops an output voltage that is proportional to the product of two input voltages.

Note that if the current i_x approaches I in magnitude, so that either i_1 or i_2 approaches zero, the base-emitter voltage across one of the left-hand pair of transistors will change by a few tenths of a volt, and the first of Eqs. 12.9 will no longer hold true. Similarly, if i_y approaches zero, the second of Eqs. 12.9 will be violated. Because of the latter restriction, and the fact that i_y can never become negative, the multiplier of Fig. 12.12 is not a true "four-quadrant" multiplier.

A modification of the circuitry of Fig. 12.12 can result in the removal of the restriction that i_y remain positive. The circuit about to be discussed is essentially that of a commercial device, a type-MC1495 or type-MC1595 integrated circuit multiplier manufactured by Motorola, Inc. Either of these devices is an excellent and inexpensive multiplier; they differ only in

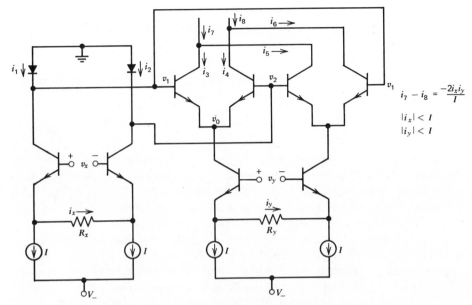

$$i_7 - i_8 = \frac{-2i_x i_y}{I}$$

$$|i_x| < I$$

$$|i_y| < I$$

Fig. 12.   13. MC1495 integrated-circuit multiplier.

the guaranteed performance tolerances. The essential circuitry is shown in Fig. 12.13. The important difference between this and the multiplier circuit of Fig. 12.12 is the addition of another differential amplifier and a pair of constant current sources.

We employ the same assumptions as in the earlier analysis. By an entirely analogous argument, Eq. 12.3 is extended to

$$\frac{i_1}{i_2} = \frac{i_4}{i_3} = \frac{i_5}{i_6} \qquad (12.10)$$

Thus, in addition to Eq. 12.6, we have the relation:

$$(i_6 - i_5) \cdot I = -(i_6 + i_5) \cdot i_x \qquad (12.11)$$

Also we have four new KCL equations:

$$i_7 = i_3 + i_5$$
$$i_8 = i_4 + i_6$$
$$i_3 + i_4 = I + i_y \qquad (12.12)$$
$$i_5 + i_6 = I - i_y$$

Our output quantity is the differential current, $i_7 - i_8$. By the first two of Eqs. 12.12 along with Eqs. 12.6 and 12.11, we see that

$$i_7 - i_8 = (i_3 - i_4) - (i_6 - i_5) = \frac{-(i_3 + i_4)i_x + (i_6 + i_5)i_x}{I}$$

$$(12.13)$$

Substitution from the third and fourth of Eqs. 12.12 leads immediately to the desired result:

$$i_7 - i_8 = \frac{-2i_x i_y}{I} \qquad (12.14)$$

The current i_y is proportional to the input voltage, v_y, by the same argument and with the same restrictions as in the case of the v_x input. A differential output voltage can be developed with the aid of external resistors and is proportional to the product of v_x and v_y. Note that the scale factor of this multiplier depends on the magnitude of the constant current source I as well as the values of these external resistors. In the MC1495 or 1595, this current is adjustable by means of an externally applied voltage. For the bias voltages and external circuitry required for the proper operation of this integrated circuit multiplier, refer to application notes AN-489 and AN-490 in the *Microelectronics Data Book*, second edition, December 1969, Motorola Semiconductor Products, Inc.†

12.7. Sampling and the sampling theorem   In many systems the need arises for measuring the instantaneous value of a signal. Even when the signal is varying in time in an unpredictable way, and is therefore carrying information, it is sometimes true that a set of samples of the signal taken at regularly spaced intervals can convey the complete information being carried. This fact, whose basis is investigated in this section, is important for a second type of multiplexing. In this multiplexing scheme, called time-domain multiplexing, samples of the various signals are interleaved in the transmission process, sent as a set of pulses, and sorted out by the receiver. As we will see it is a relatively simple matter for the receiver to generate the signal from which the samples were taken.

Let us first try to understand how a set of samples of a signal can carry the complete information available in a signal. In order for this to be true, the signal must have a finite bandwidth. Although this can never be strictly true, it can be an excellent approximation. In other words, while there is no filter that can completely attenuate signals over a continuous range of frequencies, the attenuation can be complete enough to restrict the signal, for all practical purposes, to a finite band of frequencies.

Figure 12.14 illustrates a sampling system. For simplicity, we take the spectrum of the signal to lie near zero frequency, between $f = \pm W$ to be

---

†Also available from Technical Information Center, Motorola Semiconductor Products, Inc., P.O. Box 20912, Phoenix, Ariz. 85036.

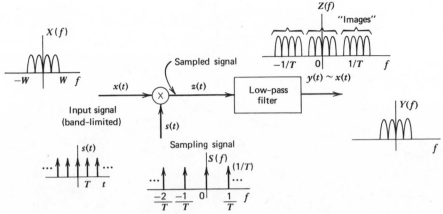

Fig. 12.   14. A sampling system, showing recovery of the original signal.

exact. In one of the problems you are asked to investigate the case when the signal is not adjacent to zero frequency. The sampling signal is a sequence of unit impulses spaced in time by T. The sampling operation is merely a multiplication of the input signal $x(t)$ by the sampling pulse train $s(t)$. In the frequency domain the multiplication is replaced by convolution; that is, the transform (or spectrum) of the signal is convolved with the transform of the sampling train to give the transform of the product wave. But we know that the transform of any periodic signal is a set of impulses at frequencies $\pm n/T$ and, in the particular case of repetitive unit impulses, the impulses in frequency are all equal in magnitude. The spectrum of the sampling train is shown as $S(f)$ in Fig. 12.14.

Convolution of the signal transform $X(f)$ with the sampling-wave transform $S(f)$ results in a set of images of the signal transform shifted to higher and lower frequencies. The *sampling theorem* states that *if the sampling interval T is less than or equal to 1/2W where W is the bandwidth of the signal, then it is possible to recover the complete signal from the set of samples.* That this is true can be seen by considering the spectrum of the sampled signal when this condition is met. The images are then found to be nonoverlapping. In fact, in order to recover the original signal from $z(t)$, one needs only to pass the sampled signal through a low-pass filter and multiply by a constant to make up for attenuation by the factor $1/T$ in the multiplier.

In real sampling systems, we must be content to multiply the signal by a short, but finite pulse rather than by an impulse. Next we investigate several methods for sampling a signal with a short pulse.

12.8. Sampling circuits   Another name for a sampling circuit is a linear gate. In this type of circuit the function is to pass a signal proportional to

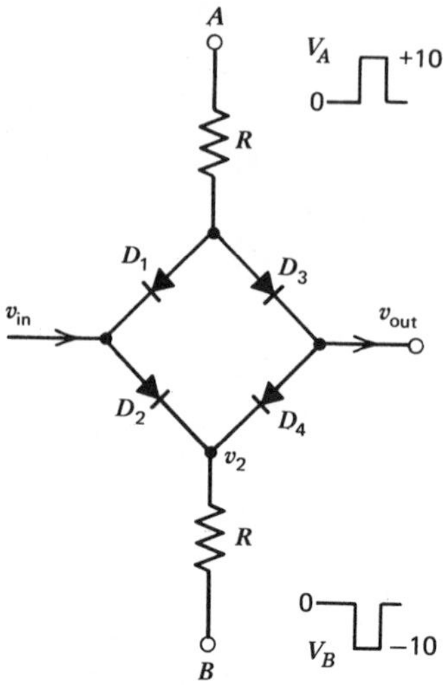

Fig. 12.   15. Diode-bridge linear gate.

the input only when a gate pulse is present. If the gate pulse is absent, the output is zero. One way to do this is with a multiplier; one input is the signal and the other is the gate pulse. A simpler way of making a linear gate is with the diode bridge gate circuit of Fig. 12.15. When the gating signals at points A and B are both zero, an input signal of either polarity is blocked by the diodes. For example, suppose $v_{in} > 0$. Then D_1 is reverse biased, and no current flows through the upper path from input to output. Diode D_2 conducts and raises the junction point between diodes D_2 and D_4 above ground potential, but because D_4 is now reverse biased no current flows to the output.

Next consider what happens when A is positively biased and B is negatively biased. To be specific, let $V_A = 10$ V and $V_B = -10$ V, and let $v_{in} = 3$ V. Since $v_{in} = 3$ V, v_2 will be less than this by about 0.7 V. Now since diodes D_3 and D_4 must be forward biased, v_{out} exceeds v_2 by the diode voltage drop of 0.7 V. The result is, that to the accuracy that the diode drops in D_2 and D_4 are equal, $v_{out} = v_{in}$. By symmetry, the same result holds for negative input voltages, except that negative signals are transmitted by diodes D_1 and D_3.

Another type of gate that can be used for signals of one polarity is the shunt transistor gate of Fig. 12.16. In this very simple gate, the bipolar transistor acts as a switch that is either open or closed. When the gate input voltage is zero or negative, the transistor does not conduct, and the input signal appears at the output, less any voltage drop in resistor R. When a sufficient positive voltage is applied to the gate input, the transistor saturates, and the output voltage is held within less than a tenth of a volt of zero.

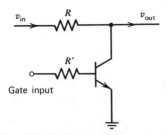

Fig. 12.   16. Shunt-transistor linear gate.

Linear gates can be fabricated as integrated circuits. An example is the novel circuitry of the Motorola MC1445/1545 gated radio-frequency amplifier, shown in Fig. 12.17. This circuit is a parallel combination of two differential amplifiers: Q_1, Q_2 and Q_3, Q_4. If the gate input voltage is greater than about 2 V or if the gate input is open circuited, Q_5 conducts the entire constant current I. In this case Q_1 and Q_2 form the differential amplifier and transistor Q_6, and therefore Q_3 and Q_4 are nonconducting. The amplified differential signal across the 1-kΩ resistors is fed through emitter followers Q_7 and Q_8, and Q_9 and Q_{10} to the output terminals. The purpose of the emitter followers is to shift the quiescent dc levels back down to near 0 V. In order to switch to the Q_3, Q_4 amplifier, the gate voltage is held at zero or a negative value.

Not only does this exceedingly useful circuit allow one to switch between two differential inputs, it also provides a voltage gain of 10 over a frequency range from dc to about 75 MHz. The switching time from one channel to the other can be as fast as a few tens of nanoseconds! The operation of this circuit is summarized by the equations:

$$
\left.\begin{aligned}
v_{0+} &= 10 \cdot (v_1 - v_2) \\
v_{0-} &= -10 \cdot (v_1 - v_2)
\end{aligned}\right\} \quad v_{\text{gate}} \geqslant 2 \text{ V.}
$$
$$
\left.\begin{aligned}
v_{0+} &= 10 \cdot (v_3 - v_4) \\
v_{0-} &= -10 \cdot (v_3 - v_4)
\end{aligned}\right\} \quad v_{\text{gate}} \leqslant 0 \text{ V.}
\tag{12.15}
$$

12.9. Sample-and-hold circuits   Unless the samples of a signal are to be interleaved with samples of other signals in a time-domain multiplexing system, the usual requirement is to make a measurement on the sampled signal, for example, to do an A/D conversion. During any such operation the sample must be held at constant voltage. Circuits that accomplish this task are called sample-and-hold circuits. An elementary form of a sample-and-hold circuit is shown in Fig. 12.18a. When the switch (e.g., the diode switch of Fig. 12.15) is closed, the capacitor charges, with a time constant RC, toward the voltage V. After the switch is opened, the capacitor holds the charge for a time determined by the leakage time

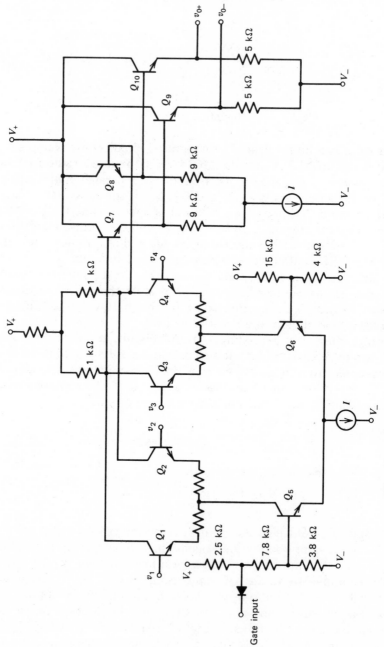

Fig. 12. 17. MC1445 integrated-circuit gated r-f amplifier.

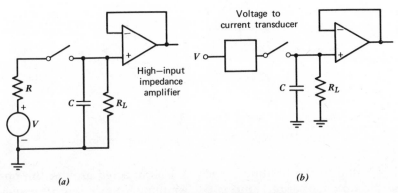

Fig. 12.   18. Sample-hold circuits.

constant, $R_L C$, where R_L is the total leakage resistance, part of which may be due to a high-impedance amplifier used to measure the capacitor voltage. The ratio of useful hold time to sample time is proportional to R_L/R. Typical values for R_L and R might be 10 MΩ and 10 Ω, so that $R_L/R = 10^6$. However, the hold-to-sample time ratio may be about two orders of magnitude less than this, depending on the error one can tolerate because of both incomplete charging of C during the sampling interval and exponential decay of the capacitor voltage during the hold period.

Another approach (Fig. 12.18b) is to charge the capacitor with a current proportional to the voltage to be sampled during the sampling interval. It is not at all necessary for the capacitor to charge to the input voltage. Although this technique has the advantage of generally shorter sampling times, it has the disadvantage that the capacitor must be completely discharged before each sample is taken, to preserve the proportionality between the input voltage and the voltage to which the capacitor charges.

Operational amplifiers can reduce some of the inaccuracies of the simple capacitor sample-and-hold. In the circuit of Fig. 12.19 a capacitor,

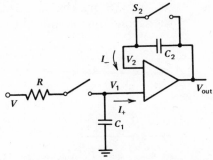

Fig. 12.   19. Op-amp sample-hold circuit.

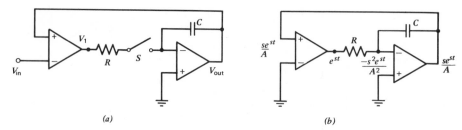

Fig. 12.   20.   Two-op-amp sample-hold circuit.

equal in value to the holding capacitor, is connected around the op-amp during the holding period. During the sampling interval, both switches are closed, and the op-amp acts as a voltage follower so that the output voltage V_{out} is very closely equal to the voltage V_1 across C_1. When the switches open, both capacitors must supply bias current to the op-amp. As C_1 discharges, the voltage V_1 falls toward zero and, of course, V_2 falls with it because of the virtual short approximation. But as C_1 is discharging, C_2 is charging, and if $I^+/C_1 = I^-/C_2$, the rate of discharge of C_1 is equal to the rate of charge of C_2, and V_{out} does not change.

Another type of op-amp sample-and-hold circuit is illustrated in Fig. 12.20. Notice that the output of the second op-amp is connected around to the + input of the first op-amp. When switch S is closed, the polarities are such that V_{out} is driven toward V_{in}. For example, if $V_{out} > V_{in}$, the output of the first op-amp is positive and current flows through R and onto the left plate of C; since this plate is at virtual ground, the output of the second op-amp falls. Now when the switch opens, the output voltage remains constant except for the discharge of C that is necessary to supply the bias current to the second op-amp.

The circuit of Fig. 12.20a introduces complications because instability can occur if the op-amps are the internally compensated type (e.g., 741) so that their gain at high frequencies is given by A/s. Suppose, as in Fig. 12.20b, that the output of op-amp number 1 is e^{st} and that its —input is grounded. Then $V_{out} = se^{st}/A$ and the voltage at the —input of the second op-amp is $-(s^2 e^{st}/A^2)$. Since the sum of the currents through R and C into the —input of the second op-amp must be zero, we have

$$\left(1 + \frac{s^2}{A^2}\right) \cdot \frac{1}{R} + \left(\frac{s}{A} + \frac{s^2}{A^2}\right) \cdot sC = 0 \qquad (12.16)$$

which can be rewritten as:

$$1 = \frac{-1}{s^3 \cdot \dfrac{RC}{A^2} + s^2 \left(\dfrac{RC}{A} + \dfrac{1}{A^2}\right)} \qquad (12.17)$$

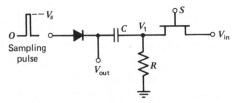

Fig. 12.   21. Fast sample-hold circuit.

It is not surprising that this system can be unstable, that is, have a solution for an s with positive real part, because even for small imaginary values of s, the phase shift introduced by the denominator is more than $180°$.

The remedy for this instability is to connect a small capacitor C_1 across R. It is left for one of the problems to show that the equation that corresponds to Eq. 12.17 has a term sRC_1 in the denominator of the right-hand side which, if C_1 is large enough, can force the magnitude of the right-hand side to become less than 1 before the phase shift reaches $180°$.

Sometimes it is necessary to obtain a sample over a very brief interval of time (~ 1 nsec) of very rapidly varying signals. The circuits we have considered up until now are inadequate for this purpose because they are basically low-pass filters while they are sampling, and therefore their outputs cannot follow very rapid changes in the inputs.

A basic circuit used to obtain samples of very rapidly varying waveforms is shown in Fig. 12.21. MOSFET switch S is closed to begin with, so that the voltage V_1 follows V_{in} very closely. In order to sample, a brief positive pulse of maximum voltage V_s is applied to the left side of the diode. Assuming an ideal diode, the capacitor charges to a voltage $V_s - V_1$ and, at some time after the sampling pulse, switch S is opened so that the right side of C is at 0 V and the left side is at $V_0 = V_s - V_1$. If V_s is known, for example, by measuring V_0 with $V_1 = 0$, a measurement of

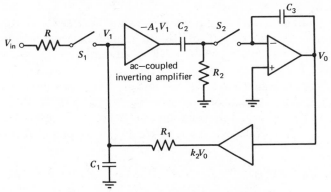

Fig. 12.   22. Another fast sample-hold circuit.

V_0 allows V_1 — the input voltage during the sampling interval — to be determined.

Another fast sample-and-hold circuit is shown in Fig. 12.22. This circuit consists of an inverting ac amplifier, A_1, an op-amp, A_2, a pair of switches, S_1, S_2, and various passive components.

To analyze the operation of this sample-hold circuit we first suppose that switch 1 is open so that the input voltage is disconnected. We also suppose that switch 2 has been open long enough that capacitor C_2 is completely discharged, and that there are nearly constant voltages V_0 and V_1 at the output and on capacitor C_1. Now switch 2 is closed, completing the feedback loop. After a few time constants $R_1 C_1$, the voltages V_1 and V_0 will have changed by amounts ΔV_1 and ΔV_0 to new steady values. The change ΔV_1 is amplified by the ac amplifier and inverted by the op-amp circuit, so that

$$\Delta V_0 = k_1 \, \Delta V_1 \tag{12.18}$$

where $k_1 = A_1 \cdot (C_2/C_3)$.

Furthermore, after C_1 charges, we must have

$$V_1 + \Delta V_1 = k_2 (V_0 + \Delta V_0) \tag{12.19}$$

If we substitute for ΔV_0 from Eq. 12.18 into Eq. 12.19 and rearrange, we find

$$V_1 - k_2 V_0 = \Delta V_1 \cdot (k_1 k_2 - 1). \tag{12.20}$$

Now suppose $k_1 k_2 > 1$, as it would be in a practical sample-hold circuit. Equation 12.20 tells us that if V_1 were equal to $k_2 V_0$ at the time S_2 was closed, then $\Delta V_1 = 0$, so that from Eq. 12.18, $\Delta V_0 = 0$, and the output voltage remains constant. On the other hand, if V_1 were a little larger than $k_2 V_0$ at the time of closure of S_2, then $\Delta V_1 > 0$; and if V_1 were a little smaller than $k_2 V_0$ at that time, then $\Delta V_1 < 0$.

Sampling is accomplished by closing switch S_1 for a brief but definite time while S_2 is open. Prior to the closing of S_1, $V_1 = k_2 V_0$; but while S_1 is closed, V_1 changes part way toward the value V_{in} as C_1 charges slightly. Let us say V_1 changes by 1/5 of the difference between V_{in} and V_1, so that $V_1 = k_2 V_0 + (1/5)(V_{in} - V_1)$. (For example, if $V_{in} = 3$ V and V_1 were initially 1 V, the value of V_1 has become $1 + (1/5)(3 - 1) = 1.4$ V at the time S_1 opens again.) Now if we choose $k_1 k_2$ properly we can make the additional ΔV_1, which occurs after S_2 is closed, such that the overall change in ΔV_1 is precisely enough to make V_1 equal to V_{in}. The condition for this to occur is $\Delta V_1 = (4/5)(V_{in} - V_1)$ and from Eq. 12.20 we have

$$k_2 V_0 + \tfrac{1}{5}(V_{in} - V_1) - k_2 V_0 = \tfrac{4}{5}(V_{in} - V_1)(k_1 k_2 - 1) \tag{12.21}$$

which gives $k_1 k_2 = 5/4$.

What the circuit has done is to: (1) sense the fact that V_1 was not equal to V_{in} during a brief sampling interval when S_1 was closed, and (2) subsequently increment V_1 by the right amount to make $V_1 = V_{in}$ during the time S_2 is closed. After V_0 has reached its desired value but before the output V_2 starts decaying toward zero, S_2 is opened again, and V_0 remains constant except for the bias current supplied by C_3.

Problems for Chapter 12

12.1. A musical tone consisting of 2 octaves ($\cos 2\pi f_0 t + \cos 4\pi f_0 t$) is multiplied by a carrier wave $\cos 2\pi f_c t$ to produce an amplitude-modulated wave. This wave is now filtered so that components at and below f_c are eliminated.

   (a) Sketch and dimension the spectrum of the resulting single-sideband (SSB) wave.

   (b) Sketch and dimension the spectrum of the product of the SSB wave of part (a) with $\cos 2\pi f_c t$ and thereby show how the original musical tone may be reproduced.

   (c) Sketch and dimension the spectrum of the product of the SSB wave with a 90° phase-shifted carrier, $\sin 2\pi f_c t$.

   (d) Sketch and dimension the spectrum of the product of the SSB wave with a frequency-shifted carrier $\cos 2\pi(f_c + \Delta f)t$. Are the original tones an octave apart?

   (e) Why is it that SSB modulation may be satisfactory for voice signals but not music signals?

12.2. A signal with the spectrum shown (Fig. P12.2) is sampled by multiplication with a set of impulses spaced by $t = T$.

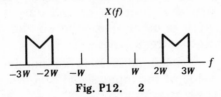

Fig. P12.   2

   (a) Find T, in terms of W, such that the spectrum of the sampled signal contains accurate images of the portion of the original spectrum between $2W$ and $3W$.

   (b) Design a system that reconstructs $x(t)$ from the samples.

12.3. An AM superheterodyne receiver may be modeled by the system in Fig. P12.3 $l(t)$ is a local sinusoidal oscillator signal that is adjusted to be 455 kHz higher than the carrier signal in $x(t)$ that is to be demodulated.

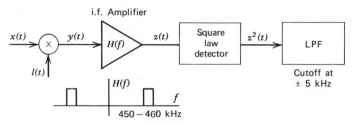

Fig. P12.   3

(a) Sketch the frequency components in the signals $y(t)$, $z(t)$, and $z^2(t)$ if $x(t)$ is a 1-MHz unmodulated carrier.

(b) Now a 1-kHz signal modulates the carrier. Again sketch the frequency components in the signals $y(t)$, $z(t)$, and $z^2(t)$.

(c) In an attempt to detect a station transmitting at 650 kHz, the local oscillator is tuned to 650 + 455 = 1105 kHz. You hear a station transmitting at 1560 kHz. Explain why by showing the spectra at different points in the system.

12.4.   A low-frequency wave is multiplied by a square wave of frequency f_s that switches between 1 and 0 in a chopper-type dc amplifier and then filtered so that its spectrum is centered on f_s. Then it is shifted back down near dc by multiplication with a square wave that is delayed in time by $1/4\, f_s$ (a quarter period), and filtered to eliminate images other than the one at dc. Show that the resultant output signal is zero, whether the original time function was an even function of time, an odd function of time, or a combination of the two.

12.5.   Show that if an arbitrary wave $x(t)$ is multiplied by a periodic wave, which can be represented by $s(t)*h(t)$ [where $s(t)$ is a periodic sequence of unit impulses with period T and $h(t)$ is arbitrary], then the spectrum of the output wave is given by $[S(f)\cdot H(f)]*X(f)$. In other words the Fourier transform of $x(t)$ is still shifted to other frequencies.

12.6.   Show that a sample-hold system (a nonlinear system), which assumes the value of the input signal $x(t)$ every T seconds, can be modeled by the system in Fig. P12.6.

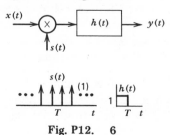

Fig. P12.   6

By considering the frequency-domain description of the system show that $y(t) \approx x(t)$ if the width of $X(f)$ is much less than $1/T$. Sketch $y(t)$ for a signal such that this condition is true.

12.7. The system in Fig. P12.7 represents the input circuitry of a *sampling oscilloscope*, which is used to display extremely rapid *periodic* signals; that is, ones whose frequency components are higher than can be amplified by the oscilloscope amplifier. The idea is to sample the fast periodic input signal once each cycle, but at successively later times in each cycle. The additional delay of the sampling wave is Δ, so that if the period is T, the sampling interval is $T + \Delta$. The input signal $x(t)$ is assumed to have no frequency components higher than W_x.

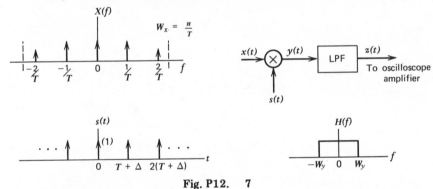

Fig. P12.   7

(a) Sketch the signals $x(t)$ and $y(t)$ in the time domain.
(b) Find Δ in terms of W_x in order that an accurate replica of $x(t)$ can be reconstructed. *Hint.* Use the sampling theorem.
(c) Show that the signal $y(t)$ has a Fourier transform that consists of groups of impulses spaced by $\Delta/T(T + \Delta)$ and that each group of impulses contains all the information in the original wave. To be specific you might take an example in which $X(f)$ contains no frequency components higher than $3/T$.
(d) If the LPF is used to pass only the impulses in $Y(f)$ that lie in the group near zero frequency, what is the lowest possible cutoff frequency W_y of the LPF? By what factor is this lower than the highest frequency present in the original signal $x(t)$?
(e) Suppose that the samples are taken, not $T + \Delta$, but $10T + \Delta$ apart so that only every tenth wave is sampled. Repeat parts (c) and (d).

Note that the function of the sampling system is to form a slowly varying version $x(at)$ of the original signal, where $a < 1$.

12.8. The FM-stereo broadcasting system works as follows. The signal $l(t)$ from the "left microphone" is added to and subtracted from the signal $r(t)$ from the "right microphone" to form sum and difference signals $s(t)$, $d(t)$. That is,

$$s(t) = l(t) + r(t)$$
$$d(t) = l(t) - r(t)$$

Both $l(t)$ and $r(t)$ and therefore $s(t)$ and $d(t)$ may be assumed to be band-limited to the audio range, that is, to about ±15 kHz. Now $D(f)$ is shifted so that the center of its spectrum lies at ±38 kHz. In addition, a sinusoidal carrier at $f_0 = 19$ kHz and an unshifted $S(f)$ are used to frequency modulate a high-frequency carrier (88 MHz $< f_c <$ 108 MHz). After detecting the signal in the FM receiver, its total spectrum $X(f)$ is as shown.

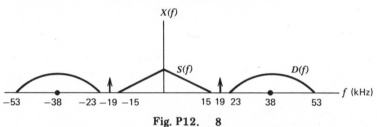

Fig. P12. 8

(a) Nonstereo FM receivers can be modeled, from this point on in the system, as low-pass filters with cutoff frequency about ±15 kHz. What is heard on a nonstereo FM radio when it receives the composite signal $X(f)$ sketched in Fig. P12.8?

(b) A stereo receiver contains circuitry to isolate $S(f)$ and to shift $D(f)$ back to zero frequency. Then $s(t)$ and $d(t)$ are added and subtracted to reproduce $l(t)$ and $r(t)$. Sketch a system that uses the 19-kHz signal to shift $D(f)$ back to dc.

(c) Why must the 19-kHz signal be broadcast and not simply be generated by an oscillator in the receiver?

(d) In actual practice $l(t)$ is obtained by squaring $\cos 2\pi f_0 t$ (where $f_0 = 19$ kHz) multiplying the result by $x(t)$, and passing the result through a LPF. Also $r(t)$ is obtained by phase shifting $\cos 2\pi f_0 t$ to obtain $\sin 2\pi f_0 t$, squaring, multiplying the result by $x(t)$ and passing the result through a LPF. Explain why this system works by sketching appropriate frequency-domain graphs.

12.9. In Fig. 12.20b, take $R = 10^3$ Ω, $A = 10^6$, $C = 10^{-7}$ F and assume a capacitor $C_1 = 10^{-9}$ F placed across R.

(a) Show that the condition for unity open loop gain can be written

$$1 = \frac{-1}{s(10^{-16}s^2 + 10^{-10}s + 10^{-6})}$$

(b) Make a Nyquist plot of the right-hand side to show that the system is marginally stable. That is, the right-hand side becomes 1 for $s = 10^5 j$.

(c) Find the effect of making $C_1 = 2 \times 10^{-9}$ F, so that the condition for unity open loop gain becomes

$$1 = \frac{-1}{s(10^{-16}s^2 + 10^{-10}s + 2 \cdot 10^{-6})}$$

Is the circuit stable or unstable? Find the angular frequency such that the loop gain is real. What is the loop gain at this frequency?

12.10. The output signal of a system is given by $y = A_1 x + A_2 x^2$ where x is the input time function, $x = \cos 2\pi f_0 t$.

(a) Find all the frequencies present in the output and sketch the Fourier transforms of $x(t)$ and $y(t)$ if $A_1 = 10, A_2 = 1$.

(b) What is the ratio of the average *power* in the second harmonic to that in the first harmonic in the output signal? (Power is proportional to the square of the amplitude for sinusoidal waves.)

(c) Now another term is added to the input, of the form $\cos 6\pi f_0 t$. Again sketch the Fourier transform of the output. Note the presence of "cross terms" that give rise to "intermodulation" frequencies.

12.11. A radar signal consists of a sinusoidal signal of frequency f_0 turned on and off periodically; the duration of the on time is τ and of the off time is $T - \tau$. Take $\tau = 10/f_0$, $T = 100/f_0$ and sketch and dimension the magnitude of the spectrum of this periodic signal.

12.12. A train of unit impulses spaced 1 msec apart is passed through an ideal LPF with a cutoff frequency of 10.5 kHz, and the resulting signal is squared as shown in Fig. P12.12. Show that the spectrum of $z(t)$ has an envelope that is triangular in shape, decreasing linearly from 1 to 20 kHz.

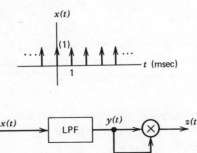

Fig. P12. 12

12.13. An amplitude modulated signal consists of a carrier wave 100 $\cos 2\pi \times 10^6 t$ plus the signal $(20 \cos 2\pi \times 10^3 t + 50 \cos 4\pi \times 10^3 t) \cdot \cos 2\pi \times 10^6 t$.

(a) Sketch and dimension the spectrum of this signal.
(b) Find the ratio of the power at the modulation frequency, 10^6 Hz, to the total power at the modulation frequencies (all but 10^6 Hz).

12.14. One form of a "ring modulator," used to produce amplitude modulation, is shown in Fig. P12.14. The carrier voltage sources $v_c(t)$ are sinusoidal at a high frequency. Assume $R_1 \ll R_2$ and that the diodes are ideal.

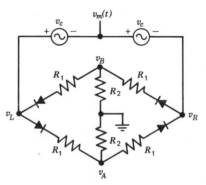

Fig. P12.   14

(a) Show that, when $v_L > v_R$, $v_A = (v_L + v_R)/2$ and $v_B = 0$, and that when $v_L < v_R$, $v_A = 0$ and $v_B = (v_L + v_R)/2$.
(b) If $v_m(t)$ is a slowly varying triangular wave with zero average value, sketch the waves at v_A and v_B.
(c) Show that the difference wave $v_A - v_B$ is a square wave with an amplitude that switches between $\pm v_m(t)$ at the carrier frequency.

12.15. Another version of a "ring modulator" (see previous problem) is shown in Fig. P12.15. This time points A and B are connected to virtual grounds at the —inputs of op-amps. Again the diodes are ideal.

(a) Sketch the current waveforms $i_A(t)$ and $i_B(t)$ if $v_m = 0$.
(b) Sketch the current waveforms if $v_m = 1$ and v_c is a sinusoidal wave of amplitude 3 V.
(c) Sketch the difference wave $i_A - i_B$ for the situation in (b).

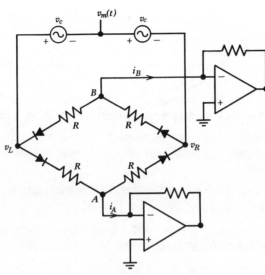

Fig. P12.   15

REFERENCES

Bleuler, E., R. O. Haxby, Eds., *Methods of Experimental Physics, Vol. 2, Electronic Methods*, Chapter 8, Academic Press, New York, 1975.

Carlson, A. B., *Communication Systems: An Introduction to Signals and Noise in Electrical Communication*, Chapters 5 and 7, McGraw-Hill, New York, 1968.

Gilbert, B., "A New Technique for Analog Multiplication," *IEEE J. Solid State Circuits, SC10*, 437 (1975).

Gilbert, B., "Precise, 4-quadrant multiplier with subnanosecond response," *IEEE J. Solid State Circuits, SC3*, 365 (1968).

Hnatek, E. R., *Applications of Linear Integrated Circuits*, Chapter 4, Wiley-Interscience, New York, 1975.

Sheingold, D. H., Ed., *Nonlinear Circuits Handbook*, Analog Devices, Norwood, Mass., 1974.

Schwartz, M., *Information Transmission, Modulation and Noise*, Chapters 3 and 4, McGraw-Hill, New York, 1959.

CHAPTER 13

Noise

13.1. Introduction In Chapters 1 and 3 you learned that random signals, or noise, may be characterized by their autocorrelation function or by their spectral density, either of which permit a calculation of the mean and variance of the random signal as well as providing a quantitative description of the rate at which the random signal takes on new values. You also learned how a random signal propagates through an LTI system.

Now we must investigate the sources of noise. Our plan is to study the *thermal noise* contributed by a resistor, the *shot noise* produced by the random emission of electrons, and the way in which the noise generated in two-port networks can be characterized. With this background it is possible to determine how the various parts of a system contribute to the overall noise at the output of a system and begin to understand the limitations in the ability of a system to detect and measure signals.

13.2. Spectral density — a review The spectral density $W_x(f)$ provides the best way of characterizing a random signal $x(t)$ because it is easy to compute the spectral density at the output of a system if we know the system impulse response (or its Fourier transform) and the spectral density of the input signal. The spectral density, a positive, real function of frequency, is defined (Eq. 3.13) as

$$W_x(f) = \lim_{\tau \to \infty} \frac{1}{\tau} |X_\tau(f)|^2 \qquad (13.1)$$

where $X_\tau(f)$ is the Fourier transform of a truncated version of the signal $x(t)$, lasting for time τ. (We use τ here rather than T to avoid confusion with the absolute temperature.) That is,

$$X_\tau(f) = \int_{-\tau/2}^{\tau/2} x(t)e^{-j2\pi ft}\, dt \qquad (13.2)$$

The average value of the square of a random signal may be computed by integrating the spectral density over all frequencies. That is (Eq. 3.14),

$$\overline{[x(t)]^2} = \int_{-\infty}^{\infty} W_x(f)\, df \qquad (13.3)$$

If we wish to compute the average of the square of the output of an LTI system whose input is a random signal with known spectral density, we must first find the spectral density $W_y(f)$ of the output from Eq. 3.19:

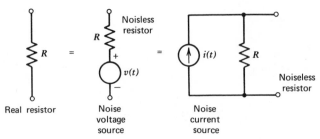

Fig. 13.   1. Noise models of a real resistor.

$$W_y(f) = W_x(f) \cdot |\, H(f)\,|^2 \tag{13.4}$$

where $H(f)$ is the Fourier transform of the system impulse response, and then integrate $W_y(f)$ over all frequencies to obtain $|\overline{y^2(t)}|$.

13.3. Thermal noise   In 1928, J. B. Johnson and H. Nyquist published articles in the journal, *Physical Review*,†‡ describing experimental and theoretical investigations of the fluctuations in voltage across a resistor caused by the random behavior of the conduction electrons in the resistor. It had become clear that a real resistor at temperature T could be represented by either of the equivalent circuits in Fig. 13.1, where the voltage generator $v(t)$ and the current generator $i(t)$ are random noise sources with zero mean, but with a nonzero mean square value that depends on both the value of the resistance and its absolute temperature. First, we use one of Nyquist's arguments to discover how the mean square value of $v(t)$ or $i(t)$ depends on the resistance.

The circuit of Fig. 13.2 consists of the parallel combination of resistors R_1 and R_2. If the resistors are at the same temperature, the second law of thermodynamics suggests that there can be no net transfer of energy on the average from one to the other. On the other hand, if R_1, for example, is hotter than R_2, then there is a net electrical transfer of energy from R_1 to R_2 that tends to bring the resistors to the same temperature. The power flow P_{12} from R_1 to R_2 is equal to the $i^2 R_2$ power developed in R_2 by the current $v_1(t)/(R_1 + R_2)$, produced by the noise voltage source $v_1(t)$. That is,

$$P_{12} = \left[\frac{v_1(t)}{R_1 + R_2}\right]^2 \cdot R_2 \tag{13.5}$$

† J. B. Johnson, "Thermal Agitation of Electricity in Conductors," *Phys. Rev.*, 32, 97—109 (1928).

‡ H. Nyquist, "Thermal Agitation of Electric Charge in Conductors," *Phys. Rev.*, 32, 110—113 (1928).

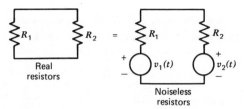

Fig. 13. 2. Circuit for showing that mean square noise voltage of a resistor is proportional to its resistance.

The power flow P_{21} in the other direction is obtained by switching subscripts.

$$P_{21} = \left[\frac{v_2(t)}{R_1 + R_2} \right]^2 \cdot R_1 \qquad (13.6)$$

Our thermodynamic argument asserts that, on the average, these powers are equal if the resistors are at the same temperature. This equality leads to the relation:

$$\frac{\overline{v_1^2(t)}}{R_1} = \frac{\overline{v_2^2(t)}}{R_2} \qquad (13.7)$$

Since Eq. 13.7 is based on the very general second law of thermo-dynamics, the quantity $\overline{v^2(t)}/R$ must be independent of the physical form of the resistor (size, construction, etc.) and independent of the resistance value. However, $\overline{v^2(t)}/R$ must be an increasing function of temperature to account for the experimental fact that energy flows from a hotter resistor to a colder one. Thus we have discovered that the mean square value $\overline{v^2(t)}$ of the noise voltage source $v(t)$ must be proportional to R. Since from Fig. 13.1 the voltage source model and the current source model must give the same open-circuit voltage across R, we have $v(t) = i(t) \cdot R$, so that $\overline{i^2(t)} = \overline{v^2(t)}/R^2$. Since $\overline{v^2(t)}$ is proportional to R, $\overline{i^2(t)}$ is inversely proportional to R.

As yet we have no way of knowing how the constant of proportionality between $\overline{v^2(t)}$ and R depends on temperature. This is one of the questions

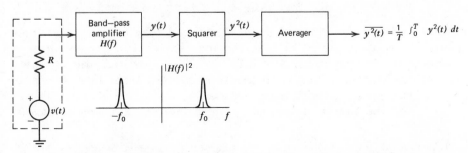

Fig. 13. 3. Measuring the spectral density of thermal noise.

that was answered by Johnson's original experiments. However, before tackling the question from a theoretical standpoint, let us first discuss the spectral density of the noise sources $v(t)$ and $i(t)$.

Johnson studied the spectral density of the thermal noise in a quite fundamental way. Figure 13.3 shows a simplified version of his measurement system with modern notation used to describe the subsystems. The band-pass amplifier has a known frequency response or system function $H(f)$ and a very high input impedance so that the input voltage is $v(t)$. The measurement of the spectral density of $v(t)$ goes as follows. From Eqs. 13.3 and 13.4 we can express the mean square value $\overline{y^2(t)}$ of the output of the amplifier as

$$\overline{y^2(t)} = \int_{-\infty}^{\infty} W_y(f)\, df = \int_{-\infty}^{\infty} W_v(f)\, |\, H(f)\, |^2\, df \qquad (13.8)$$

Now if the passband of the amplifier is very narrow, as suggested in Fig. 13.3, we can consider $W_v(f)$ to be equal to $W_v(f_0)$ over the passband. Therefore, $W_v(f_0)$ can be factored out of the integral in Eq. 13.8, and we can write

$$\overline{y^2(t)} = \int_{-\infty}^{\infty} W_v(f_0)\, |\, H(f)\, |^2\, df = W_v(f_0) \int_{-\infty}^{\infty} |\, H(f)\, |^2\, df \qquad (13.9)$$

But $\overline{y^2(t)}$ is measured directly by the circuit in Fig. 13.3, and $\int_{-\infty}^{\infty} |\, H(f)\, |^2\, df$ can be obtained from the known system function $H(f)$;[†] therefore, $W_v(f_0)$ may be computed. Furthermore, if band-pass amplifiers with various center frequencies are used, one can trace out $W_v(f)$ as a function of frequency. With this technique Johnson discovered that $W_v(f)$ was a *constant*, that is, independent of frequency, up to at least the highest frequencies for which he could obtain band-pass filters. Nyquist presented a theoretical argument in his 1928 article that showed that the independence of the spectral density of thermal noise with frequency should hold up to frequencies satisfying the relation $hf = kT$, where h is Planck's constant and k is Boltzmann's constant. At room temperature, this is a frequency of:

$$f = 1.38 \times 10^{-23} \times 300/6.63 \times 10^{-34} \doteq 6 \times 10^{12} \text{ Hz}$$

which is much higher than can be handled by an ordinary electronic system. Our next job is to find the magnitude of $W_v(f)$ and how it depends on temperature. Although Johnson answered this question experimentally, we will attack it theoretically. The theoretical results are in complete agreement with the experiments.

Consider the circuit of Fig. 13.4, which consists of a real resistor at

---

[†]Determined, for example, by sending sinusoids of different frequencies through the amplifier, or by taking the Fourier transform of its measured impulse response.

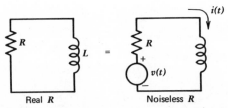

Fig. 13.   4. Circuit for theoretical determination of spectral density of resistor noise voltage source.

temperature T connected in parallel with an ideal inductor. If we replace the real resistor by the noise voltage-source model, we see that the voltage source will produce a current $i(t)$ in the circuit. At each instant of time there will be an energy $\frac{1}{2}Li^2(t)$, stored in the magnetic field surrounding the inductor. According to the general thermodynamic principle of equipartition of energy, the average value of this stored energy should be equal to $\frac{1}{2}kT$ where k is Boltzmann's constant and T is the absolute temperature. Our plan of attack is to find an expression for $i^2(t)$ in terms of $W_v(f)$, set $\frac{1}{2}Li^2(t)$ equal to $\frac{1}{2}kT$ or $i^2(t) = kT/L$ and solve for $W_v(f)$.

In order to calculate an expression for $i^2(t)$ we imagine that $i(t)$ is the output of an LTI system whose input is $v(t)$. We find the system function $H(s)$ by assuming a voltage of the form e^{st} and then finding the current in Fig. 13.5. The result is

$$H(s) = I(s) = \frac{1}{R + sL} \qquad (13.10)$$

Next we substitute $s = j2\pi f$ to obtain the Fourier transform $H(f)$:

$$H(f) = \frac{1}{R + j2\pi fL} \qquad (13.11)$$

Finally we have

$$\overline{i^2(t)} = \int_{-\infty}^{\infty} W_i(f)\,df = \int_{-\infty}^{\infty} W_v(f)\,|\,H(f)\,|^2 df = W_v \int_{-\infty}^{\infty} \left|\frac{1}{R + j2\pi fL}\right|^2 df \qquad (13.12)$$

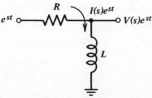

Fig. 13.   5. Finding the system function for the circuit of Fig. 13.4.

where we have used the experimental fact that $W_v(f)$ is constant. Continuing, we have

$$\overline{i^2(t)} = W_v \int_{-\infty}^{\infty} \frac{1}{R^2 + 4\pi^2 f^2 L^2} df$$

$$= \frac{W_v}{R^2} \int_{-\infty}^{\infty} \frac{df}{1 + \dfrac{4\pi^2 f^2 L^2}{R^2}} = \frac{W_v}{2\pi RL} \int_{-\infty}^{\infty} \frac{dx}{1 + x^2} \qquad (13.13)$$

where we have made the substitution $x = 2\pi f L/R$.

The integral may be evaluated by setting $x = \tan u$, $dx = \sec^2 u \, du$; the result is π. So

$$W_v = 2RL\overline{i^2(t)} = 2RkT \qquad (13.14)$$

As is expected from our earlier argument, W_v is proportional to R and is an increasing function of T.

We can also calculate the spectral density of the noise current source $i(t)$. Although we could do this from scratch as we did for W_v, it is easier to use the fact that $i(t) = v(t)/R$ in Fig. 13.1. Since $i^2(t) = v^2(t)/R^2$, we must have

$$W_i(f) = W_v(f)/R^2 = \frac{2kT}{R} \qquad (13.15)$$

Let us summarize what we have learned about thermal noise in a resistor and its effect on the circuit to which the resistor is connected. First, we know that we can represent a real resistor by either a random voltage source in series with, or a random current source in parallel with, an ideal (noiseless) resistor. The voltage and current sources are time functions with zero mean, but with mean square values that are directly proportional to the absolute temperature of the resistor. When the resistor is connected to a circuit, the output $y(t)$ of the circuit (either a voltage or a current) is a random variable whose spectral density can be computed with the help of Eq. 13.4, where $W_x(f)$ is the spectral density of either the voltage source (given by Eq. 13.14) or the current source (given by Eq. 13.15), and where $H(f)$ is the system function† of the rest of the circuit *including the resistor R*.

Example. Find the mean square output voltage $\overline{v_0^2(t)}$ of the op-amp circuit in Fig. 13.6a. Assume that the op-amp has a gain $10^6/s$, infinite input impedance, and that it contributes nothing to the noise in the circuit. The temperature is $300°$K.

---

† If the voltage source model is used, $H(f)$ is the system function for an input voltage; if the current source model is used, $H(f)$ is the system function for an input current.

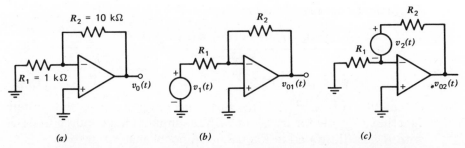

(a)   (b)   (c)

Fig. 13. · 6. Example of the use of the resistor noise-voltage model.

The output noise contributions of the two resistors are uncorrelated. That is, the mean square output $v_{01}{}^2(t)$ due to thermal noise in R_1 and the mean square output $v_{02}{}^2(t)$ due to thermal noise in R_2 add to give the total mean square output voltage $v_0{}^2(t)$. We do the problem in two parts:

(a) Calculation of $v_{01}{}^2(t)$. We plan to compute the spectral density $W_{v_{01}}(f)$ and integrate over all f to obtain the mean square output. First we must calculate $H_1(f)$, the system function when, as in Fig. 13.6b, the input is a voltage source in series with R_1, and the output is the output voltage of the op-amp. Assuming an input e^{st} and an output $V_{01}(s)e^{st}$, we note that the voltage at the inverting input of the op-amp must be $(-sV_{01}(s)e^{st})/10^6$. A nodal analysis gives

$$\frac{1 + \dfrac{sV_{01}(s)}{10^6}}{10^3} = \frac{\dfrac{-sV_{01}(s)}{10^6} - V_{01}(s)}{10^4} \tag{13.16}$$

from which we obtain

$$V_{01}(s) = \frac{-10}{1 + 11s/10^6} \tag{13.17}$$

The desired system function is obtained by substituting $s = j2\pi f$:

$$H_1(f) = V_{01}(j2\pi f) = \frac{-10}{1 + 11j2\pi f/10^6} \tag{13.18}$$

Since the spectral density of the input voltage is $W_{v1}(f) = 2kTR_1$, we find the mean square output voltage from:

$$\overline{v_{01}{}^2(t)} = \int_{-\infty}^{\infty} W_{v1}(f) \, |H_1(f)|^2 df$$

$$= \int_{-\infty}^{\infty} 2kT \cdot 10^3 \cdot \left| \frac{-10}{1 + 11j \cdot 2\pi f/10^6} \right|^2 df \tag{13.19}$$

To integrate, we make the substitution $x = 11 \cdot 2\pi f/10^6$.

$$\overline{v_{01}{}^2(t)} = \frac{2 \cdot 1.38 \cdot 10^{-23} \cdot 300 \cdot 10^5 \cdot 10^6}{22\pi} \cdot \int_{-\infty}^{\infty} \frac{dx}{1 + x^2}$$

$$= 3.8 \times 10^{-11} V^2 \qquad (13.20)$$

(b)   Calculation of $v_{02}{}^2(t)$. The difference here is that the system function $H_2(f)$ is for a system with the input voltage source in series with R_2, as illustrated in Fig. 13.6c. A nodal analysis gives

$$\frac{sV_{02}(s)}{10^6 \cdot 10^3} = \frac{-\dfrac{sV_{02}(s)}{10^6} + 1 - V_{02}(s)}{10^4} \qquad (13.21)$$

whence

and

$$V_{02}(s) = \frac{1}{1 + 11s/10^6} \qquad (13.22)$$

$$H_2(f) = V_{02}(j2\pi f) = \frac{1}{1 + 11j2\pi f/10^6} \qquad (13.23)$$

This time the spectral density of the input voltage is $W_{v2}(f) = 2kTR_2$, so

$$\overline{v_{02}{}^2(t)} = \int_{-\infty}^{\infty} W_{v2}(f) \mid H_2(f) \mid^2 df = 3.8 \times 10^{-12} V^2 \qquad (13.24)$$

The total mean square output voltage is the sum of these or about $4.2 \times 10^{-11} V^2$. The root-mean-square output voltage is the square root of this, or 6.5 μV.

13.4. Shot noise   Shot noise is caused by the random emission of electrons such as from the photocathode of a photomultiplier tube, the cathode of a vacuum tube, or from the valence band of a semiconductor. Each electron (or hole) that is released contributes a current proportional to its velocity as long as it is moving; in a vacuum tube this time would be closely the same for all emitted electrons, while in a semiconductor we would expect a distribution of lifetimes of both electrons and holes before recombination. These fluctuations in the lifetimes of electrons and holes in semiconductors contribute noise in excess of the thermal noise due to the resistance of the semiconductor device that we will not consider further. However, we consider one important application of the shot noise phenomenon, that is, the noise diode.

    A noise diode is a vacuum-tube diode with cathode and anode. It is designed with a relatively small anode-cathode spacing so that the electron transit time from cathode to anode is small, on the order of a nanosecond.

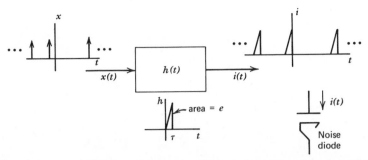

Fig. 13. 7. Current pulses produced by individual electrons in a noise diode.

Also the temperature of the cathode is kept relatively low so that the number of electrons in transit at any time is small. This condition is required to prevent correlations between the times of arrival at the anode of different electrons. That is, for the following theory to be valid, the arrival times must be completely random. Our problem is to compute the spectral density of the diode current if we are given the average diode current I_0.

The output current of the noise diode is a sequence of randomly occurring pulses, each of duration τ, the time for an electron to travel from cathode to anode. The *area* of each current pulse is just the charge transported by the electron, or e. The actual form of $h(t)$ depends on the geometry of the diode; for a planar diode, $h(t)$ is a linear ramp, which is the result of a constant acceleration of each electron by the uniform electric field between the electrodes while it is in transit. (The current in any circuit external to the diode can be shown† to be proportional to the scalar product of the instantaneous velocity vector of the electron, with the electric field vector caused by an anode-cathode potential of 1 V.)

To find the spectral density of the diode current, we imagine that we have a system, whose input is a random sequence of unit impulses and whose impulse response is the current pulse of area e and duration τ produced by the transit of one electron. Figure 13.7 illustrates this idea. The diode current is viewed in this particular way because we already know (Chapter 3, Fig. 3.9) an expression for the spectral density of a Poisson sequence of impulses. To obtain the spectral density of the diode current, we only need to use Eq. 13.4 and multiply the spectral density of the Poisson process by the square of the magnitude of the Fourier transform of $h(t)$. This calculation is illustrated in Fig. 13.8.

Since the average rate of occurrence of the Poisson impulses is I_0/e, $W_x(f)$ is the sum of an impulse of area I_0^2/e^2 at $f = 0$ plus a constant I_0/e. The Fourier transform of a short pulse of duration τ is nearly constant up

---

†See, for example, C. Hemenway, R. Henry, and M. Caulton, *Physical Electronics*, second edition, pp. 140–141, Wiley, New York.

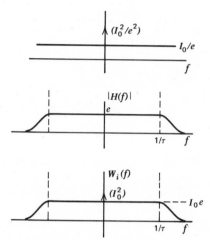

Fig. 13.   8. Finding the spectral density of shot noise.

to a frequency of the order $1/\tau$, as can be deduced from the fact that the factor $e^{-j2\pi ft}$ that multiplies $h(t)$ in the Fourier transform integral is near 1 if $ft \ll 2\pi$. Furthermore, the zero-frequency value of $H(f)$ is equal to the area under $h(t)$, or e. Thus $W_i(f) = W_x(f) \mid H(f) \mid^2$ consists of an impulse of area $I_0{}^2$ superposed on a flat spectrum of magnitude $I_0 e$, which falls to zero above a frequency $1/\tau$.

13.5. Experimental determination of noise performance of an amplifier   Any electronic system contributes noise to its output that is in excess of the output noise contributed by an input resistor. For example, as we saw in the example at the end of Section 13.3, it is a relatively simple matter to compute the mean square output of an amplifying system that is caused by thermal noise in an input resistor; all one needs to know are the value of the input resistor, its temperature, and the system function $H(f)$ of the amplifier for either a voltage or current input. The actual mean square noise at the output will be larger than this calculated value because of noise generated in the various components of the amplifier itself; resistors, transistors, etc.

One common figure of merit for the "noisiness" of an amplifier is the noise factor F, which compares the actual mean square noise output with that due to thermal noise in the input resistor (at $290°K$). If we denote by N_R the mean square output that is due to noise in the input resistor and by N_A the mean square output noise contributed by the amplifier, then, since the resistor noise and amplifier noise are uncorrelated, the actual mean square output noise must be the sum $N_R + N_A$. With these definitions, the noise factor is defined as:

$$F = \frac{N_R + N_A}{N_R} \quad (R \text{ at } 290^\circ \text{K}) \tag{13.25}$$

Note that $F = 1$ for a noise-free amplifier and $F > 1$ for any real amplifier. Remember that N_R may be calculated by:

$$N_R = \int_{-\infty}^{\infty} W_R(f) \mid H(f) \mid^2 df = \frac{2kT}{R} \int_{-\infty}^{\infty} \mid H(f) \mid^2 df \tag{13.26}$$

where, since we have explicitly used the noise current-source spectral density for the resistor, $H(f)$ is the system function for the amplifier's response to a *current* input.

One method of measuring the noise factor for an amplifier involves the use of a noise diode. The idea is to add noise to the input circuit with the noise diode in an amount such that the actual output mean square noise is doubled. If the additional output noise due to the shot noise in the input circuit is denoted by N_S, then under the doubling condition:

$$N_S = N_R + N_A \tag{13.27}$$

because the additional noise is uncorrelated with either the resistor noise or the amplifier noise. But we know how to calculate N_S. It is given by:

$$N_S = \int_{-\infty}^{\infty} W_S(f) \mid H(f) \mid^2 df = I_0 e \int_{-\infty}^{\infty} \mid H(f) \mid^2 df \tag{13.28}$$

provided that we can arrange the input circuitry so that the amplifier's response $H(f)$ to the noise diode is zero at $f = 0$ (so that the impulse at $f = 0$ in the shot noise spectral density of Fig. 13.8 does not contribute to the output). If we can also arrange that $H(f)$ is the same (for $f \neq 0$) for noise current contributed by the noise diode as for noise current contributed by the resistor, the integrals $\int_{-\infty}^{\infty} \mid H(f) \mid^2 df$ in Eqs. 13.26 and 13.28 will have the same numerical value. Therefore, substitution from Eqs. 13.26 and 13.28 gives

$$F = \frac{N_S}{N_R} = \frac{I_0 e}{2kT/R} = \frac{I_0 e R}{2kT} \approx 20\, I_0 R \tag{13.29}$$

where I_0 is in amperes and R is in ohms. Remember, in Eq. 13.29, I_0 is that dc current in the noise diode that doubles the output noise.

The way in which the noise diode is "coupled" to the amplifier circuit is shown in Fig. 13.9. The "ideal" connection is sketched in Fig. 13.9a; a constant current is supplied to the noise diode at a high enough voltage that the electron transit time in the diode is very short. A blocking capacitor prevents any constant current from flowing through R and changing the bias conditions at the input of the amplifier, but the shot-noise *fluctuations* in the diode current are passed by the capacitor. Usually C is made large enough that the time constant RC is at least a few

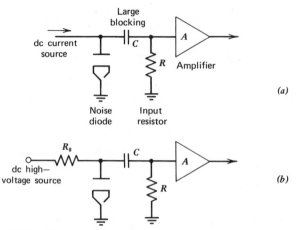

(a)

Fig. 13.   9. Noise-diode measurement of the noise produced by an amplifier.

tenths of a second. This insures that the $H(f)$ for diode current fluctuations is the same as the $H(f)$ for resistor current fluctuations, except over a negligibly small range of frequencies near zero.

The "ideal" coupling circuit is not practical because one would have to have a way of adjusting the current source magnitude to equal the diode current. A practical circuit in which the current source is replaced by a high-voltage source and series resistor is shown in Fig. 13.9b. The series resistor must either be much larger than R or else the R in Eq. 13.29 must be replaced by the parallel combination of R and R_S, because for ac signals, these resistors are effectively in parallel from the amplifier input to ground. With the coupling circuit of Fig. 13.9b the output noise with *no* shot noise contribution is found by reducing the cathode temperature to room temperature. Then the cathode temperature is raised until the output noise is doubled.

An alternate method of determining the noise factor uses the change in

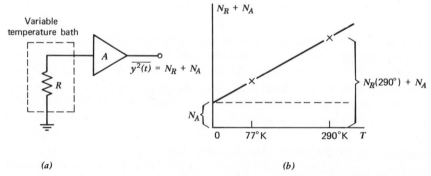

Fig. 13.   10. Variable-temperature method of measuring amplifier noise.

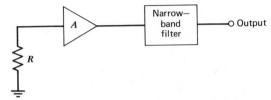

Fig. 13.   11. Measuring the spot-noise factor of an amplifier.

temperature of a resistor to change the input noise while leaving the amplifier's contribution unchanged. If the temperature of the resistor could be reduced to absolute zero, the remaining output noise would be caused by the amplifier alone. Figure 13.10 illustrates this method. The mean square output, $N_R + N_A$, is plotted as a function of the temperature of the input resistor, R. The result will be a straight line, with the intercept at $T = 0$ representing the amplifier's contribution to the mean square output. In practice, only two temperatures need to be used, such as room temperature and the boiling point ($77°K$) of a liquid nitrogen bath. The linear variation of thermal noise spectral density with temperature allows us to extrapolate the straight line to $T = 0$. The numerator and denominator of Eq. 13.25 are proportional to the distances shown on the graph in Fig. 13.10. A word of caution must be introduced concerning this method. A carbon composition resistor should not be used as the input resistor in this method because it has an appreciable percentage change in resistance between $290°K$ and $77°K$. A metal film resistor, with a very low temperature coefficient of resistance, however, is quite satisfactory for this application.

The addition of a narrow-band filter (Fig. 13.11) following the amplifier in either of the experimental setups described above allows the amplifier's noise contribution to be measured as a function of frequency. In effect we can then determine the noise factor over a narrow frequency range. Sometimes this is called the spot noise factor. The definition of spot noise factor is the same as in Eq. 13.25 but now the "system" is redefined to include the narrow-band filter.

In the literature, other quantities besides noise factor and spot noise factor are used to describe a system's noise performance. For example, the noise factor and spot noise factor are often expressed in decibels ($10 \log_{10} F$). (The coefficient is 10 rather than 20 because the noise factor is already a ratio of two powers, rather than a ratio of two currents or two voltages.) Another quantity that is used is the noise temperature of a system. The noise temperature T_N is the temperature at which the input resistor would have to be to produce a mean square output from a noisefree amplifier equal to the amplifier's contribution to the mean square output. Figure 13.12 shows how the noise temperature can be computed. The dashed sloping line represents the input resistor's

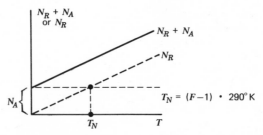

Fig. 13. 12. Definition of noise temperature of an amplifier.

contribution to the noise as a function of temperature. The temperature at which this equals the amplifier's contribution is T_N. The lower the noise contribution of the amplifier, the lower is the noise temperature.

13.6. Equivalent circuit for noise in two-ports When any real two-port network has its input either open-circuited or short-circuited, noise will appear at the output. This noise is the result of various random processes inside the network. Although thermal noise in resistors and shot noise may contribute to the output noise of a two-port, they are by no means the only random processes that can occur. Other processes may give rise to noise that has a nonuniform spectral density; a common example is so-called $1/f$ noise, which has a spectral density, as its name implies, that varies with frequency as $1/f$ down to frequencies as low as tenths or hundredths of hertz. Furthermore, the spectral density of the output noise may depend on the input termination; usually there is an input termination (complex impedance or admittance) that gives a minimum noise factor; the noise factor increases if either the real part or the imaginary part of the termination is varied in either direction from the ideal.

One way of modeling the noise performance of a two-port is to represent the real two-port by an ideal noiseless two-port preceded by a noise current generator $i_n(t)$ and a noise voltage generator $v_n(t)$ connected as shown in Fig. 13.13. Clearly this representation satisfies the requirement that there be output noise when the input is either open-circuited or

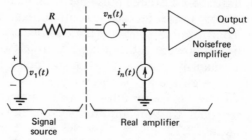

Fig. 13. 13. Modeling noise in a two-port network.

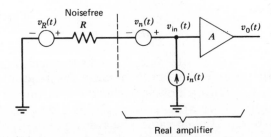

Fig. 13.   14. Comparison of noise in a two-port with thermal noise in the input resistor.

short-circuited since, with the input open, the noise current generator feeds the input and with the input shorted, the noise voltage generator feeds the input. In order to calculate the noise factor of such a circuit we must terminate it at the input with a specified impedance.

We illustrate these concepts by computing the noise factor for a particular example. Consider the signal source and amplifier in Fig. 13.13. The signal source is represented by a voltage source $v_1(t)$ in series with a source resistance R. The amplifier itself is assumed to have infinite input impedance in this example. Its noise performance is modeled by the noise voltage and noise current generators $v_n(t)$ and $i_n(t)$. Each noise generator has its own spectral density $W_v(f)$ and $W_i(f)$.

There are three noise contributions to the mean square output signal $\overline{v_0{}^2(t)}$ that would exist if the actual signal $v_1(t)$ were zero. These three contributions come from:

1.   Thermal noise in the input resistor, R.
2.   The noise voltage generator $v_n(t)$.
3.   The noise current generator $i_n(t)$.

In Fig. 13.14 we have redrawn Fig. 13.13, showing the noise voltage generator associated with R explicitly as $v_R(t)$. The signal generator is no longer present because we are interested in comparing the mean square output (or input) voltage of the amplifier in the configuration of Fig. 13.14 with the corresponding voltage if $v_n(t)$ and $i_n(t)$ were zero. The ratio of the total mean square output (or input) voltage due to all three contributors to that due to the input resistor alone, is the noise factor F. The amplifier input voltage $v_{in}(t)$ is easily seen to be given by:

$$v_{in}(t) = v_R(t) + v_n(t) + i_n(t)R \qquad (13.30)$$

because, with infinite input impedance, the entire noise current $i_n(t)$ flows through R. The square of this voltage has six terms, written out in detail in   Eq. 13.31.

$$v_{\text{in}}^2(t) = v_R^2(t) + v_n^2(t) + i_n^2(t)R^2 + 2v_R(t)v_n(t)$$
$$+ 2v_R(t)i_n(t)R + 2v_n(t)i_n(t)R \qquad (13.31)$$

When we average over time to find the mean square value, $\overline{v_{\text{in}}^2(t)}$, the fourth and fifth terms will be zero because the random processes that underlie $v_n(t)$ and $i_n(t)$ are not "correlated" with the random processes that underlie $v_R(t)$. This is another way of saying that at all the times that $v_n(t)$ or $i_n(t)$ have a particular value, the average of $v_R(t)$ *at those times* is zero. This result need not be the case for the sixth term of Eq. 13.31; the degree of correlation of $v_n(t)$ and $i_n(t)$ must be determined by experiment. Taking averages on both sides of Eq. 13.31, we have

$$\overline{v_{\text{in}}^2(t)} = \overline{v_R^2(t)} + \overline{v_n^2(t)} + \overline{i_n^2(t)}R^2 + \overline{2v_n(t)i_n(t)}R \qquad (13.32)$$

This result may be written in terms of spectral densities.

$$\overline{v_{\text{in}}^2(t)} = \int_{-\infty}^{\infty} W_R(f)\,df + \int_{-\infty}^{\infty} W_v(f)\,df + R^2 \int_{-\infty}^{\infty} W_i(f)\,df$$
$$+ 2R\epsilon \int_{-\infty}^{\infty} \sqrt{W_v(f)} \cdot \sqrt{W_i(f)}\,df \qquad (13.33)$$

where the factor ϵ in the last term denotes the amount of correlation between $v_n(t)$ and $i_n(t)$. If $\epsilon = 0$ these noise sources are completely uncorrelated, while if $\epsilon = 1$ the correlation is complete. If we restrict ourselves to a narrow band of frequencies Δf at $\pm f_0$, the total contribution to the mean square input voltage is

$$\text{spot noise} = 2kTR \cdot 2\,\Delta f + W_v(f_0) \cdot 2\,\Delta f + W_i(f_0)R^2 \cdot 2\,\Delta f$$
$$+ 2\epsilon R \sqrt{W_v(f_0)} \cdot \sqrt{W_i(f_0)} \cdot 2\,\Delta f, \qquad (13.34)$$

where we have replaced the spectral density $W_R(f)$ by $2kTR$ according to Eq. 13.14.

The total amplifier *output* noise in a narrow frequency band is proportional to the right-hand side of Eq. 13.34 and is seen to depend quadratically on the source resistance R. There is a term independent of R, a pair of terms proportional to R, and a term in R^2. Therefore, a graph of the spot noise versus R would appear as in Fig. 13.15.

The contribution of the first term in (13.34) can be reduced by cooling the resistor. (A method of determining $W_v(f_0)$, $W_i(f_0)$, and ϵ, utilizing an extension of the method outlined in Fig. 13.10 for finding the noise factor, can be developed.) The spot noise factor is determined by dividing Eq. 13.34 by the contribution of R only. Thus

$$F(f_0) = 1 + \frac{W_v(f_0) + W_i(f_0)R^2 + 2\epsilon R\sqrt{W_v(f_0)} \cdot \sqrt{W_i(f_0)}}{2kTR}$$

$$(13.35)$$

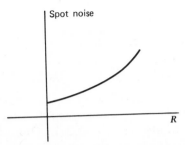

Fig. 13. 15. Total amplifier noise (at one frequency) as a function of input resistance.

This spot noise factor can be shown to be a minimum for a particular amplifier, if the source resistance is chosen to be

$$R = \sqrt{\frac{W_v(f_0)}{W_i(f_0)}} \tag{13.36}$$

However, this result is misleading because the amount of noise in the output, relative to the amount of signal — rather than the noise caused by R — will always be a minimum for $R = 0$.

Problems for Chapter 13

13.1. *Approximate derivation of the spectral density for shot noise.* This method is based on the fact that if one counts independent events, such as the number of electrons passing a point in a circuit in a fixed time interval, the different numbers have a *mean square deviation* from the average equal to the average itself. Thus, if N electrons are counted, on the average, the mean square deviation will also be given by N.

(a) Given a dc current I_0, how many electrons flow, on the average, in 1 second?

(b) If the bandpass of a low-pass filter is Δf, the effective averaging time is on the order of $1/\Delta f$. What, then, is the average number of electrons occurring during this averaging time?

(c) The output of the filter fluctuates because the number of electrons that arrive during the averaging time fluctuates. The root-mean-square fluctuation is equal to the square root of the number of electrons. Express this mean square fluctuation in terms of I_0, e, and Δf.

(d) The root-mean-square fluctuation of the *current* is obtained by multiplying the root-mean-square number by the charge on an electron and dividing by the averaging time, $1/\Delta f$. Express $i_{rm s}$ in terms of I_0, e, and Δf.

(e) Thus, show that the mean square fluctuation of the current is given by $\overline{i^2} = I_0 e \Delta f$. The missing factor of 2 was lost in the conversion from frequency response to averaging time.

13.2. A phototube acts basically as a current generator, with the magnitude of the current being proportional to the light intensity striking the cathode (Fig. P13.2a). One way of supplying a collecting bias voltage to the phototube (often used in photo-multiplier tubes, where the electrodes shown are one pair of a series) is with a supply current generator I and resistor R. The ac equivalent circuit is that in Fig. P13.2b, where i represents the ac fluctuations in current between cathode and anode. These fluctuations could be produced by a change in light intensity or by random fluctuations (shot noise) in what is otherwise a steady response to a dc light intensity.

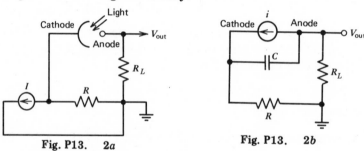

Fig. P13.   2a   Fig. P13.   2b

(a) If $R = 10^5 \ \Omega$, $R_L = 100 \ \Omega$, C = stray capacitance = 20 pF, find the system's transfer function (ratio of V_{out} to i) and plot on a Bode plot.

(b) Now a capacitor (0.05 μF) is placed across R. Repeat part (a) under this condition.

(c) If the dc phototube current is 1.0 μA., find the rms output voltage under the conditions (a) and (b).

13.3. Show that the variance of a Gaussian distribution, of the form $e^{-x^2/2\sigma^2}$ is σ^2. *Hint.* The mean square value of x is

$$\frac{\int_{-\infty}^{\infty} x^2 \, e^{-x^2/2\sigma^2} \, dx}{\int_{-\infty}^{\infty} e^{-x^2/2\sigma^2} \, dx}$$

The numerator is the derivative of the denominator with respect to some quantity. The denominator reduces to a simple integral if you use the fact that

$$\int_{-\infty}^{\infty} e^{-u^2} \, du \cdot \int_{-\infty}^{\infty} e^{-v^2} \, dv = \iint_{-\infty}^{\infty} e^{-(u^2+v^2)} \, du \, dv = \int_{0}^{2\pi} \int_{0}^{\infty} e^{-r^2} r \, dr \, d\theta.$$

13.4. An alternative method of finding the spectral density, W_v, of the noise voltage of a resistor (assuming that the spectral density is independent of frequency) is to connect it in parallel with a capacitor (see Fig. P13.4). The mean energy $\tfrac{1}{2}CV^2$ stored in the capacitor should equal $\tfrac{1}{2}kT$. Find W_v by this method.

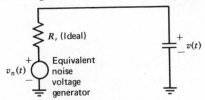

Fig. P13.   4

13.5. A series circuit consists of a resistor R, capacitor C, and inductor L. Use the voltage source noise-model for the resistor and imagine a system for which the noise voltage is the input and the current is the output.
   (a) Show that the system function is $H(f) = 1/\{R + j[2\pi fL - (1/2\pi fC)]\}$.
   (b) Write down an integral expression for $\overline{i^2(t)}$ and show that it can be reduced to

$$2kT/\pi L \int_{0}^{\infty} \frac{a \, du}{a^2 + (u - 1/u)^2}$$

   where $a = R\sqrt{C/L}$. *Hint.* Use Eq. 13.4.)
   (c) The integral in (b) can be shown to equal $\pi/2$. What is the average energy stored in the inductor? Is the result expected?

13.6. In the series RLC circuit of the previous problem, imagine a system for which the resistor noise voltage is the input and the capacitor voltage is the output.
   (a) Show that the system function is $1/j2\pi fC/\{R + j[2\pi fL - (1/2\pi fC)]\}$.
   (b) Find an integral expression for the average energy stored in the capacitor and show that it reduces to

$$\bar{E}_C = kT/\pi \int_{0}^{\infty} \frac{a \, du}{a^2 u^2 + (u^2 - 1)^2}$$

   where $a = R\sqrt{C/L}$. The integral can be shown to equal $\pi/2$ so that $\bar{E}_C = kT/2$, as expected. Note also that this integral can be put in the same form as the integral in the preceding problem by the substitution $u = 1/x$. One way to do the basic integral is to break it into an integral from 0 to 1 plus an integral from

1 to ∞, substitute $u = 1/y$ in the latter so that there is just a single integral from 0 to 1, and then let $u - (1/u) = x$. The final integration gives an inverse tangent.

13.7. A 1-MΩ resistor is connected across the input of an ideal noiseless low-pass filter, with gain (see Fig. P13.7). Find the value of the mean square output voltage, $\overline{y^2(t)}$.

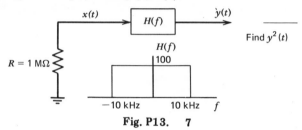

Fig. P13.   7

13.8. The *bolometer*: One method of making an absolute power measurement is to use a device, such as a thermistor ⎓⏦⎓ whose resistance varies with temperature. If the waveform whose power is to be measured has no dc component, (e.g., a white-noise waveform) then the configuration in Fig. P13.8 may be useful.

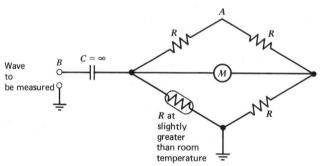

Fig. P13.   8

Suppose that with no input waveform at B, a dc voltage at A of 10 V is required to bring the temperature of the thermistor to a value where its resistance is R. This condition obtains when the meter reads zero.

Next the unknown waveform is introduced at B, and the power it expends in the bridge circuit raises the temperature of the thermistor so that the bridge is no longer balanced. But when the dc source voltage is reduced to 5 V the bridge again becomes balanced. Find the mean square voltage of the wave at B.

13.9. A resistor R is connected across the input of a two-port network whose noise contribution is representable by independent noise voltage and noise current generators as in Fig. 13.13 of the text.

(a) Show that the spot noise factor $F(f_0)$ is

$$F(f_0) = 1 + \frac{1}{2kT}\left[\frac{W_{V_n}(f_0)}{R} + W_{I_n}(f_0)R\right]$$

where $W_{V_n}(f)$ is the spectral density of the network's noise voltage source and $W_{I_n}(f)$ is the spectral density of its noise current source.

(b) Show that the noise factor is a minimum if R is chosen to be equal to $\sqrt{W_{V_n}/W_{I_n}}$. This value of F is known as the optimum noise factor.

13.10. A resistor R is connected across the input of a two-port network, with input impedance r (pure resistance). Find how the power developed in r depends on R and r. Express your answer in terms of the *spectral densities* of the noise voltage and current sources in Fig. 13.13.

13.11. Two resistors R_1 and R_2 in parallel are employed as the input to an amplifier of infinite input impedance. Show, by adding the effects of the two independent noise sources, that the noise voltage at the input of the amplifier is the same as would be generated by a single resistor across the input of value equal to the parallel combination $R_1 R_2/(R_1 + R_2)$ of the two resistors. Note that this result may be extended to any number of resistors in parallel.

13.12. A radioactive source (Poisson process) may be used to generate white noise. Consider the following system (Fig. P13.12), which is designed to produce a measurement of noise factor.

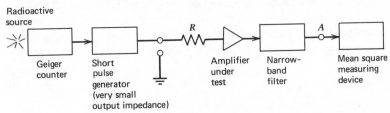

Fig. P13.   12

Suppose that, with $R = 10$ kΩ and at room temperature, the spot noise factor of the amplifier is equal to 2, so that the amplifier contributes exactly the same amount of noise as R does. Suppose that we wish to double the reading of mean square output voltage at A by introducing, on the average, 1000 rectangular pulses per second, each of width 10 μsec, from the pulse generator. What would the amplitude of the voltage of the rectangular pulses have to be?

13.13. Amplifiers using FETs in their input stages may have remarkably low spot noise figures. In fact, the noise contributed by the amplifier may be less than the *accuracy* of measuring the mean square voltage so that the usual methods of measuring noise factor become quite useless! However, the scheme illustrated in Fig. P13.13 will yield an accurate measurement.

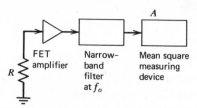

Fig. P13.   13

Step 1. Measure the mean square output in the setup pictured above. This reading is F times the amplified noise from the resistor's noise voltage source.

Step 2. Replace R by a capacitor such that its impedance (at $f = f_0$) is equal in magnitude to R. The output mean square reading is now equal to the amplifier's contribution only. If the ratio of the readings in steps 1 and 2 is 10 to 1, compute the spot noise factor of the amplifier.

13.14. Show, from the definition of noise factor in Eq. 13.25, if an input signal produces a mean-square noise voltage N_s, the noise factor can be written as the signal-to-noise ratio at the input (before N_A is added) divided by the signal-to-noise ratio at the output.

13.15. Two amplifiers are cascaded as shown in Fig. P13.15. G_1 and G_2 are the gain factors by which the mean square input signal is multiplied in passing through each amplifier. The individual noise factors of the amplifiers are defined by $F_1 = (G_1 N_R + N_{A1})G_1 N_R$

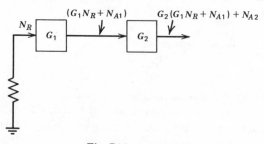

Fig. P13.   15

$F_2 = (G_2 N_R + N_{A2})/G_2 N_R$ where N_{A1} and N_{A2} are the additional mean square output signals introduced by the amplifiers. If the overall noise factor F is defined as the actual noise output divided by what the output noise would be if the amplifiers were noisefree, show that

$$F = F_1 + \frac{F_2 - 1}{G_1}$$

Thus, if the gain of the first stage is much larger than $F_2 - 1$, the second stage makes a negligible contribution to the overall noise factor.

13.16. In situations where a particular input resistor R must be used with a particular amplifier having an optimum input resistor for lowest noise figure that is different from R, it is common practice to employ a transformer (see Fig. P13.16) in order to obtain the lowest overall noise figure. As usual, the real amplifier may be modeled as an ideal *noiseless* amplifier A with a noise voltage generator v_n and a noise current generator i_n in the input circuitry. The transformer is ideal with a primary-secondary turns ratio of $1{:}n$.

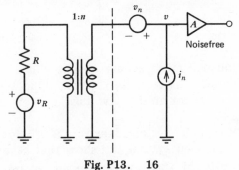

Fig. P13.   16

(a) Assume that the inductive reactance of the secondary winding is large in comparison with the reflected resistance $n^2 R$ and show that the mean square value $\overline{v_1^2}$ of the amplifier input voltage that is due to the amplifier noise is $\overline{v_n^2} + n^4 R^2 \overline{i_n^2}$ in the case where v_n and i_n are uncorrelated.

(b) Assume that the reactance across the primary winding is large in comparison with R and show that the mean square value $\overline{v^2}$ of the amplifier input voltage due to noise in the resistor is $n^2 \overline{v_R^2}$, where v_R is the noise voltage generated by the resistor.

(c) Show that the optimum turns ratio for lowest noise figure is given by $n = (\overline{v_n^2}/\overline{i_n^2}R^2)^{1/4}$.

13.17. In certain biological applications amplifiers are designed with "negative capacity" neutralization in which, in an attempt to compensate for signal degradation due to high source resistance R combined with stray capacitance C across the amplifier input, capacitive feedback is employed. The purpose of this problem is to demonstrate that, while the effect on signal (and noise in the source resistor) is as desired, the noise v_n introduced in the amplifier is enhanced by the feedback and therefore that it is best to try to reduce C rather than neutralize it. Figure P13.17 shows two systems, A and B. In Fig. P13.17b, the stray capacitance C_1 is larger than C.

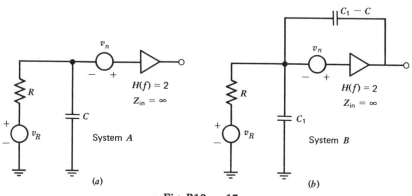

Fig. P13.   17

(a) Show that the mean square output voltage due to v_R is the same for systems A and B. [*Hint.* Find the overall system function $H'(f)$ for an input voltage $v_R = e^{st}$ and use Eqs. 13.4 and 13.14.]

(b) Find the overall system function $H''(f)$ for a noise voltage $v_n = e^{st}$ for each system and show that at angular frequency ω, the ratio of the resulting mean square output voltages $[H''_B(f)]^2 / [H''_A(f)]^2$ is

$$\frac{1 + \omega^2 R^2 (2C_1 - C)^2}{1 + \omega^2 R^2 C^2}$$

Note that this expression is greater than 1 at all frequencies, thus demonstrating that the amplifier-contributed noise is greater in B than in A.

REFERENCES

Carlson, A. B., *Communication Systems: An Introduction to Signals and Noise in Electrical Communications*, Appendix A.1, McGraw-Hill, New York, 1968.

Robinson, F. N. H., *Noise and Fluctuations in Electronic Devices and Circuits*, Clarendon Press, Oxford, 1974.

Gupta, M., "Applications of Electrical Noise," *Proc. IEEE*, 63, 996—1010 (1975).

Subcommittee on Noise, IRE Standards on Methods of Measuring Noise in Linear Twoports, 1959, *Proc. IRE*, 48, 60—68, 1960.

Subcommittee on Noise, "Representation of Noise in Linear Twoports," *Proc. IRE*, 48, 69—74 (1960).

van der Ziel, A., *Noise in Measurements*, Chapters 3 and 6. Wiley, New York, 1976.

CHAPTER 14

Noise Reduction in Electronic Systems

14.1. Introduction   In Chapter 13 we investigated two of the important sources of noise in electronic systems — thermal noise and shot noise. Here we investigate the ways in which a signal in a system is contaminated by noise and several of the ways in which the obscuration of the signal by noise can be held to a minimum.

In addition to the fundamental thermal and shot noise in a system there are always other noise sources. Some of these noise sources are external to the system itself and represent an input of electromagnetic energy in excess of the unavoidable black-body energy that is radiated by the system's surroundings. These extraneous noise sources include not only man-made interference such as that from the 60-Hz power lines, sparks from ignition systems and electric motors, and radio-frequency interference from communication systems but also natural sources of interference such as lightning and other less intense electrical discharges. Other noise sources occur in the amplifying elements of a circuit and can be attributed to the random trapping and release of electrons in energy levels and to other statistical phenomena.

An obvious and effective way to reduce the effects of external interference is to prevent its introduction into the system. Accordingly there are certain techniques of shielding and grounding that should be followed in critical applications. Once noise is in a system it is more difficult, in fact sometimes impossible, to eliminate. However, if the noise occurs primarily in one range of frequencies while the signal occupies another range of frequencies, a frequency filter will reduce the effect of noise. Finally, even if the signal and the noise occupy the same frequency range, it may happen that the signal has a definite phase relative to some sequence of events such as the cycles of a master oscillator or a set of well-defined stimulus pulses. In these situations there exist averaging techniques that literally "extract" the signal from the noise.

14.2. A time domain description of the noise problem   At the most basic level, the flow of current in electronic devices is statistical in nature because of the small but finite charge of an electron. Thus, the measurement of a current involves a counting process. The larger the current the less time it takes to measure it to a particular accuracy because larger numbers of electrons (or holes) are subject to smaller *relative* fluctuations. Whenever a current changes, the average rate of flow of electrons changes. To discover that the rate has changed one must count electrons for a long enough time to be confident that the new rate is statistically different from the old rate. If one is fortunate enough to be told the time at which the rate changed, then the problem of measuring the new rate is simply one of starting a count after the change has occurred. This is essentially what is done in the averaging techniques described in detail in later sections of this chapter. However, if the problem is to determine when the rate change occurred, then there is a fundamental limitation on the accuracy of the measurement. This fundamental limitation is based on the rate of flow of electrons in the *first stage* of the system and on the time required for these electrons to flow.

These ideas can be illustrated with the example of a photomultiplier tube for detecting light. A photomultiplier tube consists of a photosensitive cathode from which single electrons are emitted by means of the photoelectric effect and several dynodes at which electron multiplication occurs. Each photon of radiant energy in the wavelength range from the ultraviolet (say 200 nm) to the near infrared (say 1000 nm) has a certain probability of releasing an electron from the photocathode. Each electron released is attracted to the first dynode, which is maintained at a voltage of from 100 to 300 V above the cathode, and strikes the dynode with enough energy to release a small number of secondary electrons. The resulting group of electrons is accelerated toward the second dynode, where each electron in the group releases a small number of secondaries, and so on. Photomultiplier tubes have from 6 to about 15 dynodes and produce from a few thousand to a few million electrons at the final collecting electrode for each electron released from the cathode, depending on the dynode voltages and the number of dynodes.

It should be clear that the fundamental statistical limitation on the measurement of light intensity is caused by the rate at which electrons leave the cathode.† This rate might be, say, 1000/sec. The fluctuations in the number released during 1 second are then of the order of $\sqrt{N}$ or about 30. (The standard deviation or square root of the average of the square of the deviations from the average, when making repeated counts with an average number N, is $\sqrt{N}$, according to Poisson statistics.) Now after secondary-emission multiplication by, say, a factor of 10^6, the average number of electrons collected in 1 second at the anode is 10^9. However, we may *not* use Poisson statistics to determine that the fluctuations in this

---

† The "quantum efficiency" or probability of emission of an electron by an incident photon is an important property of a photomultiplier tube. For a fixed incident light flux, a higher quantum efficiency means that a larger number of electrons is released per unit time.

number are on the order of $\sqrt{10^9}$ or 3×10^4. Instead, these fluctuations are approximately the same *relative* size, about 3%, as those at the first dynode; the absolute fluctuation is therefore about 3×10^7 electrons at the anode. Actually the fluctuations are slightly larger than this because of fluctuations in the numbers of secondary electrons released by each primary electron in the secondary emission multiplication process. This contribution to the fluctuations in the final number of electrons collected at the anode is called secondary emission noise.

Another source of noise in photomultiplier tubes results from the fact that the electrons collected at the anode include some that were simply boiled off (thermionic emission) of the cathode and dynodes, without benefit of an incident photon. This "dark noise" depends on the area of the cathode and its temperature. However, the temperature of a photomultiplier tube also affects its gain so that there is usually a broad optimum range of temperature at which the tube should be operated for lowest dark-noise to signal ratio.

Note that in the preceding example we assumed that we measured the anode current for 1 second. There is no way of determining the average current to better than about 3% for that particular level of light flux in a time shorter than 1 second. If one attempts to measure the current in, for example, 0.01 sec, the average number of electrons arriving at the first dynode in that time, 10, would be subject to fluctuations on the order of $\sqrt{10}$, or 30% of the mean current.

If we try to determine the *time* at which a change in current at the anode occurs we are limited not only by the counting statistics, that is, the limitation on accuracy in a determination of the number of electrons that arrive in any given short interval, but also on the fluctuation in travel time of electrons from cathode to anode. Thus even if there is a copious supply of electrons when the light flux changes, say, from small to large, the transit time spread (on the order of a few nanoseconds) provides a lower limit to the time resolution of the device.

While we are working on this example, let us include the effect of external interference by supposing that the high-voltage supply from which the dynode voltages are derived is fluctuating at a 60-Hz rate and causes a corresponding 60-Hz variation in the average rate at which electrons reach the anode. If we are to measure a change in average anode current that is due to a change in the average light flux striking the photocathode, we must be careful to average over a time that is long compared to 1/60 sec in order to average out these 60-Hz fluctuations. This averaging may be accomplished even though the changes in light intensity occur at a frequency higher than 60 Hz. For example, if the light intensity is oscillating with a period of 0.001 sec, we can arrange to measure the anode current only during the 0.0005-second intervals when the intensity is high or during the 0.0005-second intervals when the intensity is low, provided that we measure for enough intervals to average out the 60-Hz variation. This technique for eliminating noise, called phase-sensitive detection, was discussed at the end of Chapter 1.

14.3. Shielding, grounding, and decoupling   The purpose of this section is to describe some of the ways interference can enter a system and the remedies for preventing the problems. Figure 14.1*a* illustrates a system with several errors in wiring that can allow interference to enter. The system is a common one: a high-gain amplifier with input transducer represented by the resistor R is connected to an indicating device, in this case, an oscilloscope. The signal source, which is in series with its resistance R, has been turned off so that we can concentrate on the interference pathways. The high-gain amplifier is powered by a dc voltage supply. The chassis of both the dc supply and the oscilloscope are connected to ground through their safety ground connections, the third wire in their 60-Hz power cords.

The first difficulty with this system is that unwanted signals can be coupled to the input of the amplifier by stray capacitance. Any point in

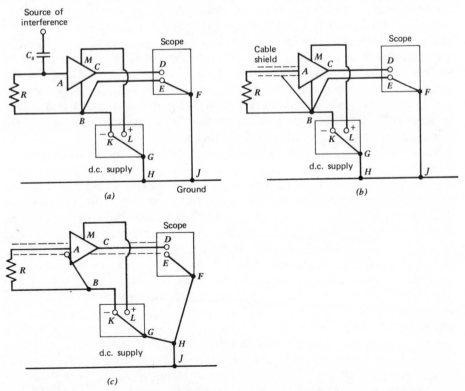

Fig. 14.   1. Typical system showing sources of interference and remedies.

the environment of this system that has an ac voltage relative to point B will be coupled by stray capacitance to point A. This is represented in the figure by the capacitor, C_s. Assuming that the source of interference is a unit voltage source e^{st}, the voltage that appears at the amplifier input has a magnitude given by

$$V_{\text{stray}\,C} = \frac{Z_T}{\dfrac{1}{sC} + Z_T} \tag{14.1}$$

where Z_T is the total impedance between points A and B, including the parallel combination of the transducer impedance R, the amplifier input impedance, and any wiring capacitance.

As an example, suppose that the parallel impedance is purely resistive, with a magnitude of $100\,\text{k}\Omega$ and the stray capacitance is $10\,\text{pF}$. A 60-Hz signal $(s = j \cdot 2\pi \cdot 60 = 377j)$ of magnitude $1\,\text{V}$ will result in $10^5/[(j/377 \times 10^{-11}) + 10^5]$ or approximately $-j \cdot 377\,\mu\text{V}$ at A. Since the situation worsens at higher frequencies and, of course, at larger amplitudes of interfering sources, one must take every precaution to eliminate capacitive coupling to the input. The remedy in this case is quite simple; we need only to surround the wire leading from the transducer R to the amplifier input A with a conducting shield maintained at the potential of point B, leaving as small an end exposed for the purpose of connecting to the transducer as possible. Often it may be possible and desirable to enclose the transducer itself in the shield. The currents induced by the interference source and stray capacitance then flow harmlessly through the shield and do not develop a potential difference between A and B that is amplified by the high-gain amplifier. The system, corrected for capacitive coupling, is redrawn in Fig. 14.1b.

Another way for interference signals to enter the input circuitry is by means of magnetic-field coupling. Here the magnetic flux ϕ, produced by a current-carrying wire, links the circuit loop between A and B and acts as a voltage source $-(d\phi/dt)$ in series with the transducer. The magnitude of this induced voltage is $M(di/dt)$, where M is the mutual inductance between the interfering circuit and the input circuit loop between A and B, and i is the current in the interfering circuit. Usually this mutual inductance coupling is insignificant, but if found to be present, it can be reduced by decreasing the areas of both the interfering circuit loop and the input circuit loop, by changing the orientations of these loops so that the magnetic flux no longer links the input loop, and by increasing the separation between the loops. At frequencies in excess of a megahertz, a highly conducting shield placed around the entire input loop, including the transducer, can virtually eliminate the magnetic coupling, because the eddy currents induced in the shield produce magnetic fields that almost exactly cancel the interfering field inside the shield.

Two subtle problems exist with regard to the grounding of the system of Fig. 14.1*a*. The purpose of the safety ground wires *GH* and *FJ* is to prevent dangerous shocks to an experimenter in case a conducting path is accidentally established between the ac power line and the chassis of an instrument. Although the safety grounds are necessary to prevent accidents, they can be the source of undesirable interference at 60 Hz and higher harmonics of 60 Hz, particularly 180 Hz. Often the 60- or 180-Hz interference can be traced to an ac potential between the points, *J* and *H*, at which the safety grounds are connected to the safety ground distribution system of the building. An appreciable fraction of this ac potential may exist between points *B* and *E*. This voltage adds to the amplified signal voltage between *C* and *B* and appears on the scope. Another way that this 60-Hz signal can enter the system is by way of a voltage drop between points *B* and *K* caused by current flowing in the path *JFEBKGH*. A dc power supply maintains a constant voltage between its output terminals, *L,K* in this example. Any 60-Hz voltage existing between *K* and *B* will also exist between *L* and *B* and between *M* and *B*. Unless the amplifier is carefully designed to prevent power supply variations from appearing (sometimes even amplified) at its output, some 60- or 180-Hz signal will appear between *C* and *B*.

Both of the problems described in the preceding paragraph may be classified as ground-loop problems. The remedy for the first is to return the safety ground wires to the same point in the building ground system. Even though this is done (see Fig. 14.1*c*), magnetic coupling may induce a current in the loop *FEBKGH*. This current is eliminated by removing the connection between points *B* and *E*, which can be done while retaining a desirable shield around the signal-carrying wire *CD*, by connecting the shield to either *B* or *E* *only*.

Another interference problem arises when more than one load circuit is to be driven by a single dc supply voltage. Figure 14.2*a* shows a power supply connected to two loads. The current drawn by each load will vary with the signal being processed by that load circuit. For example, if one load is a complex logic circuit with several inputs, the changes in load circuit will be sudden steps in current caused by the switching on and off of logic gates and flip-flops. Without proper decoupling circuits, the current changes in that load circuit can appear as voltage changes at the other load. For example, suppose the current changes by 100 mA in load 1. If the effective output resistance R_0 of the power supply is a realistic 10 mΩ, then the actual supply voltage to load 2 will change by 1 mV. This can cause an undesirable signal to appear in the signal circuitry within load 2, especially if it is a high-gain amplifier. What is worse is that the power-supply output impedance usually increases as the frequency increases because of the degradation of the frequency response of the feedback circuits used to maintain a constant voltage. Thus for *pulses* of current, the signal coupled into load 2 may be of the order of volts!

A simple remedy exists for this coupling type of interference: simply

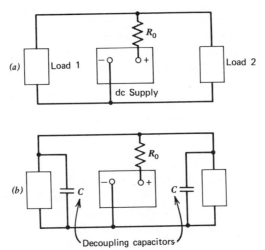

Fig. 14.   2. Decoupling one load circuit from another.

place a capacitor across the load terminals (Fig. 14.2b). This often needs to be no larger than 0.01 or 0.001 μF. However, in some cases where heavy currents are involved, the decoupling capacitor may have to be many microfarads. In such cases an electrolytic capacitor of the appropriate value should be paralleled by a high-quality, nonelectrolytic capacitor of lower value. The reason for the smaller capacitor is that the electrolytic capacitor may have considerable inductance and resistance associated with it; at higher frequencies, where the impedance is supposed to become very small, it may act as a resistor or inductor.

One final situation involving shielding must be mentioned here. If feedback exists between the output and input of a high-gain amplifier, it may be possible for the system to oscillate. We investigated the conditions for oscillations in considerable depth in Chapter 5. The point to be made here is that the feedback pathways may not be at all obvious. The feedback may occur by capacitive coupling between output and input or by inductive coupling. An even more common pathway is through the power supply. Here the large current changes in the high-level stages cause changes in the supply voltage seen by the low-level stages. The remedies for these ills are the same as for the corresponding interference problems: shield both input and output leads, reduce the areas of and increase the distance between input and output loops, prevent output currents from flowing in any part of the input loop, and decouple the power supply at the input stages. Often it is helpful to introduce a resistor or inductor in series with the power supply lead to an input stage as shown in Fig. 14.3. This resistor or inductor in conjunction with the decoupling capacitor constitutes a low-pass filter that prevents feedback of the high-frequency currents that can be the cause of oscillations.

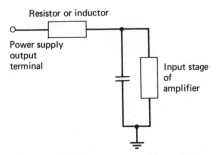

Fig. 14.   3. Decoupling a sensitive input stage.

14.4. Noise limiting by reduction of bandwidth

In Chapter 13 we learned that both thermal and shot noise are wideband noise sources. As a result, the mean square noise current or voltage at the output of a system depends not only on the gain of the system but also on its bandwidth. The smaller the bandwidth, the smaller the fluctuations in the output in response to a noise signal at the input. This effect of bandwidth on noise may be understood in the time domain by considering the impulse response of an entire system. We have learned that the duration of a system's impulse response is inversely proportional to its bandwidth. The effect of reducing the bandwidth of a system by a factor of 2 is to lengthen the impulse response duration by a factor of 2. Then at any chosen time, twice as many events, for example, photons detected by a photomultiplier tube, will still be contributing to the output signal by means of their impulse response. Because of Poisson statistics and the $\sqrt{N}$ law, the root-mean-square fluctuation or standard deviation increases by a factor of $\sqrt{2}$ while the mean output signal increases by a factor of 2. Thus, in the system with narrower bandwidth and longer impulse response, the relative noise in the output is smaller by a factor of $\sqrt{2}$.

If a signal occupies a certain bandwidth then it is wise to limit the frequency response of a system designed to detect and measure that signal to the frequency band occupied by the signal. For example, the signal might be a 1000-Hz sinusoidal current from a photomultiplier tube whose amplitude is proportional to the light intensity passing through a mechanical "chopper" operating at 1000 Hz. If the problem is to measure the amplitude of the 1000-Hz output current very accurately, then one should employ a narrow-band filter centered on 1000 Hz in order to reduce the fluctuations in output signal that are due to shot noise. We have seen examples of low-pass and band-pass filters for the audio frequencies in Chapter 8. At frequencies of 100 kHz to a few hundred megahertz, passive tuned circuits utilizing inductors and capacitors must replace those active filters. Figure 14.4 illustrates a basic tuned circuit consisting of ideal passive components. If the input signal is a unit

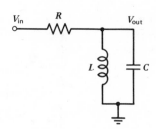

Fig. 14.   4. A passive tuned circuit or band-pass filter.

exponential voltage, e^{st}, the output voltage is readily shown, by a nodal analysis, to be

$$v_{out} = \frac{sL/R}{sL/R + s^2 LC + 1} \cdot e^{st} \qquad (14.2)$$

We know that a sinusoidal signal of frequency f consists of the sum of two exponential signals with $s = \pm j2\pi f$. Each of these exponential signals will have a maximum amplitude of 1 if $s^2 LC + 1 = 0$, which occurs at a frequency $f = 1/2\pi\sqrt{LC}$. Inductors with values on the order of a microhenry are easily formed from a few turns of wire. When such an inductor is paralleled by a 100-pF capacitor, for example, the resonant frequency is about 15 MHz.

On occasion one requires a filter (band-stop filter) that stops a particular frequency while allowing lower and higher frequencies to pass. It may be important to block a frequency component of the noise even though the signal itself has no component at that frequency. For example, if there is appreciable 60- or 180-Hz interference, a circuit may be driven into a nonlinear region of operation so that a signal in the kilohertz or megahertz range is distorted or even blocked. In this type of situation, the interference must be eliminated or reduced in the early stages of a system.

One example of a band-stop filter is shown in Fig. 14.5. A unit exponential input voltage produces an output voltage:

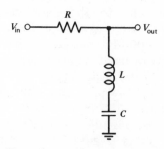

Fig. 14.   5. Band-stop filter.

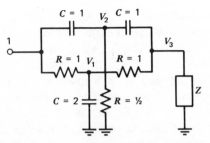

Fig. 14.   6. Twin-tee band-stop filter.

$$v_{out} = \frac{s^2 LC + 1}{s^2 LC + sRC + 1} \cdot e^{st} \tag{14.3}$$

Thus, at the critical frequency $f = 1/2\pi\sqrt{LC}$, the output voltage is zero. At audio frequencies and below, the inductors necessary to make a band-stop filter of the type shown in Fig. 14.5 require iron cores and are bulky, heavy, and expensive. Another type of filter, known as the twin-tee filter, is an excellent choice at low frequencies because it is constructed from resistors and capacitors only. A twin-tee filter is shown in Fig. 14.6. The analysis follows.

First, choose the units of s to be $1/RC$ rad/sec, so that we can set $R = 1$ and $C = 1$. Also, assume that the input signal is the usual unit exponential voltage, e^{st}. Now we write a nodal equation for the three nodes with the unknown voltages. These are:

for node V_1: $1 = V_1 \cdot (2 + 2s) - V_3$

for node V_2: $s = V_2 \cdot (2 + 2s) - sV_3$ (14.4)

for node V_3: $V_2 s + V_1 = V_3 \cdot (s + 1 + 1/Z)$

These equations are solved in straightforward fashion by determinants for V_3. The result is

$$V_3 = \frac{s^2 + 1}{2(s + 1)(s + 1 + 1/Z) - (s^2 + 1)} \tag{14.5}$$

Hence, V_3 is zero if $s = \pm j$; that is, for sinusoidal signals with frequency $f = 1/2\pi$ frequency units or $f = 1/2\pi RC$ Hz. Furthermore, at low frequencies, as $|s| \to 0$, V_3 approaches $Z/(Z + 2)$ and at high frequencies, or $|s| \gg 1$, V_3 approaches $sZ/(sZ + 2)$. These limiting values can be found directly, by noting that at high frequencies, current flows from input to output through the upper capacitors with the resistors acting as open circuits, while at low frequencies current flows from input to output through the lower resistors, with the capacitors acting as open circuits.

14.5. Extracting a signal from noise when its frequency and phase are known If a signal is repetitive it can be recovered from random noise simply by measuring it many times. Two examples of repetitive signals are those that are periodic with known phase relative to a periodic reference signal and those that repeat at a known time. In the first case the instrument used to measure the amplitude of the periodic signal buried in noise is known as a lock-in amplifier. In the second case the instrument used to measure the signal at one or many points in time after a reference time is known as a signal averager.

The heart of a lock-in amplifier is a phase-sensitive detector, which was discussed in detail at the end of Chapter 1. Recall that in a phase-sensitive detector, a signal (plus noise) is multiplied by a reference wave that is alternately +1 and −1 (in some systems the reference wave is a sinusoid), and the resulting signal is smoothed by a low-pass filter of very narrow bandwidth; for example, from 0.01 to 1 Hz wide. The phase-sensitive detector is, in effect, a very narrow-band filter whose response depends on the phase of the input sinusoid relative to the reference wave.

Most lock-in amplifiers contain a provision for shifting the phase of the reference wave, whether it be a sine wave or a square wave. The purpose of such a phase shifter is to allow the experimenter to search for components of the signal at the reference frequency but with a different phase. The reference wave is usually derived from the exciting signal in an experiment. For example, if the experiment consists of a measurement of the current in a crystal in response to a change in light intensity falling on the crystal, the reference wave will have some definite phase relative to the chopped light beam. To find the phase of the photocurrent with respect to the light beam, one simply varies the phase shift of the reference wave until the output of the low-pass filter reads a maximum positive value. At this point the phase-shifted rererence wave and the signal are in phase.

The degree to which a lock-in amplifier can pull a signal out of noise is determined primarily by the bandwidth or averaging time of the final low-pass filter, since it is this element that limits the frequency bands from which the noise contributes to the output. However, one must always be careful that the combined signal plus noise at any point in the amplifying stages of a lock-in amplifier are not so large as to drive a stage into a nonlinear response. If this should occur, then noise at different frequencies will be mixed together at the signal frequency, resulting in more than just the expected thermal or shot noise.

A phase-sensitive detector can be made by replacing the low-pass filter with an integrator, which integrates the output of the multiplier for a definite length of time. Since this type of system has the advantage that the final output of the integrator is a definite number, which can be read, for example, with an accurate digital voltmeter, the experimenter need

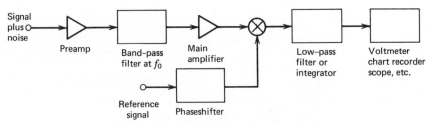

Fig. 14.   7. A lock-in amplifier system.

not approximate the average of the slowly varying output signal of a low-pass filter. On the other hand, repetition of the integration process will give a slightly different number each time it is done. These fluctuations in integrator output are a manifestation of that fact that one can never completely eliminate the noise in a measurement. The use of an integrator in a phase-sensitive detector reduces the measurement process with a lock-in amplifier to the same form as for simple physical measurements such as mass or length. One simply obtains a single number in each measurement that is subject to statistical fluctuations (noise) whose magnitude is measured by the standard deviation of the measurements.

Commercial lock-in amplifiers usually contain a great deal more circuitry than just a phase-sensitive detector. Figure 14.7 illustrates a system that contains a wide-band preamplifier, a band-pass filter centered on the chopping frequency, a main amplifier, and a phase shifter for the reference wave. The purpose of the amplification is to make the signal large enough to give an output reading on a display device such as a scope, or chart recorder. The purpose of the band-pass filter is to limit the overall size of the signal plus noise entering the amplifier so that the amplifier is not driven to nonlinearity.

14.6. Extracting signal from noise when the time of occurrence is known   In many situations the response to a brief signal such as a short pulse is a fairly complex waveform that is contaminated with noise. Two examples are the electrical signal on an electrode embedded in the visual cortex of the brain of an animal following a flash of light falling on the eyes and the return echo of a radar signal sent to the moon or to the planet Venus. The stimulus signal can be repeated many times; the problem is to find the average response to the stimulus.

This problem can be solved with various degrees of sophistication. Figure 14.8 shows a simple system that samples the signal at the same time after each stimulus pulse. The output of the sampler, $y(t)$, is a train of pulses with amplitudes that vary because of noise; their average value is proportional to the signal at a particular time after the stimulus. The

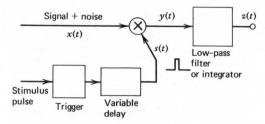

Fig. 14.   8. Sampling a repetitive signal to reduce the effect of noise.

effects of noise tend to cancel since, by definition, the noise is completely independent of the signal. As in the case of the lock-in amplifier, there will be fluctuations in the output of the low-pass filter or the integrator that are a direct result of the noise superimposed on the signal.

Another, more powerful form of signal averager is shown in Fig. 14.9, in very schematic form. This is primarily a digital system. After each stimulus pulse a sampling pulse train is produced consisting of as many as several hundred pulses spaced by as little as a few microseconds. Each sampling pulse creates a sample of the signal plus noise. The amplitude of the sample is immediately converted into a digital signal, which is added to the contents of a memory register. The sampling pulses are also fed to an address counter that controls which of the memory registers is to be updated. The read/write pulse generator produces signals to the memory that either read out the contents of a memory register to the adder or write the result of addition into the proper memory register. Each time the stimulus pulse occurs, samples of the response are taken, converted to digital form, and added to the memory register associated with that particular time. When this system is completed with a set of controls to clear the memory, vary the number and interval of the sampling pulses, etc., and a readout or display device, it becomes an extremely powerful

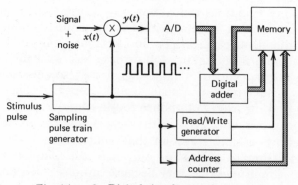

Fig. 14.   9. Digital signal-averaging system.

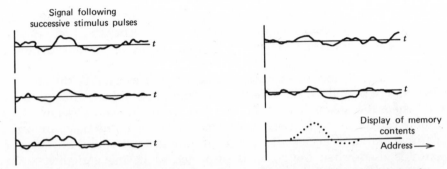

Signal following
successive stimulus pulses

Display of memory
contents

Address ——>

Fig. 14.   10. Noisy responses to successive stimuli and the averaged signal.

tool for extracting signal from noise. Figure 14.10 illustrates the way in which several signals are sampled and summed to give a relatively smooth representation of the average response to successive stimulus pulses.

Average response computers of the type shown in Fig. 14.9 are complex pieces of equipment. Price tags on commercial instruments have been on the order of five to ten thousand dollars. However, the advent of microprocessors and semiconductor memories is certain to bring enormous price reductions in the near future.

14.7. Finding hidden periodicities in a noisy signal   A very powerful technique for analyzing noisy signals for periodic components is known as autocorrelation. Because this process is quite complex, it is usually performed by an essentially digital system. An autocorrelator is an instrument that computes the autocorrelation function of a signal $x(t)$, which we defined in Chapter 1 (Eq. 1.13) as:

$$R(\tau) = \lim_{T \to \infty} \frac{1}{2T} \int_{-T}^{T} x(t + \tau) \cdot x(t) \, dt \qquad (14.6)$$

To compute the value of the autocorrelation function at any particular value of its argument τ, one needs to integrate the product of the signal $x(t)$ and an earlier version of the signal $x(t + \tau)$. A digital autocorrelator does this by storing digitized samples of the signal and then performing digital multiplications and additions to achieve the integral. These calculations must be repeated for each value of the argument τ.

The value of the autocorrelation function at zero argument is clearly the mean square value of the signal during the integration time T. For a periodic signal, $R(\tau)$ itself is periodic, because a time shift of $x(t + \tau)$ by an integral number of periods will give the same integrand and therefore the same value of the integral. Furthermore as we saw in Chapter 1, for a random signal such as noise, the autocorrelation function approaches zero for large arguments, because there is just as much likelihood for the product of two widely separated samples of the noise to be positive as negative. Thus, when the signal contains periodic components, the autocorrelation function exhibits peaks at $\tau = nT_i$ where n is any integer and T_i are the periods of the various periodic components that comprise $x(t)$.

To summarize the three basic techniques for extending the time of measurement to reduce the effect of random noise, a lock-in amplifier is used when the signal is periodic and has a definite phase relative to a periodic reference wave; a signal averager is used when the signal follows a repetitive but not necessarily periodic stimulus pulse; and an autocorrelator is used when the signal is suspected of having periodicities but when the periods are unknown.

Problems for Chapter 14

14.1. The input circuitry of a certain instrumentation amplifier may be modeled as a 10-MΩ resistance in parallel with a 100-pF capacitance. The 320-V peak-to-peak 60-Hz power lines are observed to produce a 1-V peak-to-peak signal at the amplifier input.
  (a) Find the stray capacitance, C_s, between the power line and the amplifier input terminal.
  (b) If the amplitude of the 180-Hz component of the power line signal is 1% of the 60-Hz amplitude, that is, 3.2-V peak-to-peak, find the amplitude of the 180-Hz voltage at the input terminal.

14.2. A semiconductor radiation detector is a reverse-biased diode in which hole-electron pairs produced by the absorption of a photon near the junction give rise to a burst of current as they are swept across the junction by the electric field there. A typical biasing circuit is shown in Fig. P14.2. Included is the noise current source associated with the resistor. Since the reverse leakage current also gives rise to shot noise, the mean square fluctuations in current seen by the current amplifier are the sum of the mean-square fluctuations due to shot noise and the mean square fluctuations due to resistor noise current. Suppose that the amplifier is noisefree and

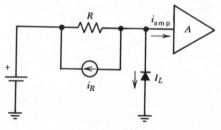

Fig. P14. 2

has an impulse response of duration 10 nsec (bandwidth $\approx 10^8$ Hz) and that the detector leakage current is 10 nA.

(a) Find the mean square noise current in the amplifier due to shot noise.

(b) Show that the resistor noise current increases the amplifier mean-square noise current by the factor $1 + 2kT/eI_L R$ over the shot noise contribution.

(c) If the resistor is at room temperature how large must it be in order to increase the noise by less than 10% of the leakage-current contribution?

(d) In silicon, approximately one charge is produced for each 3.5 eV of energy of a photon absorbed near the junction. The number of photons so produced is, of course, subject to fluctuations according to Poisson statistics. If 700 keV photons are entirely absorbed and the associated current pulses last about 1 nsec, which noise source gives the greater fluctuation in amplitude of the output current pulses: (1) Poisson fluctuations in the number of carriers produced by the photon, or (2) fluctuations due to shot noise.

14.3. In Fig. P14.3, two amplifiers are supplied by the same power supply, which has an output impedance of $0.1\ \Omega$. Each amplifier may be modeled by a 1-kΩ load resistor as far as the power supply is concerned. Suddenly the load current i_1 drawn by amplifier 1 changes by 10 mA for 1 msec. If the decoupling resistors R and capacitors C are 100 Ω and 0.1 μF, respectively, use an approximate frequency-domain analysis to estimate the maximum change in supply voltage V_2 at the other amplifier. (*Hint.* Find the system function for an input i_1 and an output V_2.)

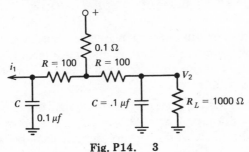

Fig. P14.   3

14.4. Design a twin-tee filter to reject 60-Hz signals (See Fig. 14.6.)

(a) Choose $R = 1$ kΩ and find the correct value of C.

(b) Make a pole-zero plot for the system function $H(s)$ if $Z \gg 1$ kΩ.

(c) Make a linear graph of $|H(j2\pi f)|$ versus f from 0 to 200 Hz.

14.5. A white noise signal $x(t)$ may be modeled as a large number of sinusoidal signals of frequencies $f_n = n \, \Delta f$ with amplitudes and phases that vary slowly and randomly during times of the order of $1/\Delta f$. In the frequency domain, then, a white noise signal may be represented as a sequence of real and imaginary impulses with areas A_n and B_n, respectively. Of course, since the white noise signal is real, $A_n = A_{-n}$ and $B_n = B^*_{-n}$ (see Fig. P14.5). We wish to compare the noise output of two systems when the input is the white noise source. The system $H_1(f)$ is an ideal *band-pass* filter with passband of width W centered on f_c, the frequency of the cosine multiplying signal. The system $H_2(f)$ is an ideal low-pass filter of cutoff frequency $W/2$. In each case $W > \Delta f$.

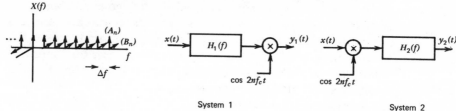

System 1   System 2

Fig. P14.   5

(a) Which of the A_n and B_n in the spectrum of $x(t)$ contribute to the final noise output in $y_1(t)$ and $y_2(t)$?

(b) Show that the noise output is the same in the two systems.

14.6. Consider the two systems of the previous problem, except that the multiplying cosine wave is replaced by a symmetrical square wave of period $T = 1/f_c$.

(a) Describe the output of the two systems.

(b) Compare the noise output of system (2) with a third system, similar to (2) but with a filter such as $H_1(f)$ preceding the multiplier.

(c) Why do lock-in amplifier (phase sensitive detector) systems generally include a *prefilter* that eliminates the third and higher harmonics of the signal to be detected before the multiplication process occurs?

14.7. Suppose in a digital averager the least significant bit corresponds to an analog voltage of 1 mV. This problem concerns the ability of the averager to detect and measure signals that have amplitudes less than 1 mV in the presence of noise whose rms level is several millivolts. The A-to-D converter acts like a multiple-threshold device, for example, if the signal + noise is between −2.5 and −1.5 mV the digital representation is −2 mV; if the signal + noise is between −1.5 and −0.5 mV the digital representation is −1 mV, etc.

(a) If we assume a Gaussian distribution of analog values for the noise alone, what is the *expected* result for the sum of $N = 100$ measurements of noise only.

(b) If the signal is 0.1 mV, about how many times out of 100 measurements will the presence of the signal cause the next higher threshold to be exceeded than the threshold that would have been exceeded with noise only.

(c) What is the expected result for the sum of 100 measurements of signal plus noise?

(d) In order for the measurement of the signal amplitude to be a good one, the expected sum in part (c) should be larger than the standard deviation of the sum. For example, if the rms noise level is 2 mV, then we expect about $\sqrt{N} \cdot 2$ mV for the standard deviation of the sum, or 20 mV in the situation outlined above. [Note that we are neglecting the contribution to the standard deviation from the uncertainty of the number calculated in part (b).] How many measurements, N, should be taken in order that the expectation value of the sum is 3 times the expected standard deviation, for 2 mV noise and a 0.1-mV signal?

14.8. In the digital averager described in the preceding problem, if the signal amplitude is such that the probability is a that any measurement will give a digital reading that is 1 mV higher than it would have been with no signal present, then

(a) Show that the expected value of the sum of N measurements of signal + noise is given by $\bar{y} = \sum_{m=0}^{N} m\, p(m)$; where $p(m) = N!\, a^m\, (1-a)^{N-m}/m!(N-m)!$.

(b) Sum this series and show that the result is Na, as expected.

(c) The standard deviation of a large number of measurements of y is given by σ, where $\sigma^2 = \overline{(y - \bar{y})^2} = \overline{y^2} - \bar{y}^2$.

Show that $\overline{y^2} = \sum_{m=0}^{N} m^2\, p(m)$ and that $\sigma^2 = Na(1-a)$. [Note that this standard deviation is a measure of the uncertainty that was neglected in part (d) of the previous problem. A good estimate of the total standard deviation can be obtained from the square root of the sum of the squares of the two standard deviations.]

REFERENCES

Jones, B., *Circuit Electronics for Scientists*, Chapter 12, Addison Wesley, Reading, Mass., 1974.

Morrison, R., *Grounding and Shielding Techniques in Instrumentation*, Chapters 3, 4, and 5, Wiley, New York, 1967.

Ott, H. W., *Noise Reduction Techniques in Electronic Systems*. Wiley-Interscience, New York, 1976.

CHAPTER 15

Discrete Systems

15.1. Introduction   In Chapters 1 to 5 we developed methods of handling LTI systems, in which the input and output signals are, in general, continuous functions of time. Then we saw in Chapter 12 that under certain circumstances a sequence of *samples* of a continuous signal can contain all of the information in the original continuous signal because the original signal can be recovered by passing the sequence of samples through an ideal low-pass filter. Now it should be clear that certain operations might be performed on the sampled data while it is in discrete form; for example, the sequence of input values might be converted to digital form by an A-to-D converter, stored in a computer, filtered or otherwise altered by the computer, and then either used directly or converted back into a continuous signal by a D-to-A converter and low-pass filter.

The primary advantage of modifying data while it is in discrete form is that a *computer program* replaces electronic hardware; to change the type of "filtering" one needs only to change the program. Another advantage is that once the input sequence has been stored, there is no longer the realizability restriction. Now the output sequence can be a function of "future" values of the input sequence as well as past values. On the other hand there are disadvantages to these discrete system techniques, which are sometimes subtle. For example, just as the inability to completely band-limit a signal results in errors when the continuous signal is reconstructed, so do we encounter errors in the reconstruction of a continuous signal from a *filtered* version of an incompletely band-limited input signal. Also digitizing errors (the fact that a digital version of a data sample is accurate to only within ± half the value of the least significant bit) adds to the noise in the signal. Another obvious disadvantage of discrete systems is that the output may be considerably delayed relative to the input because of the time required for computing.

Our treatment of discrete systems depends heavily on analogies with the continuous systems development in earlier chapters. We begin with the discrete impulse response and the discrete version of convolution, then look for special input sequences called eigensequences. This leads us into a frequency-domain approach, involving z transforms (the discrete analog of Laplace transforms) and discrete Fourier transforms (DFT). In this connection we describe one of the Fast Fourier Transform algorithms, a

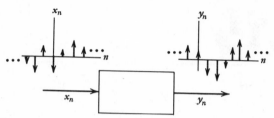

Fig. 15.   1. A discrete system; both input and output are defined at discrete times only.

method for rapid computation of the DFT on a computer. Throughout this chapter remember that the discrete analog of a continuous time function $x(t)$ is a set of values; that is, a sequence x_n, in which an index variable, here n, which can take on only integer values, replaces the continuous independent variable t.

15.2. The discrete impulse response and convolution
As in our development of continuous systems, the key concept is that of a system (Fig. 15.1) with an input sequence x_n and an output sequence y_n. In order to proceed we must limit our discrete system to be LTI. That is, if two input sequences individually give two output sequences, then the sum of the input sequences must give the sum of the corresponding output sequences (linearity), and a delayed version, x_{n-m}, of any input sequence x_n gives a delayed version y_{n-m} of the output sequence.

We can characterize the system by its impulse response h_n, which is the output sequence when the input sequence is the discrete version of a unit impulse, that is, the sequence $x_n = |\delta_{n0},|$ where δ_{nj} is the Kronecker delta.†

Since the output of the system is h_n when the input is δ_{n0}, the linearity property ensures that the output is ah_n when the input is $a\delta_{n0}$ and that the output is the delayed sequence h_{n-m} when the input is a delayed impulse δ_{nm} (see Fig. 15.2). It follows that a general input sequence x_n gives rise to the output sequence:

$$y_n = \sum_{j=-\infty}^{\infty} x_j h_{n-j} \tag{15.1}$$

To understand the right-hand side of this equation, which is the discrete analog of the convolution integral, consider that the zeroth member x_0 of the input sequence elicits the output sequence $x_0 h_n$, the first member of the input sequence x_1 elicits the output sequence $x_1 h_{n-1}$, and so on. To

---

† The Kronecker delta is a widely used symbol in mathematics and physics. Defined for integer values of its subscripts, it has the value 1 if $n = j$ and 0 if $n \neq j$. Hence, δ_{n0} represents a sequence, defined for all integers n, which is 1 at $n = 0$ and 0 for all other n.

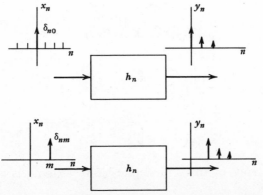

Fig. 15.   2. Impulse response of a discrete system.

obtain the total output sequence we simply sum over the various members of the input sequence.

An alternate formula for the convolution sum is found by substituting $m = n - j$, $j = n - m$ in Eq. 15.1. Now instead of summing over the j index we must sum over the m index so that we have

$$y_n = \sum_{m=-\infty}^{\infty} x_{n-m} h_m \qquad (15.2)$$

Equation 15.2 may be understood from a slightly different point of view than Eq. 15.1. A particular value of the output sequence, say y_3, is made up of the immediate response $x_3 h_0$ to the present input value x_3, plus the delayed response $x_2 h_1$ to the earlier input x_2, plus the still more delayed response $x_1 h_2$ to the still earlier input x_1, and so on. The sum in Eq. 15.2 reflects these various contributions to each output value y_n.

Example. Compute the output sequence when the input sequence and the impulse response are given in the table below.

n	x_n	h_n
< 0	0	0
0	2	1
1	2	1/2
2	2	1/4
3	2	0
$\geqslant 4$	0	0

When the number of nonzero values of both x_n and h_n is limited, the computation can be organized much as a multiplication, as shown in the scheme below.

$$
\begin{array}{ccccccc}
x_3 & 2 & 2 & 2 & 2 & x_0 \\
h_2 & 0 & \tfrac{1}{4} & \tfrac{1}{2} & 1 & h_0 \\
\hline
 & 2 & 2 & 2 & 2 \\
 & 1 & 1 & 1 & 1 \\
\tfrac{1}{2} & \tfrac{1}{2} & \tfrac{1}{2} & \tfrac{1}{2} \\
\hline
0 & \tfrac{1}{2} & 1\tfrac{1}{2} & 3\tfrac{1}{2} & 3\tfrac{1}{2} & 3 & 2 \quad y_0
\end{array}
$$

The bottom row in this "multiplication" scheme is the desired output sequence y_n.

Another way to organize the work is to form polynomials from the input and impulse sequences and then multiply them. Thus we have

$$[2 + (2/z) + (2/z^2) + (2/z^3)] \cdot [1 + (\tfrac{1}{2}/z) + (\tfrac{1}{4}/z^2)] =$$
$$2 + (3/z) + (3\tfrac{1}{2}/z^2) + (3\tfrac{1}{2}/z^3) + (1\tfrac{1}{2}/z^4) + (\tfrac{1}{2}/z^5)$$

By convention, the sequence values for *positive* values of the index are associated with *negative* powers of z and vice versa. The output sequence y_n is simply read off as the coefficients of the y-polynomial. Note that in both schemes we have performed the simple operation of multiplication to obtain the output. This example is a premonition of things to come.

One important application of discrete convolution is in smoothing noisy data. For example, a simple "running average" of an input sequence is obtained by convolving it with an impulse response consisting of some finite, usually small, number N of sequence values $1/N$. The continuous time analog of this process, convolution with a rectangular pulse, can only be approximated; to obtain an LTI system with even an approximately rectangular impulse response requires a considerable investment in hardware. On the other hand, the discrete "running averager" is implemented by writing a program that adds the previous N values of the input sequence and multiplies by the constant $1/N$. Furthermore, the impulse response of the discrete running averager can be altered simply by changing the program to one in which each previous sequence value is multiplied by some number before being added. To effect a similar change in a continuous-time system would require a complete redesign of the hardware.

15.3. Eigensequences and the discrete "system function" Next we ask, as we did in Chapter 2 for continuous-time LTI systems, whether there are special input sequences such that the output of the system is the same sequence except for multiplication of each member of the sequence by the same constant. By analogy with the continuous-time case, where the eigenfunctions are e^{st} defined for all t, we are led to try geometric sequences z^n, defined for all n, from $-\infty$ to $+\infty$.

To find the output response to the input sequence $x_n = z^n$ we substitute into the second form of the convolution sum, Eq. 15.2:

$$y_n = \sum_{m=-\infty}^{\infty} z^{n-m} h_m \tag{15.3}$$

Now we factor z^n from each term in the sum to obtain

$$y_n = z^n \sum_{-\infty}^{\infty} z^{-m} h_m = z^n H(z) \tag{15.4}$$

where

$$H(z) \equiv \sum_{-\infty}^{\infty} z^{-m} h_m \tag{15.5}$$

provided z is a number such that the sum converges. We see that the output sequence is indeed the same geometric sequence as the input sequence except for multiplication of each term in the output sequence by the constant $H(z)$.

Example. Find an expression for the system function of a discrete system whose impulse response is the "step" function $h_n = 1$, $n \geqslant 0$ and is zero for $n < 0$.

By direct substitution into Eq. 15.5 we find

$$H(z) = \sum_{0}^{\infty} z^{-m} \cdot 1 = 1 + \frac{1}{z} + \frac{1}{z^2} \cdots = 1 + \frac{1}{z} \cdot H(z) \tag{15.6}$$

Next we solve for $H(z)$ to obtain

$$H(z) = \frac{1}{1 - 1/z} \tag{15.7}$$

Since the sum of a geometric series converges if the terms decrease in magnitude as n increases, the result in (15.7) is valid only in the range $|z| > 1$. Other than this restriction, z can be any complex number.

Example. Find the system function for the exponentially decreasing impulse response sequence:

$$h_n = \begin{cases} c^n & n \geqslant 0 \qquad |c| < 1 \\ 0 & n < 0 \end{cases} \tag{15.8}$$

We substitute into the defining equation for $H(z)$:

$$H(z) = \sum_0^\infty z^{-m} c^m = \sum_0^\infty \left(\frac{z}{c}\right)^{-m} = \frac{1}{1 - \dfrac{c}{z}} \quad , \quad |z| > |c| \quad (15.9)$$

The summation has been performed as in the example above.

The eigensequence concept provides a powerful insight into the operation of discrete LTI systems.

15.4. The z transform   In the previous section we called the sum in Eq. 15.5 the discrete system function. A more common name is the z transform;[†] that is, $H(z) = \sum_{-\infty}^\infty z^{-m} h_m$ is the z *transform* of the discrete impulse response h_m. The z transform has the importance for discrete systems that the Laplace transform has for continuous-time systems. For example, the central idea that the Laplace transform of the output of an LTI system is the product of the Laplace transforms of the input and the impulse response has a direct analogy in discrete systems. That is, *the z transform of the output sequence of a discrete LTI system is the product of the z transforms of the input sequence and the impulse-response sequence*, provided that these two z transforms have at least one common value of z for which they exist. This theorem is easily proved by substituting the convolution sum of Eq. 15.2 into the defining equation for the z-transform of the output sequence. Thus

$$Y(z) = \sum_{n=-\infty}^\infty z^{-n} y_n = \sum_{n=-\infty}^\infty z^{-n} \sum_{m=-\infty}^\infty x_{n-m} h_m \qquad (15.10)$$

Now we introduce a new variable $l = n - m$, $n = l + m$ and rearrange the sum to obtain

$$Y(z) = \sum_{l=-\infty}^\infty \sum_{m=-\infty}^\infty z^{-l-m} x_l h_m = \left(\sum_{l=-\infty}^\infty z^{-l} x_l\right) \cdot \left(\sum_{m=-\infty}^\infty z^{-m} h_m\right)$$

$$(15.11)$$

Finally we recognize the two sums in parentheses as the z transforms of x_n and h_n so that

$$Y(z) = X(z) \cdot H(z) \qquad (15.12)$$

The perceptive reader will recognize that our second method of

---

[†] What we have defined by Eq. 15.5 is the *bilateral* z transform, because the sum extends over all integer values of n. For certain purposes, such as solving linear difference equations with given initial conditions, it is more desirable to work with the unilateral z transform, in which the sum runs from 0 to ∞. Of course if the sequences in question are zero for all negative values of n, the bilateral and unilateral z transforms are the same function of z.

computing the convolution sum in the example of Section 15.2, that is, forming a polynomial, was simply a use of the z transform method. The polynomials that were multiplied together were the z transforms of the input and impulse response sequences, while the product polynomial was the z transform of the output.

Just as it was possible to construct tables of Laplace transforms for various continous time functions, so it is possible to construct a table of z transforms for various sequences. If the table is sufficiently rich we can use the following stepwise procedure to find the output sequence of a discrete system given the input and impulse response sequences.

1. Look up the z transform of the input sequence x_n and the impulse response sequence h_n.

2. Multiply these two z transforms and manipulate into a form $Y(z)$ that is recognizable in the table.

3. Look up the sequence y_n whose z transform is $Y(z)$.

A short table of z transforms appears in Table 15.1.

We have already proved entries 4 and 5 by actually performing the sum, Entry 6 is easily proved by writing down the sum in entry 4:

$$\sum_{n=0}^{\infty} e^{-\alpha n} z^{-n} = \frac{1}{1 - \frac{e^{-\alpha}}{z}} \tag{15.13}$$

and then taking the negative derivative of both sides with respect to the parameter α. The resulting sum is recognizable as the z transform of ne^{-an}, and the evaluation of the sum gives the closed form expression that appears in the table.

One type of system that is handled advantageously by the z-transform technique is the discrete system with feedback such as is shown in Fig. 15.3. Although it is possible, in principle, to find the output sequence if we know x_n and h_n in this feedback system, by an iterative technique,† it is more instructive to take z transforms of each sequence to arrive at:

$$Y(z) = H(z) \cdot [X(z) - Y(z)] \tag{15.14}$$

†For example, assuming $y_n = 0$, $x_n = 0$, and $h_n = 0$ for $n < 0$, we find y_0 from the equation $y_0 = (x_0 - y_0)h_0$; then we find y_1 from:

$$y_1 = (x_1 - y_1)h_0 + (x_0 - y_0)h_1$$

y_2 from:

$$y_2 = (x_2 - y_2)h_0 + (x_1 - y_1)h_1 + (x_0 - y_0)h_2$$

and so on. Unless h_n is a relatively short sequence, this method becomes tedious for large n.

TABLE 15.1 A Short Table of z Transforms

Sequence	z Transform	Comment
1. x_n	$X(z) = \sum_{-\infty}^{\infty} x_n z^{-n}$	Definition of bilateral z transform
2. $ax_n + by_n$	$aX(z) + bY(z)$	Linearity
3. x_{n-m}	$z^{-m} X(z)$	Delay theorem
4. $\left.\begin{array}{c} e^{-\alpha n} \\ 0 \end{array}\right\} \begin{array}{c} n \geq 0 \\ n < 0 \end{array}$	$\dfrac{1}{1 - \dfrac{e^{-\alpha}}{z}} \quad ; \quad |z| > e^{-\alpha}$	A different way of writing Eq. 15.9
5. $\left.\begin{array}{c} 1 \\ 0 \end{array}\right\} \begin{array}{c} n \geq 0 \\ n < 0 \end{array}$	$\dfrac{1}{1 - \dfrac{1}{z}} \quad ; \quad |z| > 1$	Special case of number 4
6. $\left.\begin{array}{c} ne^{-\alpha n} \\ 0 \end{array}\right\} \begin{array}{c} n \geq 0 \\ n < 0 \end{array}$	$\dfrac{1}{\left(1 - \dfrac{e^{-\alpha}}{z}\right)^2} \cdot \dfrac{e^{-\alpha}}{z} \quad ; \quad |z| > e^{-\alpha}$	
7. $\left.\begin{array}{c} n \\ 0 \end{array}\right\} \begin{array}{c} n \geq 0 \\ n < 0 \end{array}$	$\dfrac{1}{z\left(1 - \dfrac{1}{z}\right)^2} \quad ; \quad |z| > 1$	Special case of number 6

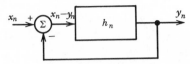

Fig. 15.   3. A discrete system with feedback.

or

$$Y(z) = \frac{H(z) \cdot X(z)}{1 + H(z)} \tag{15.15}$$

Example 1. Find the output of the feedback system of Fig. 15.3 if the input is a single unit "impulse" at $n = 0$ (i.e., $x_n = \delta_{n0}$) and if the system is an inverting delay circuit (i.e., $h_n = -\delta_{n1}$).

By direct substitution into the defining sum for the z transform, we find $X(z) = 1$ and $H(z) = -(1/z)$. Then Eq. 15.15 gives

$$Y(z) = \frac{-\dfrac{1}{z} \cdot 1}{1 - \dfrac{1}{z}} \tag{15.16}$$

which we recognize, from entries 3 and 5 in Table 15.1, as a sequence of negative unit impulses, beginning at $n = 1$ rather than $n = 0$ because of the "delay factor" $1/z$. That is,

$$y_n = \begin{array}{ll} -1 & , \quad n \geqslant 1 \\ 0 & , \quad n < 1 \end{array} \tag{15.17}$$

That Eq. 15.16 correctly gives the z transform of the geometric sequence in Eq. 15.17 can be verified by performing the sum: $-(1/z) - (1/z^2) - (1/z^3) \ldots$. Of course the final output sequence in this elementary example could have been found easily by convolution; the input impulse at $n = 0$ gives rise to an inverted output pulse at $n = 1$, which is inverted by the subtractor and causes, in turn, another inverted output pulse at $n = 2$, and so on. Figure 15.4 illustrates the output of this system.

Example 2   Find the output of the feedback system of Fig. 15.3 if the input is the sequence

$$x_n = \begin{array}{ll} 1 & , \quad n \geqslant 0 \\ 0 & , \quad n < 0 \end{array}$$

and the impulse response is the sequence

$$h_n = \begin{array}{ll} e^{-\alpha n} & , \quad n \geqslant 0 \\ 0 & , \quad n < 0 \end{array}$$

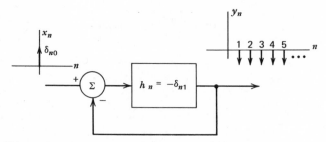

Fig. 15.   4. Example of determining the output of a discrete system with feedback.

The z transforms of x_n and h_n are given by entries 5 and 4 of Table 15.1. Thus, from Eq. 15.15,

$$Y(z) = \cfrac{1}{\left(1 - \cfrac{1}{z}\right)\left(1 - \cfrac{e^{-\alpha}}{z}\right)\left(1 + \cfrac{1}{1 - \cfrac{1}{z}}\right)} = \cfrac{1}{\left(1 - \cfrac{e^{-\alpha}}{z}\right)\left(2 - \cfrac{1}{z}\right)}$$

$$= \cfrac{1}{2\left(1 - \cfrac{e^{-\alpha}}{z}\right)\left(1 - \cfrac{\frac{1}{2}}{z}\right)} \tag{15.18}$$

This result may be further simplified by a partial fraction expansion (provided $e^{-\alpha} \neq \frac{1}{2}$) to obtain

$$Y(z) = \cfrac{\frac{1}{2}\left(1 - \cfrac{e^{\alpha}}{2}\right)^{-1}}{1 - \cfrac{e^{-\alpha}}{z}} - \cfrac{\cfrac{e^{\alpha}}{4}\left(1 - \cfrac{e^{\alpha}}{2}\right)^{-1}}{1 - \cfrac{\frac{1}{2}}{z}} \tag{15.19}$$

From entry 4 of Table 15.1, we note that y_n is the sum of two sequences:

$$y_n = \frac{1}{2}\left(1 - \frac{e^{\alpha}}{2}\right)^{-1} e^{-\alpha n} - \frac{e^{\alpha}}{4}\left(1 - \frac{e^{\alpha}}{2}\right)^{-1} \cdot \left(\frac{1}{2}\right)^{n} \qquad \begin{matrix} n \geqslant 0 \\ 0 \qquad n < 0 \end{matrix} \tag{15.20}$$

Example 3. The feedback system of Fig. 15.5a represents the way in which a mortgage loan is paid off. Note that the subtractor has been replaced by an adder. The system output y_n represents the amount owed just *after* the nth monthly payment. The amount owed is calculated by multiplying the amount owed after the previous monthly payment by 1.005 (the interest is at 6% per year or $\frac{1}{2}$% per month in this example) and then subtracting the payment. The input function represents both the original amount borrowed, a positive amount at $n = 0$, and the succeeding monthly payments, a negative amount, $-p$, each month.

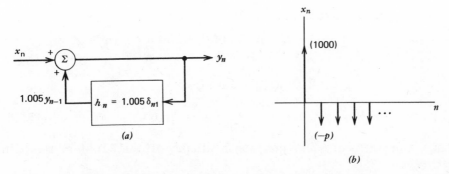

Fig. 15.   5. Feedback model for paying off a mortgage.

The equation satisfied by the system is

$$y_n = x_n + 1.005\, y_{n-1} \qquad (15.21)$$

To make the example specific, suppose that $\$1000$ is borrowed at $n = 0$, and that the amount owed is to be exactly zero after 20 years; that is, $y_{240} = 0$. The problem is to find the monthly payment p that retires the mortgage in 20 years.

The input sequence x_n may be represented as in Fig. 15.5b. The actual amount borrowed is $x_0 = 1000$; the payments are represented by $x_1 = x_2 = \ldots = -p$. The z transform of x_n is $1000 + p - p/[1-(1/z)]$ where $1000 + p$ is the z transform of the value $x_0 = 1000 + p$, and $-p/[1-(1/z)]$ is the z transform of

$$x^n - x_0 \big|= \begin{array}{cc} -p & , \quad n \geqslant 0 \\ 0 & , \quad n < 0 \end{array}$$

Now we take the z transform of each term in Eq. 15.21.

$$Y(z) = 1000 + p - \frac{p}{1 - 1/z} + \frac{1.005}{z} \cdot Y(z) \qquad (15.22)$$

Here we have used entry 3 of Table 15.1. The next step is to solve for $Y(z)$:

$$Y(z) = \frac{1000 + p}{1 - \dfrac{1.005}{z}} - \frac{p}{\left(1 - \dfrac{1}{z}\right)\left(1 - \dfrac{1.005}{z}\right)} \qquad (15.23)$$

Now we expand the second term on the right by partial fractions to obtain, after a little rearrangement,

$$Y(z) = \frac{1000 - 200p}{1 - \dfrac{1.005}{z}} + \frac{200\,p}{1 - \dfrac{1}{z}} \qquad (15.24)$$

and from entries 4 and 5 of Table 15.1, we find

$$y_n = (1000 - 200\,p)(1.005)^n + 200\,p \quad , \quad n \geqslant 0$$
$$0 \quad , \quad n = 0 \qquad (15.25)$$

Finally, by setting $n = 240$, $y_{240} = 0$ we find

$$p = \frac{5}{1 - 1.005^{-240}} = 7.164 \qquad (15.26)$$

Thus, a payment of $\$7.17$ per month will pay off this $\$1000$ mortgage in 20 years.

15.5. Stability in discrete feedback systems   As in the case of the continuous feedback systems of Chapter 5 we must worry about instability in our discrete feedback systems. Reasoning by analogy with the continuous time case, we can set the input equal to 0 for all n, assume that an eigensequence z^n (for all n) exists in the system, and then find the critical values of z that are permitted by the system. If one or more of these critical values has a magnitude greater than or equal to 1, then that eigensequence grows without bound or remains constant and we say that the system is unstable.

Example 1. Investigate the system of example 1 of Section 15.4 for stability.

With $x_n = 0$ the system reduces to that shown in Fig. 15.6. Since we found earlier that the z transform for a simple inverting delay is $-(1/z)$ we see that in order to have a signal z^n "running around" the loop, we must have $-z^n \cdot (-1/z) = z^n$ or $z = 1$. Since the magnitude of this critical z is 1, the system is classified as unstable.

Example 2. Investigate the system of example 2 of Section 15.4 for stability.

Here the impulse response is

$$e^{-\alpha n} \quad , \quad n \geqslant 0$$
$$0 \quad , \quad n < 0$$

so the system function is $1(1 - e^{-a}/z)$. In order to have an eigensequence "running around" we must have

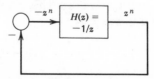

Fig. 15.   6. An unstable discrete feedback system.

$$\frac{-1}{1-\dfrac{e^{-\alpha}}{z}} = 1 \qquad \text{or} \qquad z = \frac{e^{-\alpha}}{2}$$

If the real part of α is greater than -0.693, $|z|$ is less than 1 and the system is stable.

15.6. The discrete Fourier transform   Often, in the computer analysis of sampled data, it is desirable to compute something similar to the Fourier transform of a signal. Let us suppose that the signal consists of N samples. Now the Fourier transform of a finite number of impulses is a continuous function of frequency, and it is not possible to compute a continuous function for all possible values of its arguments. What we need to do is to compute something similar to the coefficients of a Fourier *series*; that is, a function that is defined for only discrete values of frequency.

Since a Fourier series can be obtained only for a periodic signal, we arbitrarily imagine that our set of N samples is repeated with period $T = N$. The discrete Fourier transform (DFT) is defined as the set of Fourier series coefficients of this *periodic* set of samples.

Figure 15.7 illustrates these ideas. Part (a) of the figure shows a set of N samples of a time signal. The time scale is arbitrarily chosen so that the interval between the samples is 1. To prepare for the calculation of a Fourier series we form a periodic function by repeating the signal as in Fig. 15.7b. The period of this new signal is N. In Chapter 3 we learned that to evaluate the Fourier series coefficients, here denoted by A_n, we

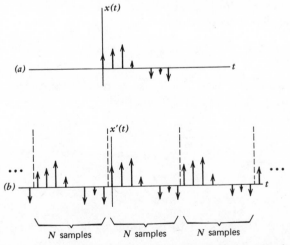

Fig. 15.   7. Forming a discrete periodic signal from a set of samples.

must multiply the time function by $e^{-j2\pi nt/T}$ and integrate over a period. Thus, from Eq. 3.4 we have

$$A_n = \frac{1}{T} \int_0^T x(t)e^{-j2\pi nt/T}\, dt = \frac{1}{T} \int_0^T \sum_{l=0}^{N-1} x_l u_0(t-l)\, e^{-j2\pi nt/N}\, dt$$

$$(15.27)$$

Since $T = N$ and because the function $x(t)$ consists of equally spaced impulses, this integral reduces to a sum:

$$A_n = \frac{1}{N} \sum_{l=0}^{N-1} x_l e^{-j2\pi nl/N} \qquad (15.28)$$

As always, integrating over a unit impulse in time gives a value that is equal to the rest of the integrand evaluated at that value of t at which the impulse occurs. There are N impulses; therefore N terms in the sum.

For what values of n should we calculate A_n? Notice that if we replace n by $n + N$ in Eq. 15.28, we have $A_{n+N} = A_n$; that is, A_n is periodic, with period N. In other words, as a consequence of the impulsive nature of $x(t)$, the Fourier series contains only N independent coefficients; these are repeated over and over as n increases. To summarize, *the DFT of a set of N data samples itself has N values*. These values of A_n are, in general, complex.

By analogy with Fourier series as applied to continuous periodic time functions we should expect to be able to recover the original $x(t)$ by an equation analogous to Eq. 3.2. That is,

$$x(t) = \sum_{n=-\infty}^{\infty} A_n e^{j2\pi nt/T} \qquad (15.29)$$

Now, however, $x(t)$ exists only at integral values of t, and furthermore we have only n independent values of A_n. Thus it is reasonable to expect that Eq. 15.29 should be replaced by:

$$x_k = \sum_{n=0}^{N-1} A_n e^{j2\pi nk/N} \qquad (15.30)$$

where we have again used $T = N$. If we insert the expression, Eq. 15.28, for A_n into the sum in Eq. 15.30, we find

$$x_k = \sum_{n=0}^{N-1} \left[\frac{1}{N} \sum_{l=0}^{N-1} x_l e^{-j2\pi nl/N} \right] e^{j2\pi nk/N}$$

$$= \frac{1}{N} \sum_{l=0}^{N-1} x_l \sum_{n=0}^{N-1} e^{j2\pi n(k-l)/N} \qquad (15.31)$$

The sum over n can be seen to be equal to N if $k = l$ and to zero if $k \neq l$. That is,

$$\sum_{n=0}^{N-1} e^{j2\pi n(k-l)/N} = N\delta_{kl} \qquad (15.32)$$

To understand this, realize that the left-hand side is the sum of N complex numbers each of magnitude 1 but with phase angles proportional to n. When $k = l$ all of the phase angles are zero so that the sum is N, but when $k \neq l$ the phase angles are such that the N complex numbers add to zero.[†]

If we substitute from Eq. 15.32 into 15.31 the result is

$$x_k = \frac{1}{N} \sum_{l=0}^{N-1} x_l \cdot N\delta_{kl} = \sum_{l=0}^{N-1} x_l \delta_{kl} = x_k \qquad (15.33)$$

which is what we set out to show. In other words the sequences x_k and A_n form a DFT pair; either sequence can be found from the other by using Eqs. 15.28 or 15.30.

Next, let us see what to expect for the DFT when the sequence x_k is a set of N samples of a sinusoidal wave. In general the period of this sinusoid wave is different from N; we denote it by T_1. In the following discussion we refer to the sketches of time signals and their transforms in Fig. 15.8. To simplify the diagrams we plot only the magnitudes of the complex Fourier transform functions.

The function whose DFT is to be found is $x(t)$ or x_k. This function can be written as the product of three time functions as shown in Fig. 15.8: $x_1(t)$, a continuous sine wave defined for all time; $x_2(t)$, a window function that limits $x(t)$ to the time interval between $t = 0$ and $t = N$; $x_3(t)$, a sampling function that causes $x(t)$ to be defined only at integral values of t. The ordinary Fourier transform of $x(t)$, *not the DFT* is, of course, the triple convolution of the three Fourier transforms, $X_1(f)$, $X_2(f)$, and $X_3(f)$. The transform $X_1(f)$ is a pair of spikes at $\pm(1/T_1)$, the transform $X_2(f)$ of the window function is a continuous function that goes to zero at nonzero integral multiples of $1/N$, and the transform $X_3(f)$ of the sampling function is a periodic set of impulses with unit spacing along the frequency axis. Therefore, the Fourier transform $X(f)$ is the sum of a set of images of $X_2(f)$ centered on $\pm(1/T_1)$, $1 \pm (1/T_1)$, $2 \pm (1/T_1)$, etc.

[†] For a formal proof for the case when $k - l = i \neq 0$, notice that

$$\sum_{n=0}^{N-1} e^{j2\pi \frac{ni}{N}} \equiv \sum_{n=0}^{\infty} e^{j2\pi \frac{ni}{N}} - \sum_{n=N}^{\infty} e^{j2\pi \frac{ni}{N}}$$

The substitution $m = n - N$ allows us to rewrite the second sum on the right:

$$\sum_{n=N}^{\infty} e^{j2\pi \frac{ni}{N}} = \sum_{m=0}^{\infty} e^{j2\pi \frac{(m+N)i}{N}} = e^{j2\pi i} \sum_{m=0}^{\infty} e^{j2\pi mi/N} = 1 \cdot \sum_{m=0}^{\infty} e^{j2\pi mi/N}$$

Since the two sums on the right are the same, the right-hand side is zero.

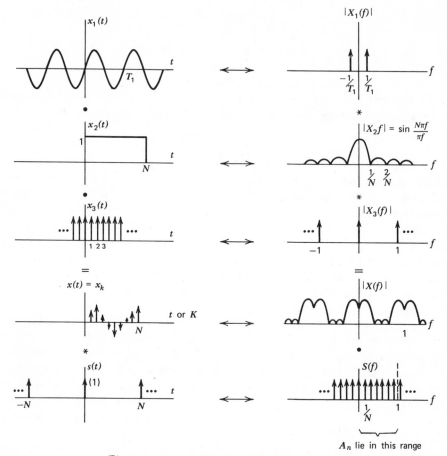

Fig. 15. 8. Finding the DFT of a sine wave.

To find the DFT of $x(t)$, recall that the DFT is the first N coefficients of the Fourier transform of a periodically extended version of $x(t)$; that is, $x(t)$ repeated at intervals of N along the t axis. Hence, we must convolve $x(t)$ with the set of impulses $s(t)$ and take the Fourier transform. But this Fourier transform is the *product* of our previously determined $X(f)$ and the transform $S(f)$, which is a set of impulses spaced at intervals of $1/N$ along the f axis. The DFT is the first N of the product impulses, beginning with the one at $f = 0$.

Several important properties are illustrated by the product $|X(f)| \cdot |S(f)|$. First, if $1/T_1$ is an integral multiple of $1/N$ (i.e., N a multiple of T_1) then the zeroes of $X(f)$ lie precisely on top of the impulses in $|S(f)|$. In this special case, only two of the coefficients in the DFT are nonzero; these lie at $f = 1/T_1$ and $1 - 1/T_1$ and they coincide with the peaks of $X(f)$. On the other hand, if N is not a multiple of T_1, then the impulses in

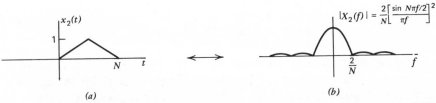

Fig. 15.   9. A window function used to reduce leakage in the computation of a DFT.

$| S(f) |$ are displaced from the peak and zeros of $X(f)$, resulting in several significant nonzero coefficients in the DFT. This important phenonenon is called *leakage*; the user of the DFT must be aware of it or else he or she will misinterpret nonzero coefficients in the DFT as being due to extra frequency components that are not actually present in the continuous time function $x_1(t)$.

A commonly applied method of reducing leakage effects is to modify the window function $x_2(t)$ so that the side lobes of $X_2(f)$ are smaller in comparison to the central peak. One such window and the magnitude of its Fourier transform are shown in Fig. 15.9. For this window the peak of the first lobe is $4/9\pi^2$ times the central maximum. For comparison purposes the peak of the first side lobe of the transform of the rectangular window function of Fig. 15.8 is $2/3\pi$ times the central maximum. On the other hand the main lobe is twice as wide for the triangular window function than for the rectangular one.

Another property of the DFT that is illustrated by the product of $| X(f) | \cdot | S(f) |$ in Fig. 15.8 is symmetry around the middle of the interval $0 < f < 1$. As a result of this symmetry, the pattern of DFT coefficients can be interpreted unambiguously only if $1/T_1 < \frac{1}{2}$ or if $T_1 > 2$. This is really a restatement of the sampling theorem; that is, the original continuous wave must be sampled at least as often as twice per cycle of the highest frequency present in the wave.

Many applications of DFT techniques can be found in the literature. For example, an important one is in an optical technique called Fourier Transform Spectroscopy. The objective is to determine spectra, that is, the intensity of a light source, such as a collection of hot atoms, as a function of frequency. Fourier Transform Spectroscopy employs a Michelson interferometer with one fixed mirror and one moveable mirror. A graph of intensity versus position of the moveable mirror (an interferogram) can be shown to be the Fourier transform of the required spectrum. In practice, the DFT of a sampled version of the interferogram is obtained rather than the ordinary Fourier transform, and the actual spectrum is deduced from the DFT.

Another application of the DFT is in electronic recognition and synthesis of speech. Different sounds can be detected by the relative content of several frequency components. The coming era of verbal communication between human beings and computers is certain to depend heavily upon DFT analysis of voice signals.

15.7. The Fast Fourier Transform   The discrete Fourier transform, discussed in the previous section and defined by Eq. 15.28, appears to require the computation of the sum of N products for each of the N DFT coefficients, A_n, or a total of N^2 multiplications. Even for a fast computer that performs a multiplication in, say, 10 μsec, a 1000-sample DFT would require 10 seconds to do the multiplications, a 10,000 sample DFT would require 1000 seconds, and so on. The Fast Fourier Transform (FFT) is a computational algorithm that reduces the number of required multiplications to approximately $N \log_2 N$, so that the time for multiplications is reduced by a factor of approximately $N/\log_2 N$. For $N = 1000$ this is a factor of ≈ 100; for $N = 10,000$ this is a factor of ≈ 700.

The FFT algorithm achieves its enormous speed advantage because certain key products and sums of products appear over and over again in the computations and need to be computed only once if the entire computation is properly organized. The computation of the FFT is most

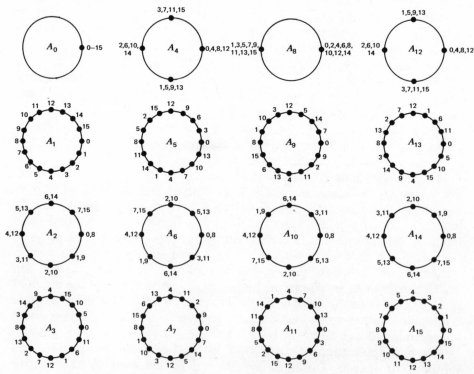

Fig. 15.   10. Computation of DFT coefficients for $N = 16$.

easily accomplished if N is some integral power of 2, through similar algorithms can be found for certain other values of N.

Why the savings can be accomplished is illustrated in Fig. 15.10, for the special case of $N = 16$. The computation for each of the 16 DFT coefficients is illustrated by writing the 16 sample values, x_l, around a unit circle. The position of the sample value on the circle indicates the amount by which the phase angle of the sample value must be shifted before adding. For example, to calculate A_0, the phase shift for every x_l is zero. For A_1, x_0 is shifted by zero, x_1 by $-\pi/8$, x_2 by $-\pi/4$, etc. For A_2, x_0 is shifted by zero, x_1 by $-\pi/4$, x_2 by $-\pi/2$, and so on.

The first key simplification illustrated by Fig. 15.10 is that sample values whose indices differ by 8, such as x_0 and x_8, or x_1 and x_9, always are multiplied by complex numbers that differ in phase by $0°$ or $180°$. The first step, then, in the computation of the FFT is to phase shift x_8 by 0 and $-\pi$ and add to x_0 and to do the same for x_1 and x_9, x_2 and x_{10}, through x_7 and x_{15}. In other words, we calculate $x_n + x_{n+8}(0)$ and $x_n + x_{n+8}(8)$, for $n = 0$ to 7 where the number in parentheses indicates the amount of phase shift in multiples of $-\pi/8$ rad. Furthermore, since only these 16 sums are used in the remainder of the calculation, the same 16 computer memory locations that held x_0 to x_{15} can hold the sums. This useful feature of the FFT algorithm applies also to the remaining steps of the calculation and is called the "in place" property of the algorithm.

This first step of calculating $x_n + x_{n+8}(0)$ and $x_n + x_{n+8}(8)$ is illustrated in the signal flow graph of Fig. 15.11. The first column lists the original sample values in what is called bit-reversed order;[†] the second column lists the sums and differences. Each node in a horizontal row represents a single memory location, whose contents are changed at each major step (column) in the program by combining its previous contents with the contents of one other location. The signal flow graph represents the fact that after step 1, location 0 contains $x_0 + x_8(0)$, location 1 contains $x_0 + x_8(8)$, location 2 contains $x_4 + x_{12}(0)$, location 3 contains $x_4 + x_{12}(8)$, and so forth. The number written next to the line from one node to another indicates the amount of phase shift applied to that number before adding it to the other number.

To understand step 2 of the algorithm, look at the ways in which the four values x_0, x_4, x_8, and x_{12} are combined in the 16 DFT coefficients.

[†] Bit-reversed order is obtained by writing the binary representation of a subscript backwards. Thus x_2 is *fourth* in the list because $2_{10} = 0010_2$ and, when the bits are reversed, $0100_2 = 4_{10}$. As another example, x_7 is fourteenth because $7_{10} = 0111_2$ and $1110_2 = 14_{10}$.

There are only four ways in which these appear:

$$x_0 + x_4 \qquad + x_8 \qquad + x_{12} \qquad = (x_0 + x_8) + (x_4 + x_{12})$$
$$x_0 + x_4(4) \quad + x_8(8) + x_{12}(12) = [x_0 + x_8(8)] + [x_4 + x_{12}(8)](4)$$
$$x_0 + x_4(8) \quad + x_8(0) + x_{12}(8) \quad = (x_0 + x_8) + (x_4 + x_{12})(8)$$
$$x_0 + x_4(12) + x_8(8) + x_{12}(4) \quad = [x_0 + x_8(8)] + [x_4 + x_{12}(8)](12)$$

$$(15.34)$$

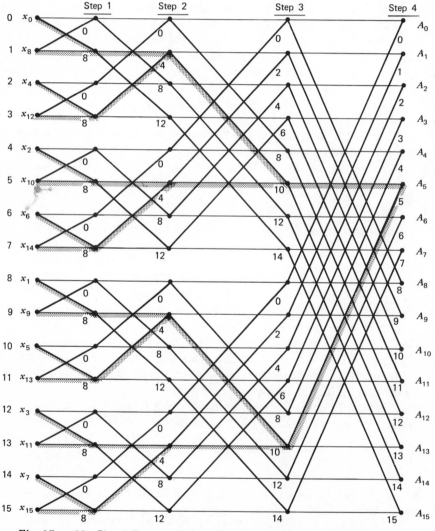

Fig. 15. 11. Signal flow graph for a Fast-Fourier-Transform (FFT) algorithm.

On the right-hand side of each equation we have rearranged the left-hand side in a way that shows how the quantities calculated in step 1 can be phase shifted and added to obtain the four desired combinations of x_0, x_4, x_8, and x_{12}. For example, the right-hand side of the last equation means that $x_4 + x_{12}(8)$, which had been stored in location 3, is phase shifted by 12 times ($-\pi/8$), before being added to $x_0 + x_8(8)$, which had been stored in location 1. The result is placed in location 3. The first four nodes in the step 2 column of the signal flow graph in Fig. 15.11 summarize the four operations of Eq. 15.34.

The four values x_1, x_5, x_9, and x_{13} are combined in the same four ways as x_0, x_4, x_8, and x_{12}, and so are x_2, x_6, x_{10}, and x_{14} and x_3, x_7, x_{11}, and x_{15}. These combinations complete step 2 of the algorithm and are illustrated in the signal flow graph.

In step 3 we form all possible combinations of the eight values $x_0, x_2, x_4, x_6, x_8, x_{10}, x_{12}$, and x_{14}. There are eight such combinations, which differ in the amount of phase shift between adjacent terms in the sum. It should be clear by now that the previously computed combinations of x_0, x_4, x_8, and x_{12} can be combined with the previously computed combinations of x_2, x_6, x_{10}, and x_{14} to give the necessary eight combinations. Again the signal flow graph indicates the amount of phase shift required before addition and also where each result is to be stored. The various combinations of $x_1, x_3, x_5, x_7, x_9, x_{11}, x_{13}$, and x_{15} are obtained in identical fashion.

Finally, step 4 completes the algorithm. Here the eight previously computed combinations of even-subscripted values are combined with the eight combinations of odd-subscripted values by phase shifting the latter by different amounts before adding to the former.

To check that the algorithm described by the signal flow graph is correct, look at the calculation of one of the values, say A_5. The "route" taken through the algorithm by each of the 16 original sample values is shown shaded. The total phase shift of each of the sample values, modulo 16, should give the phase shifts shown for A_5 in Fig. 15.10. Thus we have

$$A_5 = x_0(0 + 0 + 0 + 0) + x_1(0 + 0 + 0 + 5) + x_2(0 + 0 + 10 + 0)$$
$$+ x_3(0 + 0 + 10 + 5) + x_4(0 + 4 + 0 + 0) + x_5(0 + 4 + 0 + 5)$$
$$+ x_6(0 + 4 + 10 + 0) + x_7(0 + 4 + 10 + 5) + x_8(8 + 0 + 0 + 0)$$
$$+ x_9(8 + 0 + 0 + 5) + x_{10}(8 + 0 + 10 + 0) + x_{11}(8 + 0 + 10 + 5)$$
$$+ x_{12}(8 + 4 + 0 + 0) + x_{13}(8 + 4 + 0 + 5) + x_{14}(8 + 4 + 10 + 0)$$
$$+ x_{15}(8 + 4 + 10 + 5)$$
$$= x_0(0) + x_1(5) + x_2(10) + x_3(15) + x_4(4) + x_5(9) + x_6(14) + x_7(3)$$
$$+ x_8(8) + x_9(13) + x_{10}(2) + x_{11}(7) + x_{12}(12) + x_{13}(1) + x_{14}(6)$$
$$+ x_{15}(11) \tag{15.35}$$

We have not included a formal "proof" here of the FFT algorithm. Such proofs may be found in more specialized books and journal articles.[†] Instead, our purpose has been to illustrate the algorithm in such a way that the reader could extend it to other values of N that are integral powers of 2. In addition, the signal flow graph illustrates that there are only $N \log_2 N$ multiplications required because there are $\log_2 N$ steps each with N multiplications.

An extensive literature on the FFT has developed since it was originally described by J. Cooley and J. Tukey in 1965.[‡] For an excellent bibliography, consult the article by R. C. Singleton.[§]

Problems for Chapter 15

15.1. Sketch the sequence j^n, defined for all integral values of n, from $-\infty$ to ∞. This sequence is sent through a system with a discrete impulse response consisting of unit values at $n = 0,1,2$ and zero for all other n. Find and sketch the output. Is it true that the output is just a constant times the original sequence? If so, is the constant equal to $H(j)$?

15.2. Find the z transform of the alternating sequence

$$\begin{cases} (-1)^n ; & n \geqslant 0. \\ 0 ; & n < 0. \end{cases}$$

15.3. Show that the z transform of the sequence

$$\begin{cases} \dfrac{a^n}{n!} ; & n \geqslant 0 \\ 0 ; & n < 0 \end{cases}$$

is $e^{a/z}$. Sketch the sequence for the first few values of n, for the case $a = 1$.

15.4. If the sequence x_n has a z transform $X(z)$, show from the definition of the z transform that the z transform of $e^{-\alpha n} \cdot x_n$ is $X(ze^\alpha)$.

[†]For example, G-AE Subcommittee on Measurement Concepts, "What is the Fast Fourier Transform," *IEEE Trans. Audio and Electroacoustics*, AU-15, 45—55 (June 1967).

[‡]J. W. Cooley and J. W. Tukey, "An Algorithm for Machine Calculation of Complex Fourier Series," *Math. Computation*, 19, 297—301 (April 1965).

[§]R. C. Singleton, "A Short Bibliography on the Fast Fourier Transform," *IEEE Trans. Audio and Electroacoustics*, AU17, 166—167 (June 1969).

15.5. In the feedback system shown in Fig. P15.5 show that the system is stable provided $a < \pi$.

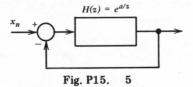

$$H(z) = e^{a/z}$$

Fig. P15.   5

15.6. In the feedback system of Fig. 15.3, the input is a sequence of unit values beginning at $n = 0$.

$$x_n = \begin{array}{l} 1, \ n \geqslant 0 \\ 0, \ n < 0 \end{array}$$

and the impulse response is the sequence

$$h_n = \begin{array}{l} (\tfrac{1}{2})^n, \ n \geqslant 0 \\ 0, \quad n < 0 \end{array}$$

Find and sketch the output sequence, y_n. (Note that this problem is a special case of Example 2 of section 15.4.)

15.7. Find the monthly payment required to retire a \$20,000 mortgage in 30 years (360 payments) at an annual interest rate of 6%.

15.8. An auto loan of \$3000 at 12% annual interest on the unpaid balance is to be paid off by 36 equal monthly payments. Find the total amount of interest paid by the borrower.

15.9. A certain discrete system with input x_n and output y_n is described by $y_n = K\, y_{n-1} - x_n$.
(a) Find and sketch the discrete impulse response of the system.
(b) Find the system function $H(z)$.
(c) Draw a block diagram of the system, using elementary blocks.

15.10. A discrete analog of a resonant system has a realizable impulse response $e^{-an} \cos \omega_0 n$ (for $n \geqslant 0$).
(a) Show that the system function is given by

$$H(z) = \frac{z^2 - ze^{-a}\cos \omega_0}{z^2 - 2ze^{-a}\cos \omega_0 + e^{-2a}}$$

(*Hint*. Use number 4 of Table 15.1.)
(b) Show that this "digital filter" can be implemented by the

system (or computer program) represented in Fig. P15.10. Note that the computation of y_n involves only the present value and one earlier value of x_n and two earlier values of y_n.

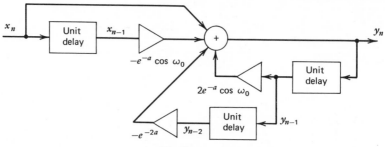

Fig. P15.   10

(c) Now the discrete signal $x_n = \cos \omega n$ is sent into this filter. Show that this input can be written as a linear combination of two eigensequences and find the value of z for each of the eigensequences.

(d) Sketch the impulse response, the input sequence, and the output sequence as functions of n for the following parameter values: $a = 0.693$, $\omega_0 = \pi$, $\omega = 3\pi/4$. (*Hint.* Each eigensequence is multiplied by a complex constant.)

(e) Repeat part (d) for the same filter parameters but $\omega = \omega_0 = \pi$. In comparing the output sequences y_n for parts (d) and (e) do you notice a resonance phenomenon? Which parameter should be changed to obtain a sharper "resonance"?

15.11. In Eq. 15.32 sketch the terms on the left-hand side on a complex number diagram for the special case $N = 7, k = 4, l = 1$, and thus show that the sum is zero.

15.12. Construct graphs that are analogous to Figs. 15.10 and 15.11 for a Fast Fourier Transform algorithm with $N = 8$.

15.13. Use the circle diagrams of Fig. 15.10 or the signal flow graph of Fig. 15.11 to sketch the FFT coefficients A_n as functions of n for the following simple cases.
(a) All $x_k = 1$ ($k = 0$ to 15)
(b) $x_1 = 1$; all others $= 0$
(c) $x_{15} = 1$; all others $= 0$
(d) $x_1 = x_{15} = \frac{1}{2}$; all others $= 0$
(e) $x_5 = 1$; all others $= 0$
(f) $x_{11} = 1$; all others $= 0$

REFERENCES

Bogner, R. E., and A. C. Constantinides, Eds., *Introduction to Digital Filtering*, Wiley-Interscience, New York, 1975.

Childers, D., and A. Durling, *Digital Filtering and Signal Processing*, West Publishing Co., 1975.

Gold, B., and C. M. Rader, *Digital Processing of Signals*, McGraw-Hill, New York, 1969.

Jury, E. I., *Theory and Applications of the z-Transform*, Kreiger Publishing Co., Huntingdon, New York, 1973.

ANSWERS TO PROBLEMS

Chapter 1

1.1 $y(t)$ = straight line segments joining the points: $(0,0)$, $(1,1)$, $(2,-2)$, $(3,1)$, $(4,0)$ and $y(t) = 0$ elsewhere.

1.3 $y_1(t)$ contains segments joining: $(0,0)$, $(1,-1)$, $(3,1)$, $(4,0)$
$y_2(t)$ contains segments joining: $(0,0)$, $(0.5,-0.5)$, $(1,0)$, $(1.5,1.5)$, $(2,1)$, $(2.5,-0.5)$, $(3,-1)$, $(4,0)$
$y_3(t)$ contains segments joining: $(0,0)$, $(1,-\sqrt{2})$, $(2,\sqrt{2})$, $(3,0)$
The response to the $x(t)$ of problem 1.2 is recognized by the fact that it has the highest peak value $[y(2) = 2]$.

1.5 Same answer as for 1.4.

1.7 Output is a triangular pulse with vertices: $(0,0)$, $(\Delta, AB/\Delta)$, $(2\Delta, 0)$. The area is AB as expected.

1.9
$$y(t) = \begin{cases} 0 & t \leqslant 0 \\ (1 - e^{-bt})/b & 0 \leqslant t \leqslant a \\ (e^{ab} - 1)e^{-bt}/b & t \geqslant a \end{cases}$$

1.11 Just convolve the actual mass distribution $P(m)$ with the response $P_0(m)$ to an impulsive distribution.

1.2 Output = $-y(t)$ where $y(t)$ is answer to problem 1.1.

1.4
$$y(t) = \begin{cases} 0 & t \leqslant 0 \quad t > T \\ \dfrac{T}{2\pi}\left(1 - \cos\dfrac{2\pi t}{T}\right) & 0 < t < T \end{cases}$$

1.6 Since any input wave is a superposition of impulses, and each response to an impulse averages to zero, the average response to any number of input impulses is zero.

1.8 (a) Step response = $u_{-1}(t) \cdot (1 - \cos 2\pi t/T)$.
(b) Move the boom quickly to the halfway point, wait $T/2$ for the load to swing through half a cycle, and finally move the boom quickly to a point above the load.

1.10 (b) Decay rate
$$= \begin{cases} 0 & t \leqslant 0 \\ R_0(1 - e^{-\lambda t}) & 0 \leqslant t \leqslant T_0 \\ R_0 e^{-\lambda t}(e^{\lambda T_0} - 1) & t \geqslant T_0 \end{cases}$$
(c) Do two consecutive convolutions.
(d) Decay rate
$$= \begin{cases} 0 & t \leqslant 0 \\ \dfrac{N_0 \lambda_1 \lambda_2}{\lambda_2 - \lambda_1}(e^{-\lambda_1 t} - e^{-\lambda_2 t}) & t \geqslant 0 \end{cases}$$

1.12 The sum in the problem statement is the infinite series expansion of e^a. When multiplied by the e^{-a} factor, the result is 1.

1.13 $\bar{y} = aAT/2$
$\Delta y^2 = \text{variance} = aA^2T/3$
$\overline{y^2} = aA^2T/3 + a^2A^2T^2/4$

1.14 $y = \sum\limits_{n_1=0}^{\infty} \sum\limits_{n_2=0}^{\infty} p(n_1)p(n_2)(n_1A$
$+ n_2B) = a(A + B)$

$\overline{y^2} = \sum\limits_{n_1=0}^{\infty} \sum\limits_{n_2=0}^{\infty} p(n_1)p(n_2)(n_1^2A^2$
$+ n_2^2B^2 + 2n_1n_2AB)$
$= a^2(A + B)^2 + a(A^2 + B^2)$

1.15 $\dfrac{A^2}{2} \cos 2\pi\tau/T_1$

1.16 $(a/b^2 - 1/b^3 + e^{-ab}/b^3)/T + a^2/b^2T^2$

Chapter 2

2.1 (a) $e^t/3$.
(b) e^{-t}.
(c) Integral does not converge.
Re(s) > -2 for convergence.

2.2 $e^{-s\Delta}$

2.3 $Y(0) = H(0) \cdot X(0)$

2.4 $u_{-1}(t)(1 - e^{-bt})/b - u_{-1}(t$
$- a)[1 - e^{-b(t-a)}]/b$

2.6 $u_{-1}(t)(\omega e^{-at} - \omega \cos \omega t$
$+ a \sin \omega t)/(\omega^2 + a^2)$

Chapter 3

3.1 (a) $x_n = 0, n$ even; $x_n = 1, n$ odd.

3.2 (a) Two impulses of area $\frac{1}{2}$ at $f = \pm\omega/2\pi$.
(b) Impulse of area $1/(4 + 2j)$ at $f = \omega/2\pi$ and of area $1/(4 - 2j)$ at $f = -\omega/2\pi$.
(c) $\cos (t - 26.6°)/\sqrt{5}$.
(d) No.

3.3 (a) Impulses of area $-j/2$ at $f = \omega/2\pi$ and of area $j/2$ at $f = -\omega/2\pi$.
(b) Impulses of area $-j/(4 + 2j)$ at $f = \omega/2\pi$ and of area $j/(4 - 2j)$ at $f = -\omega/2\pi$.
(c) $\sin (t - 26.6°)/\sqrt{5}$.
(d) No.

3.4 $x_n = \sin (n\pi/2)/n\pi$

3.5 $T = 4$

3.6 Output is a 3-kHz sine wave.

3.10 Impulses: area $= \frac{1}{8}$ at $\pm 3f$, area $= \frac{3}{8}$ at $\pm f$. The transform is real, as expected.

3.11 Impulses at frequencies $f = n/T$ with areas $(\sin n\pi/3)/n\pi$. Those for $n = \pm 3, \pm 6$, etc. are missing.

3.12 (a) Real, impulsive, even, nonperiodic in f.
(b) $\frac{1}{2}$.
(c) $2/\pi$.
(d) $2u_0(f)/\pi, 2u_0(f \pm \frac{1}{2})/3\pi,$
$-2u_0(f \pm 1)/15\pi$.

3.13 (a) $u_0(t + a) - u_0(t - a)$
(b) $P(f) = e^{j2\pi fa} - e^{-j2\pi fa}$
(c) $P(f) = j2\pi fQ(f)$
(d) $Q(f) = \dfrac{\sin 2\pi fa}{\pi f}$

3.14 $X(f) = \dfrac{\sin^2 \pi f}{\pi^2 f^2}$

Transform of periodic triangular

wave $= \displaystyle\sum_{n=-\infty}^{\infty} u_0(f - n/2) \cdot \dfrac{\sin^2 \pi f}{2\pi^2 f^2}$

3.15 $\dfrac{e^{-j\pi f a}}{b + j2\pi f} \left[\displaystyle\sum_{n=-\infty}^{\infty} \dfrac{u_0(f - n/T) \sin \pi f a}{\pi f T} \right]$

Chapter 4

4.1 $h_i = R_1 + R_2 R_3/(R_2 + R_3)$,
$h_f = -R_2/(R_2 + R_3) = -h_r$
$h_o = 1/(R_2 + R_3)$

4.3 (a) $s^2 R C^2/(1 + 2sRC)$
(b) $sRC/(1 + sRC)$
(c) $(1 + 2sRC)/sC(1 + sRC)$

4.5 $I = 15/4,\ V_1 = 11/4,\ V_2 = 9/2,$
$V_3 = 17/4.$

4.7 $L_{eff} = 1/4\ H$
4.8 (a) $(2s + 1)/(3s + 2).$
(b) Zero at $-\tfrac{1}{2}$, pole at $-\tfrac{2}{3}$.
(c) $H(s) = \dfrac{2s}{3s + 2} + \dfrac{1}{3s + 2} = HPF +$

LPF.
(d) $H(s) \longrightarrow \tfrac{1}{2}$ for small ω, $\tfrac{2}{3}$ for
large ω.

4.2 $h_i = R_A R_B/(R_A + R_B)$,
$h_f = -R_A/(R_A + R_B) = -h_r$
$h_o = 1/R_C + 1/(R_A + R_B)$

4.4 $h_i = R(2 + sRC)/(1 + sRC)$
$-h_f = h_r = 1/(1 + sRC)$
$h_o = sC/(1 + sRC)$

4.6 (a) $r + R(1 + \beta)$
(c) $h_i = r,\ h_f = -1 - \beta$
$h_r = 1,\ h_o = 1/R.$

Chapter 5

5.1 (a) $\pm 2\sqrt{2}j.$
(b) 0.14 at $-98.2°.$
(c) Starts at $\tfrac{1}{4}$, crosses $180°$ line at
-0.028, approaches 0 along $90°$
line.

5.3 (a) Stable.
(b) $s = -55 \pm j \cdot 3.2 \cdot 10^3.$

5.5 Stable for $0 < A < 0.25,\ A > 1.60.$
5.8 Stable for $A \leqslant 2.$

5.2 (a) $A < 1.5$ for stability.
(b) 0.11 Hz.

5.4 (a) $-0.5 < A < 4$
(b) 0.28 Hz

5.6 0.55 Hz, $A = 8.$
5.9 (a) $V_{out} = V_{cc}(1 - e^{-t/R_1 C});\ t =$
$1.10\ R_1 C.$
(b) $t = R_1 R_2(2R_1 - R_2)C/(R_1$
$+ R_2)(R_1 - 2R_2)$ provided
$R_1 > 2R_2.$
(c) $0.69\ R_1 C.$
(d) $3.77.$

Chapter 6

6.1 ≈ 1 V

6.2 Minority carrier flow from base to
emitter, which does not contribute
to collector current, is thereby
kept small compared with minority
carrier flow from emitter through
base into collector.

6.3 Current is small. Case (1): most voltage drop occurs across BE junction.

6.4 (a) $I_2 + I_C' = \alpha_F I_E', I_3 + I_2 = I_1,$
$I_E' = I_1 + \alpha_R I_C'.$
(b) $I_C' = 2.11$ mA, $I_E' = 3.29$ mA, $I_3 = 0.54$ mA.
(c) Solve Eq. 6.5 for the V's.
(d) BE junction is more forward biased than BC junction.

6.5 $I_C = -0.70$ mA, $I_E = 0.71$ mA, $I_B = -0.01$ mA, $V_{EB} = 0.29$V, $V_{CB} = -2.9$ V.

Chapter 7

7.1 486 mV, 535 mV.

7.3 (a) 4.5 mA
(b) 4.27 mA
(c) Gain = 0.98

7.4 (b) Pole at $s = -10^8$, zero at $s = 10^{10}$.

7.5 (b) Gain = $-(0.002 + sC)/(5.15 \times 10^{-4} + 0.005sC)$
(c) 164 μF

7.8 Q_1 on: $V_E = 1.96$ V, $V_{B2} = 0.98$ V, $I_{C1} = 3.84$ mA.
Q_2 on: $V_E = 2.5$ V, $V_{B2} \approx 3$ V, $I_{C2} = 5.0$ mA.

Chapter 8

8.1 102, 1.02 MΩ.
8.3 $-10^7/[s^2 RC + s(1 + 10^7 RC) + 10]$, marginally stable for all C.
8.4 (a) $A = 10^5, a = 10.$
(b) $1/(101 + 10^{-4}s).$
(c) $s = -10^4, -10^6.$
8.7 $10^{-4}s$
8.13 (a) $V_0 = 0$
(b) -1.02 V
8.15 (b) $R_L = R$
(c) R_1

8.2 $V_0 = R_1 A V_1 /(R_1 + R_2 + R_1 A)$

8.5 $(1 + 10^{-5}s + 10^{-11}s^2)/(10^{-5} + 10^{-6}s + 2.2 \times 10^{-10}s^2 + 10^{-17}s^3)$

8.12 $V_0 = V_{os}(1 + R_2 /R_1) + I_- R_2.$
8.14 $v_B = -i_A /sC, 106$ kΩ.

Chapter 9

9.1 (b) $Q_1 = \overline{A \cdot B + C}$

9.2 Circuit uses six inverters, three 2-input NANDs.

9.3 Sum = $A\overline{B}\overline{C} + \overline{A}B\overline{C} + \overline{A}\overline{B}C + \overline{A}\overline{B}\overline{C}$
Carryout = $AB + AC + BC$

9.6 (a) $A + BC$
(b) $A\overline{D} + B\overline{D} + \overline{A}BCD$

9.7 $y_0 = \overline{A_0 B_0}, y_1 = \overline{A_1 B_1}, y_2 = \overline{A_2 B_2},$
$y_4 = \overline{EO_B}.$

9.8 $A = B = 1, C = 0$ $A = 0$
for   20 nsec $< t <$ 80 nsec

9.9 Inputs 0, 1, 2, 3 = $\overline{D}$; 4, 5 = 0;
6 = D; 7 = 1.

9.10 Inputs 0, 1, 2, 3 = 0; 4 = D;
5, 6, 7 = 1.

9.11 $V_A = Q_2 + \overline{Q}_4 Q_1 + Q_4 \overline{Q}_1;$
$V_B = \overline{Q}_4 Q_2 \overline{Q}_1 + Q_4 Q_2 Q_1.$

9.12 For example, $V_A = \overline{Q}_2 Q_1 + Q_2 \overline{Q}_1;$
$V_B = \overline{Q}_4 \overline{Q}_2 + Q_4 Q_2.$

9.13 $A_n' = \overline{E}\overline{A}_n + EA$

Chapter 10

10.1 (a) Both outputs eventually become 0 and remain 0.

  (b) Both outputs eventually become 1 and remain 1.

10.4 Negative-edge triggered.

10.7 $J_1 = K_1 = 1; J_2 = K_2 = Q_1;$
$J_4 = \overline{Q}_8 Q_2 Q_1; J_8 = Q_4 Q_2 Q_1;$
$K_4 = K_8 = Q_2 Q_1.$

10.2 (a) Gates 1 to 7 are at logic 1,0,1,1,1,0,1, respectively.

  (b) Gates 1 to 7: 1,0,1,0,0,1,1. (Critical race at gate 7.)

10.6 $D_4 = Q_2' Q_1' + Q_4' \overline{Q}_1';$
$D_2 = Q_2' \overline{Q}_1' + \overline{Q}_4' \overline{Q}_2' Q_1'; D_1 = \overline{Q}_1'.$

Chapter 11

11.4 $V_{out} = k[V_1 + 2V_2 + 4V_4 + 8(1 + x)V_8]$

11.6 (a) 127, -128.

11.5 (a) $R_N = 1\ k\Omega$

  (b) $R_N = 9\ k\Omega$

  (c) $R_N = 15\ k\Omega$

11.7 (a) Sawtooth wave: rise time $= \tau$, fall time $= \tau \cdot (I_0 R - V_{in})/V_{in}$.

  (d) Frequency is too low to measure accurately in a reasonable time.

Chapter 12

12.1 (a) Impulses at $-f_c - 2f_o$, $-f_c - f_o, f_c + f_o, f_c + 2f_o$.

  (b) Impulses at $\pm 2f_o, \pm f_o$, plus others at $|f| > 2f_c$ that are rejected by a LPF.

  (c) All impulses now have imaginary area.

  (d) Impulses at $\pm(f_o - \Delta f_c)$, $\pm(2f_o - \Delta f_c)$; not an octave apart.

  (e) Music signals require preservation of frequency ratios.

12.3 (a) $y(t)$ contains impulses at ± 2455 and ± 455 kHz; $z^2(t)$ contains impulses at 0 and ± 910 kHz.

  (b) $y(t), z(t),$ and $z^2(t)$ contain additional impulses 1 kHz above those in (a).

  (c) Spectrum near 1560 kHz is shifted down to passband of i.f. amplifier.

12.8 (a) $s(t)$.

  (c) Correct phase must be maintained.

12.10 (a) Impulses at $f = 0, \pm f_o, \pm 2f_o$.

  (b) 400.

12.13 6.9

12.2 (a) $T = n/5W; n = 1, 2, 3, 4, 5;$ or $T \leqslant 1/6W;$ or $1/4W < T < 1/3W;$ or $1/T = 1/2W.$

  (b) Pass sampled wave through band-pass filter to pass $2\ kHz < |f| < 3\ kHz$.

12.7 (b) $\Delta = 1/2W_x$

  (d) $W_y = W_x \Delta/(T + \Delta)$

12.9 (c) Stable, $\omega = 1.4 \times 10^5$, loop gain $= 1/2$.

12.11 Groups of impulses $f_o/100$ apart; groups are spaced by f_o.

Chapter 13

13.1 (a) I_0/e
    (b) $I_0/e\ \Delta f$
    (c) $\sqrt{I_0/e\ \Delta f}$
    (d) $\sqrt{I_0e\ \Delta f}$

13.5 $kT/2$, yes.

13.8 $v_B^2 = 19\ V^2$

13.11 The spectral densities add to give $2kTR_1R_2/(R_1 + R_2)$ at the amplifier input.

13.13 1.1.

13.2 (a) $-100/(1 + 2 \times 10^{-6}s)$.
    (b) Approx. $-100/(1 + 2 \times 10^{-9}s)$.
    (c) $0.02\ \mu V$, $0.64\ \mu V$.

13.7 $1.66 \times 10^{-9}\ V^2$ at $T = 300°K$

13.10 $2kTr\ \Delta f/(R + r)$ where Δf is the effective bandwidth.

13.12 $41\ \mu V$

13.17 (a) $H'(f) = 2/(1 + j2\pi fRC)$.
    (b) $H_A''(f) = 2$,
    $H_B''(f) = 2[1 + R(2C_1 - C)j2\pi f]/(1 + RCj2\pi f)$.

Chapter 14

14.1 (a) 0.9 pF
    (b) 6.1 mV

14.3 System function $\doteq 100/$
    $(1100 + 0.02s + 10^{-10}s^2)$
    ΔV_2 (Max) ≈ 0.9 V

14.6 (a) System 2 now includes noise from $x(t)$ that was near $f = \pm 3f_c, \pm 5f_c$, etc.
    (b) System 3 eliminates this extra noise.
    (c) Prefilter also removes noise near these odd harmonic frequencies.

14.2 (a) $3.2 \times 10^{-19}A^2$
    (c) $R \geqslant 52\ M\Omega$
    (d) Poisson fluctuations:

    (d) Poisson fluctuations:

14.4 (a) $2.65\ \mu F$

14.5 (a) Only those within $\pm W/2$ of f_c

14.7 (a) Zero
    (b) 10
    (c) 10 mV
    (d) 3600

Chapter 15

15.1 Output sequence = $-j$ times input sequence. Yes, H(j) = $-j$.

15.6 $y_0 = 1, y_1 = 3/2, y_2 = 7/4, \ldots$

15.8 $587.15

15.10 (c) $z = e^{j\omega}, e^{-j\omega}$.
    (d) Eigensequences are multiplied by $1.36\ e^{\pm j/2}$.
    (e) Decrease α.

15.13 (a) $A_0 = 1$, all others $= 0$.
    (b) $A_n = (e^{-j\pi n/8})/16$
    (c) $A_n = (e^{j\pi n/8})/16$
    (d) $A_n = (\cos j\pi n/8)/16$
    (e) $A_n = (e^{-j5\pi n/8})/16$
    (f) $A_n = (e^{-j \cdot 11\pi n/8})/16$

15.2 $(1 + 1/z)^{-1}$

15.7 $119.91

15.9 (a) $y_n = -K^n$
    (b) $-1/(1 - K/z)$

INDEX